2

From Hot Air to Action?

Climate Change, Compliance and the Future of International Environmental Law

by

Meinhard Doelle

THOMSON
CARSWELL

Library and Archives Canada Cataloguing in Publication

Dolle, Meinhard, 1964-
From hot air to action? : climate change, compliance and the future of international environmental law / Meinhard Dolle.

Includes bibliographical references and index.
ISBN 0-459-24277-6

1. Environmental law, International. 2. Climatic changes. I. Title.

K3585.D64 2005 344.04'6 C2005-906548-6

Composition: Computer Composition of Canada Inc.

THOMSON CARSWELL

One Corporate Plaza
2075 Kennedy Road
Toronto, Ontario
M1T 3V4

Customer Service:
Toronto: 1-416-609-3800
Elsewhere in Canada/U.S. 1-800-387-5164
Fax 1-416-298-5082
www.carswell.com
E-mail: carswell.orders@thompson.com

Acknowledgments

This book would not have been possible without the support from family, friends and colleges. At the risk of leaving out countless valuable contributions, I would like to especially acknowledge the support of Dr. David VanderZwaag, Dr. Aldo Chircop, Dr. Douglas Johnston, and Dr. Jutta Brunnée who have offered guidance and support throughout this project with patience, wisdom, and incredible insight. All remaining shortcomings and errors are of course, mine alone.

Dedication

This book is dedicated to my family. To my parents, for teaching me to always question, to think critically, and to see things from different perspectives. To my partner, Wendy, for her patience, love and support during this four year long journey. Now it is your turn again. And to my children, Klara, Alida, and Nikola, may this remind you that learning is one of the great joys in life.

Preface

This book considers the evolution, current state, and future prognosis of the global climate change regime under the umbrella of the United Nations Framework Convention on Climate Change. The focus of the book is on State compliance with the Kyoto Protocol. Compliance is considered from the perspective of the internal compliance regime developed under the Kyoto Protocol as well as a select set of potential external international law influences. The book concludes with an assessment of the level of compliance to be expected and its potential influence on the future of the climate change regime. Implications for international environmental law more generally are also considered.

To provide the necessary context, the state of international environmental law generally and the climate change regime under the UNFCCC more specifically is assessed. This is followed by an assessment of likely levels of compliance with Kyoto considering a number of internal and external influences. The focus of internal factors is on an assessment of the compliance system under the Kyoto Protocol within a theoretical context of rational choice versus norm-based compliance theories.

The consideration of external influences includes that of links between the climate change regime and the World Trade Organization, the UN Law of the Sea Convention, International Human Rights Norms, and Multilateral Environmental Agreements. These influences are considered both in terms of their potential contribution to compliance with Kyoto and, more generally, their possible influences on the future evolution of the climate change regime. The book concludes with an assessment of the impact of the Kyoto Protocol on compliance, on the future of the climate change regime, and on the future of international environmental law more generally.

Contents

Chapter 8 **Linking Climate Change and Other MEAs: From Fragmentation to Integration?**

PART III: THE FUTURE

Chapter 9 **The Future of the Climate Change Regime**

Introduction: The Climate Change Regime in Context

In December 1997, the Parties to the United Nations Framework Convention on Climate Change (UNFCCC)[1] successfully negotiated the Kyoto Protocol.[2] It was at the time and still is the first and only international agreement for binding emission reductions designed to mitigate human-induced climate change. Most States signed the Protocol, but initially few ratified it. Seven years later, the Kyoto Protocol finally came into force in February 2005 with the ratification by the Russian Federation.

Ratification by Russia had become essential after the United States declared in 2001 that it was not prepared to join the Kyoto process. It required States, representing 55% of developed country emissions, to ratify in order for the Protocol to come into force.[3] With the US representing 36.1% and Russia representing 17.4%, the Protocol could not come into force until at least one of them ratified. Russia deposited its ratification in November 2004, clearing the way for Kyoto to come into force on February 16, 2005.

There has been considerable debate over Kyoto since 1997. When the Kyoto Protocol was signed in December of 1997, it was heralded by some as a landmark in negotiations on climate change[4] and multilateral environmental agreements more generally. It included, for the first time, binding targets for the reduction of greenhouse gases. This is particularly noteworthy given that the Protocol com-

---

1 United Nations Framework Convention on Climate Change, Intergovernmental Negotiating Committee for a Framework Convention on Climate Change OR, 5th Sess., Annex, UN Doc. A/AC.237/18 (PartII)/Add.1 (1992), 31 I.L.M. 849, online: UNFCCC <http://unfccc.int/resource/docs/a/18p2a01.pdf> [UNFCCC or The Framework Convention].

2 Conference of the Parties to the Framework Convention on Climate Change: Kyoto Protocol, 10 December 1997, U.N. Doc. FCCC/CP/1997/L.7/add. 1, 37 I.L.M. 22 (1998), [hereinafter the Kyoto Protocol].

3 See The Kyoto Protocol, *supra* note 2, art. 25. The U.S. represents 36.1% of Annex I emissions, Russia about 17.4%, Japan about 9%, Australia 2%, Canada 3%, and most of the rest split between the EU and Eastern European countries. This means, once the Protocol was ratified by the EU and Eastern Europe, Russia and Japan were key to the Kyoto Protocol coming into force. The data used here is based on information in the first national communications of Parties under the UNFCCC, submitted on or before December 11, 1997, and used in the negotiations on Article 25. For the status of ratification, see http://unfccc.int/.

4 See Roger Ballentine, "Kyoto an Important First Step in Fighting Warming (Global Warming)" (2000) 17:3 The Environmental Forum 39, and Thomas Richichi, "Although storm clouds threatened throughout the global warming conference in Kyoto, the conferees reached an agreement on greenhouse gas emissions" (1997) 20 Nat'l L.J. B4.

mitted Parties to the pursuit of an environmental objective that would have an impact on almost every aspect of life.

Kyoto was seen as fundamentally different from other agreements, such as the Montreal Protocol on ozone-depleting substances[5], in that it tackled an environmental issue for which there were no obvious technological fixes. It was considered unlikely that the climate change issue could be fully addressed without confronting the emerging reality that development had become unsustainable. Compliance with these targets was expected to create significant net short-term economic costs for developed countries. In addition, the Kyoto Protocol was negotiated at a time when the scientific debate on the results of anthropogenic emissions was ongoing,[6] indicating some recognition of the precautionary principle introduced in the Rio Declaration.[7]

---

5 Montreal Protocol on Substances that Deplete the Ozone Layer, 16 September 1987, amended at London on 29 June 1990, amended at Copenhagen on 25 November 1992, amended at Vienna in 1995, amended at Montreal on 17 September 1997, and amended at Beijing on 3 December 1999, 1522 U.N.T.S. 3, Can. T.S. 1989 No. 42, 26 I.L.M. 1550 (entered into force 1 January 1989), online: United Nations Environment Programme <http://www.unep.org/ozone/pdf/Montreal-Protocol2000.pdf>.

6 At the same time, climate change had even by 1997 been studied more than most other environmental issues. There is general recognition that our climate system is so complex that accurate predictions on the precise impact of GHG emissions will not be possible for the foreseeable future in spite of an incredible amount of scientific research on this issue. For a comparison of the state of the science between 1997 and 2001, compare the Second and Third Assessment Report Series of the IPCC: IPCC Second Assessment Climate Change 1995: A Report of the Intergovernmental Panel on Climate Change (Cambridge, England: Cambridge University Press, 1996); John T. Houghton *et al*, eds., *Climate Change 1995: The Science of Climate Change: Contribution of Working Group I to the Second Assessment of the Intergovernmental Panel on Climate Change* (Cambridge, England: Cambridge University Press, 1996); Robert T. Watson, Marufu C. Zinyowera & Richard H. Moss, eds., *Climate Change 1995: Impacts, Adaptations and Mitigation of Climate Change: Scientific-Technical Analyses: Contribution of Working Group II to the Second Assessment of the Intergovernmental Panel on Climate Change* (Cambridge, England: Cambridge University Press, 1996); James P. Bruce, Hoesung Lee & Erik F. Haites, eds., *Climate Change 1995: Economic and Social Dimensions of Climate Change: Contribution of Working Group III to the Second Assessment of the Intergovernmental Panel on Climate Change* (Cambridge, England: Cambridge University Press, 1996); Robert T. Watson *et al*, eds., *Climate Change 2001: Synthesis Report: A contribution of Working Groups I, II and III to the Third Assessment Report of the Intergovernmental Panel on Climate Change* (Cambridge, England: Cambridge University Press, 2002); John T. Houghton *et al*, eds., *Climate Change 2001: The Scientific Basis: Contribution of Working Group I to the Third Assessment Report of the Intergovernmental Panel on Climate Change* (IPCC) (Cambridge, England: Cambridge University Press, 2002); James J. McCarthy *et al*, eds., *Climate Change 2001: Impacts, Adaptation & Vulnerability: Contribution of Working Group II to the Third Assessment Report of the Intergovernmental Panel on Climate Change* (IPCC) (Cambridge, England: Cambridge University Press, 2002); Bert Metz *et al*, eds., *Climate Change 2001: Mitigation: Contribution of Working Group III to the Third Assessment Report of the Intergovernmental Panel on Climate Change* (IPCC) (Cambridge, England: Cambridge University Press, 2002).

7 *Rio Declaration on Environment and Development*, UN CEDOR, Annex, Agenda Item 21, UN Doc. A/CONF.151/26/Rev.1 (1992) at principle 15 [Rio Declaration].

Others have been more critical of the Protocol, especially with respect to the relatively modest targets set, and the flexibility mechanisms included that will allow countries to offset emissions in their own countries by funding reductions elsewhere.[8] Much of the criticism relates to what has taken place since December 1997. First of all, the Montreal Protocol had set a precedent in international environmental law of quick ratification and amendments in response to new scientific information to make the agreement more effective in reaching its environmental objective.[9] This has not happened with the Kyoto Protocol. On the contrary, it took over seven years for it to even come into force.[10] Furthermore, the agreement reached in Kyoto is now seen as having been significantly watered down since 1997,[11] while the science on the cause and effect of climate change has become stronger.[12]

In short, criticisms have ranged from concerns that Kyoto is an inadequate response to a serious threat to nature and to human survival, to the view that Kyoto will cause unnecessary economic and social harm. The entry into force of the Kyoto Protocol provides an opportunity to reflect on these debates over its significance. Kyoto is now destined to be the focal point of the effort of the international community to cooperatively address climate change. It is also likely to have a significant influence over domestic action on climate change, particularly in high emission, developed countries such as Canada, the European Union, and Japan.[13]

---

[8] See, for example David Victor, *The Collapse of the Kyoto Protocol, and the Struggle to Slow Global Warming*, (Princeton, Princeton University Press, 2001), and Chris Rolfe, "Kyoto Protocol to the United Nations Framework Convention on Climate Change: A Guide to the Protocol and Analysis of its Effectiveness" (1998), online: West Coast Environmental Law Association <http:www.wcel.org/wcelpub/1998/12152.html>.

[9] See Elizabeth R. DeSombre, "The Experience of the Montreal Protocol: Particularly Remarkable, and Remarkably Particular" (2001) 19 UCLA J. Envtl. L. & Pol'y 49. See also Richard E. Benedick, *Ozone Diplomacy: New Directions in Safeguarding the Planet 2d ed.* (Cambridge: Harvard University Press 1998), and Winfried Lang, "Is the Ozone Depletion Regime a Model for an Emerging Regime on Global Warming?" (1991) 9 UCLA J. Envtl. L. & Pol'y 161.

[10] See The Kyoto Protocol, *supra* note 2, art. 25 for the ratification formula for the entry into force of the Protocol.

[11] For copies of a variety of position papers by the Climate Action Network and its member organizations, criticizing decisions made on how Kyoto would be implemented, see the CAN website: <http://www.climatenetwork.org/>. One issue of particular note in this regard has been the use of land use change and forestry activity under Article 3.4 of the Kyoto Protocol. Other areas of contention have been the rules for emissions trading and the Clean Development Mechanism.

[12] See discussion of the science of climate change in section 2 of Chapter 2.

[13] See, for example, section 7 of Chapter 8 for an assessment of Canada's Kyoto implementation plan. For an overview of the current EU approach to climate change, see "Winning the Battle Against Global Climate Change", Commission of the European Communities, Brussels, 9 February 2005 {SEC(2005) 180} online: <http://72.14.207.104/search?q=cache:crDpXngSfK8J:europa.eu.int/comm/environment/climat/pdf/comm_en_050209.pdf/>.

One part of the debate over Kyoto has been based on concerns that we do not know enough about the science of climate change to justify the level of effort Kyoto requires. This concern has lost considerable credibility over the past decade. There is now overwhelming scientific evidence that climate change is taking place and that is caused mainly by the release of greenhouse gases from human activity. The debate about the science is therefore not considered here.[14] The more dominant and credible debate in recent years has been over the adequacy of the Kyoto Protocol in addressing climate change on the one hand, and the level of effort and potential long-term influence of Kyoto on global cooperation on climate change on the other hand. Beyond this, the debate has extended to the impact of Kyoto on the future of international environmental law and compliance in international law.[15]

With the entry into force of the Kyoto Protocol, the climate change regime is at a key juncture in its evolution. Negotiations on commitments beyond 2012 have been under way informally for some time, but are expected to intensify between 2005 and 2008. In recent years, the science on climate change is pointing more and more convincingly to the need for significant further reductions in the near future. Estimates for reductions needed by 2050 to meet the objective of preventing dangerous interference with the global climate system range from 50 to 80% below 1990 levels of emissions.[16]

In the meantime, even the modest Kyoto obligations on developed States to reduce GHG emissions by about 5% by 2012 are expected to have considerable economic and social consequences. Parties with emission reductions obligations are now entering the final stages of domestic emission reduction efforts to meet those obligations. This raises important questions about the level of compliance

---

[14] See, for example, Petr Chylek & Glen Lesins, eds., 1st International Conference on Global Warming and the Next Ice Age, (Conference Proceedings, Halifax, Nova Scotia, Canada, 19-24 August 2001) (Halifax: Dalhousie University, 2001). See also Henry R. Linden, "CO(2) does not pollute: but Kyoto's demise won't end debate" Public Utilities Fortnightly 139:10 (May 15, 2001) 22, and Bruce Pardy "The Kyoto Protocol: Bad News for the Global Environment" (2004) 14 JELP 27.

[15] See, for example, Jutta Brunnée, "The Kyoto Protocol: A Testing Ground for Compliance Theories?" (2003) 63:2 Heidelberg J. of Int'l. L. 255.

[16] See, for example, Government of Canada, Climate Change Plan for Canada, (Ottawa: Government of Canada, 2002) The plan was further supplemented with Government of Canada Project Green, Moving Forward on Climate Change, A Plan for Honouring our Kyoto Commitment (Ottawa, Government of Canada, 2005) at 3, both available at <http://www.climatechange.gc.ca>. The Council of the European Union on October 17, 2002 at its 2,457th meeting in Luxembourg adopted conclusions in preparation for COP 8 that included a recognition that reductions of 70% below 1990 levels are likely to be required. For a copy of the press release, see http://ue.eu.int/ueDocs/cms_Data/docs/pressData/en/envir/72808.pdf. See also Department of Trade and Industry (UK), Energy White Paper: Our Energy Future–Creating a Low Carbon Economy (Norwich: The Stationary Office, 2003), online: DTI Energy Group <http://www.dti.gov.uk/energy/whitepaper/ourenergyfuture.pdf>.

with the Kyoto obligations to be expected and the future of the regime more generally.[17]

With every passing year since the negotiation of the UNFCCC in 1992, the stakes have continued to rise. On the one hand, it is becoming clearer that achieving emission reductions will be difficult. Emissions in most countries have continued to rise since 1990, in some developed States by as much as 20%. On the other hand, it is also becoming clearer that inaction is not an option. The science on climate change has developed significantly over time. With it the level of certainty has grown that human activities are the primary cause of global climate change; that those changes will be largely harmful and that they may become irreversible.

It is at this key stage in the development of the climate change regime, that this book asks the following questions to determine the state of the regime and the prognosis for its future development:

- How effective is the compliance system under the Kyoto Protocol likely to be in encouraging States to meet their obligations at a potentially high short term economic and social cost?
- What influence are other key international regimes, norms, and obligations likely to have on State compliance with the Kyoto obligations?
- What influence are other key international regimes, norms, and obligations likely to have on the future evolutions of the climate change regime?
- What influence is compliance with the Kyoto obligations likely to have on the future of the climate change regime?
- What influence is the climate change regime likely to have on the future of international environmental law?

These questions are addressed in three parts. Part I provides an assessment of the climate change regime, starting in Chapter 2 with an overview of the state of the science, and an analysis of the heart of the regime, the Kyoto Protocol. In this context, Chapter 2 considers how far the Kyoto Protocol takes the international community substantively toward an effective climate change regime. This is followed in Chapters 3 and 4 with a detailed assessment of the compliance regime under Kyoto, based on the assumption that at this stage of the development, compliance with the first commitment period obligations under the Kyoto Protocol is an essential first step toward an effective climate change regime.

---

[17] See section 2 of Chapter 4 for a more detailed discussion of the importance of compliance with the Kyoto obligations for the future of the climate change regime, for compliance in international law, and for international environmental law more generally.

In Part II, the focus shifts from an assessment of the regime itself to a consideration of some key external influences on the regime generally and compliance with obligations in the Kyoto Protocol more specifically. This part considers the relationship between the climate change regime and a number of other potential international law influences on state behaviour. They include the World Trade Organization (WTO), the United Nations Convention on the Law of the Sea (UNCLOS), International Human Rights norms, and Multilateral Environmental Agreements (MEAs) such as the Convention on Biological Diversity (CBD).

Finally, Part III takes a look forward. Based on the conclusions reached in Part II, it considers the future of the climate change regime, compliance, and the potential influence of the climate change regime on international environmental law more generally. Specifically, Chapter 9 assesses proposals for the development of the regime post-2012, and considers the main challenges ahead in moving toward effective GHG emission reductions globally. The Epilogue provides some initial thoughts on how the climate change regime may influence international environmental law more generally. Will MEAs continue to play a central role in international environmental law? How do States ensure that an integrated approach to environmental challenges is taken? How can international law begin to keep up with the accelerating rate of environmental degradation?

The international climate change regime, of course, has not evolved in a vacuum. Rather, it is a product of decades of efforts toward more effective international cooperation on environmental issues. Given the profile of the climate change regime, and the general recognition that global cooperation on environmental issues has not been effective to date in solving a series of global, regional and local environmental challenges,[18] it is not surprising that the success or failure of the climate change regime is expected to have a significant effect on the future evolution of international environmental law.

It has become clear over the past three decades that the traditional concept of state sovereignty is fundamentally incompatible with many environmental objectives, given that more and more impacts of decisions made by States are felt outside state borders in terms of environmental degradation of some form.[19] To put it another way, only by convincing States to give up state sovereignty when it comes to the right to make decisions that have environmental impact beyond State borders, can we hope to address many of the environmental challenges we currently face.

---

[18] See Robert T. Watson, *Millennium Ecosystem Assessment Synthesis Report*, pre-publication draft (Washington: Island Press, 23 March 2005) [due for publication in September 2005] online: <http://www.millenniumassessment.org/en/index.aspx>.

[19] *Ibid.* at 56, 114.

Various attempts to generically pierce through state sovereignty through some form of or world governance of the environment have failed to date.[20] Other efforts have included reliance on customary international law principles of state responsibility, soft law agreements, single issue treaties and international regimes tasked with coordinating international efforts on particular issues. With respect to climate change, the international community has followed a recent trend toward a regime-based approach to international cooperation.[21]

This book is built on the premise that environmental problems with global impacts such as climate change can only be solved through global cooperation. If it is generally recognized that the global community stands to benefit from addressing the problem, and that the global community is seriously threatened if the problem cannot be solved, one would expect States to be motivated to give up some sovereignty to address climate change, as long as they are assured that others will do the same.[22]

The level of assurance that other States will cooperate and the impact of that sense of assurance on the development and implementation of international environmental law is thought to depend on two basic factors, acceptance of commitments and obligations in agreements (throughout negotiation and implementation), and the effectiveness of compliance measures to ensure that States meet their obligations and fulfill their commitments.

The first seeks to align, in some way, self-interest with the global interest, and address issues of fairness of relative obligations in an overall effort to ensure all Parties are sufficiently motivated to comply with commitments taken on. The second seeks to ensure that the house of cards that is built on the motivation of each party does not collapse when, either motivation or intervening circumstances pose a challenge to a party's compliance and thus threaten to alter the interconnection of self and global interests that lead to the global commitment to address the challenge in question.

In this regard, from an international environmental law perspective, all eyes are on Kyoto at the moment, particularly with respect to compliance. The obli-

---

[20] For an assessment of global governance options, see: J. Hierlmeier, "UNEP: Retrospect and Prospect – Options for Reforming the Global Environmental Governance Regime" (2002) 14 Geo. Int'l Envtl. L. Rev. 767, and M. A. Drumbl, "Northern Economic Obligation, Southern Moral Entitlement, and International Environmental Governance" (2002) 27 Colum. J Envtl. L. 363.

[21] For a detailed analysis of the emergence of regimes in international law, see, for example, S. D. Krasner, ed., *International Regimes*, (Cornell University Press, 1983).

[22] In fact, this may be recognized as gaining rather than giving up sovereignty, in that international cooperation may be the only way a given State can gain control over the mitigation of climate change impacts within its boundaries.

gations under Kyoto are set, and the Protocol has entered into force. The equity of those obligations will undoubtedly be the subject of ongoing debates for years to come, and with it the question of how effective the substance of Kyoto has been at motivating States to ratify and implement their commitments. What is left is the question of whether States will meet their commitments and obligations.[23] The stakes, environmentally, economically and in terms of equity, are arguably higher than they have been for any other multilateral environmental agreement (MEA).

If regime building is to become an effective alternative or comlement to world governance, States need to know that compliance will be high even when the stakes are high. We have learned from the Montreal Protocol on Ozone Depleting Substances that compliance with MEA obligations can be difficult, but the economic stakes are considerably lower than is the case with climate change.[24] Kyoto is the testing ground for a new level of cooperation on environmental issues. It will give us further insight into the extent to which States can rely on others to take environmental obligations seriously, and where environmental concerns rank in priority.

It is to this end that this book considers the effectiveness of various compliance tools generally, the compliance system under the Kyoto Protocol more specifically, and other opportunities in international law to motivate States to take action on climate change. The consideration of these various influences on compliance with Kyoto are considered in two broad categories; internal and external. Internal influences refer to those within the climate change regime; external to those outside the regime.

In the first Chapter, the historical context within which the climate change regime evolved is briefly explored. Specifically, the development of Multilateral Environmental Agreements, soft law principles and customary law at the international level are considered as major influences on the climate change regime.

---

[23] See Section 1 of Chapter 3 for a more detailed discussion of the terms, obligations and commitments and how they are used here. See section 1 of Chapter 4 for a discussion of the importance of compliance for the future of the climate change regime.

[24] See Elizabeth R. DeSombre, *supra* note 9.

PART I

THE STATE OF THE CLIMATE CHANGE REGIME

In this Part, the current state of the climate change regime is assessed. Chapter 2 focuses on the substantive obligations imposed on States through a combination of the UNFCCC, the Kyoto Protocol, and a number of subsequent agreements on how to implement the Protocol for the 2008 to 2012 commitment period. Chapter 3 sets the stage for an assessment of the compliance system under the Kyoto Protocol, by providing the theoretical foundation on compliance in international law. This is followed in Chapter 4 with a detailed analysis of the compliance system under Kyoto.

1

The International Environmental Law Context

1. INTRODUCTION

For any thorough analysis of the climate change regime and its role in international environmental law,[1] it must be considered in the context of the evolution of international environmental law. A comprehensive history of environmental law generally or international environmental law more specifically is not warranted here. A brief overview of this evolution over the course of the 20th century, however, should help to ensure the climate change regime and the issue of compliance are considered in this historical context. It will further facilitate the assessment of the importance of the climate change regime for the future evolution of international environmental law in the Epilogue. The focus in this Chapter is therefore on providing a brief history of international environmental law, and to consider the particular influence of the ozone layer depletion (OLD) regime on the climate change regime.

The start of the modern day environmental movement and with it the birth of environmental law in North America is generally considered to be the publication of *Silent Spring* by Rachel Carson.[2] Concerns about the environmental impact of human activity were raised in Europe around the same time and similarly led to development of domestic environmental laws in Western European countries. While both the quantity and the nature of legal activity with respect to environmental issues have evolved significantly domestically and internationally since the publication of *Silent Spring*, the roots of international environmental law actually go back much further.[3] The following is a brief overview of its evolution at the international level in three stages, international environmental law before

---

[1] For a general overview of the history of international environmental law, see Philippe Sands, ed., *Greening International Law* (New York: New Press, 1994), and Edith Brown Weiss, "International Environmental Law, Contemporary Issues and the Emergence of a New World Order" (1993) 81 Geo. L. J. 675.

[2] Rachel Carson, *Silent Spring* (London: H. Hamilton, 1963). For a history of the environmental movement and the evolution of human impact on nature, see:J. R. McNeil, *Something New Under the Sun:An Environmental History of the Twentieth-Century World* (New York: W. W. Norton & Co., 2000).

[3] See Edith Brown Weiss *et al*, *International Environmental Law: Basic Instruments and References* (Salem, NH: Butterworth, 1992).The second edition covers instruments negotiated during the period from 1992 until 2000.

the Stockholm Conference on the Human Environment in 1972, from Stockholm to Rio[4] in 1992, and post-Rio.

2. PRE-STOCKHOLM

At the beginning of the 20th century, there were relatively few international agreements on environmental issues in place, and those that dealt with environmental protection in some form were concerned with very specific and short-term human benefits to be preserved or protected through limited international cooperation. These agreements, negotiated in most cases on a bilateral, sometimes regional basis, were entered into in the context of an acceptance of unrestrained national sovereignty over natural resources. Substantively, they tended to deal with shared waterways and other instances where Parties identified a need to protect national interests in certain natural resources of commercial importance. Examples include agreements dealing with the Rhine River, bilateral fisheries issues, and fur seals.[5] These agreements tended to address access to resources rather than concern about the protection of the resources from pollution or other human interference.

In the early 20th century, some agreements started to address conservation issues, even if in the specific context of a motivation to protect commercially important species from immediate harm. A typical example would be the 1902 *Convention for the Protection of Birds Useful to Agriculture*.[6] The Convention only dealt with birds which directly preyed on insects that were considered to be a threat to agriculture. There is no indication of a broader consideration of the importance of preserving a healthy habitat for these birds, or the relationship between ecosystem health and the objective of preserving the birds for their pest control function.

The 1930s and 1940s brought a gradual consideration of broader conservation issues in the context of the negotiation of international environmental agreements. This period saw the negotiation of agreements on the preservation of fauna and flora in their natural state, on nature and wildlife preservation, and more specific

---

[4] *Rio Declaration on Environment and Development*, UN CEDOR, Annex, Agenda Item 21, UN Doc. A/CONF.151/26/Rev.1 (1992) [hereinafter Rio Declaration].

[5] *Treaty Between the United States and Great Britain Providing for the Preservation and Protection of Fur Seals*, 7 February 1911, United States-Great Britain, 37 Stat. 1538.

[6] *Convention for the Protection of Birds Useful to Agriculture*, 19 March 1902, 102 B.F.S.P. 969, 191 Cons. T.S. 91(entered into force May 11, 1907).

agreements such as the International Whaling Convention[7] and other agreements on fisheries and land based species.[8]

The creation of the United Nations (UN) in 1945 marked a turning point in international law. The full impact of the United Nations on international environmental law, however, would not be felt for some time to come. The environment was not a focus of the United Nations during this period from 1945 until the late 1960s. At the same time, a number of organizations created under the UN umbrella did have some limited mandate to address environmental issues. They included the Food and Agriculture Organization (FAO), the United Nations Educational, Scientific and Cultural Organization (UNESCO), the World Health Organization (WHO), the International Maritime Organization (IMO) and the General Agreement on Tariffs and Trade (GATT).[9]

The UN did hold its initial conference on the environment as early as 1949, even if only in the limited context of conservation and utilization of resources. This was followed by a conference initiated by the UN General Assembly in 1954 on the Conservation of the Living Resources of the Sea leading to the 1958 Geneva Conventions. Other issues with environmental implications addressed through international treaties during this period included atomic energy, nuclear arms, and agreements on oil pollution, high seas fishing and wetlands.[10]

The focus during this period of international environmental law was to react to specific issues of concern to individual or groups of nations. Efforts tended to be isolated, reactive, and lacking consistent structure and approach. There is little indication that environmental issues addressed were considered in any broader context. Little attention was paid to linkages among environmental issues, trends that might suggest underlying problems in the direction taken by humankind, questions about overall limits to the ability of nature to absorb the impact of human activity, north/south equity issues, or intergenerational equity considerations.

---

[7] *Convention between the United States of America and other Powers for the Regulation of Whaling*, 24 September 1931, 49 U.S. Stat. 3079, 155 L.N.T.S. 349.

[8] See, for example, Douglas M. Johnston, "International Environmental Law: Recent Developments and Canadian Contributions" in R. St. J. Macdonald, Gerald L. Morris, and Douglas M. Johnston, eds., *Canadian Perspectives on International Law and Organization*, (Toronto: University of Toronto Press, 1974).See also Douglas Johnston, ed., *The Environmental Law of the Sea* (Gland, Switzerland: IUCN, 1981).

[9] For a general discussion of the role of these international institutions in international environmental law, see Patricia Birnie *et al*, *International Law & the Environment*, 2nd ed. (Oxford: Oxford University Press, 2002), 34-38.

[10] *Treaty on the Non-Proliferation of Nuclear Weapons*, 1 July 1968, 21 U.S.T. 483, (entered into force 5 March 1970), *Agreement for Cooperation in Dealing with Pollution of the North Sea by Oil*, 9 June 1969, 704 U.N.T.S. 3, *Convention on Wetlands of International Importance Especially as Waterfowl Habitat* (RAMSAR), 2 February 1971, 996 U.N.T.S. 245, 11 I.L.M. 963.

3. FROM STOCKHOLM[11] TO RIO

The year 1972 is generally recognized as the starting point for contemporary international environmental law. Not only was it the year of the United Nations Stockholm Conference on the Human Environment, but a number of issue-specific treaties were negotiated at or around the same time.[12] More importantly, Stockholm marked the beginning of an explosion of international treaty making to address specific environmental issues for which international cooperation was deemed essential. Inspired by the birth of environmental movements in North America and Europe in the 1960s, and facilitated through the United Nations, the 1972 Conference on the Human Environment was the first global conference to consider environmental issues from a broader perspective.

The conference produced a number of non-binding instruments, and led to the creation of the United Nations Environment Programme (UNEP). It also provided the context for the negotiation of a number of binding multilateral treaties, such as CITES,[13] the London Ocean Dumping Convention, the World Heritage Convention, and the first regional seas convention.[14] Most of the treaties between 1972 and 1992 were still stand alone treaties. This period, however, did mark the beginning of the framework convention-protocol approach which has become so dominant in international environmental law in the last two decades. The first such agreement was the *Barcelona Convention for the Protection of the Mediterranean Sea Against Pollution* signed in 1976.[15]

In addition to the sheer volume of international agreements dealing with environmental matters inspired by the Stockholm Conference, the 1972 conference is considered a turning point in the development of international environmental law in two other respects, the creation of an international structure to champion

---

[11] *Declaration of the United Nations Conference on the Human Environment*, 16 June 1972, UN GAOR, U.N. Doc. A/CONF.48/14/Rev.1 (1973), 11 I.L.M. 1416, online: United Nations Environment Programme <http://www.unep.org/Documents/Default.asp?DocumentID=97&ArticleID=1503>, [hereinafter Stockholm Declaration]. See also: *Report on the United Nations Conference on the Human Environment at Stockholm: Final Documents*, 16 June 1972, 11 I.L.M. 1416.

[12] Most notably, the *Convention on International Trade in Endangered Species of Wild Fauna and Flora* (CITES), 3 March 1973, 27 U.S.T. 1087, and the *Convention on the Prevention of Marine Pollution by Dumping of Wastes and Other Matter*, 29 December 1972, 26 U.S.T. 2403, 1046 U.N.T.S. 120.

[13] *Ibid.*

[14] See Edith Brown Weiss, *supra* note 3, for a chronology of significant international environmental agreements and instruments.

[15] *Barcelona Convention for the Protection of the Mediterranean Sea Against Pollution*, 16 February 1976, 15 I.L.M. 290.For a recent article on the convention, see Jamie Benedickson, "The Great Lakes and the Mediterranean Sea: Ecosystem-Management and Sustainability in the Context of Economic Integration" (2004) 14 J. Envtl. L. & Prac. 107.

environmental issues in the form of UNEP, and the first attempt at an overall (non-binding) substantive context for the development of international environmental law. The substantive context was provided through the Stockholm Declaration in the form of twenty-six guiding principles and supplemented through an action plan with over a hundred recommendations for international action on a range of environmental issues.[16]

The focus of the twenty-six guiding principles is anthropocentric, and it is very much based on the concept of state sovereignty over the use of nature within its borders, assuming that areas of incompatibility of state sovereignty and state responsibility for transboundary harm will be limited and manageable through specific agreements on an issue by issue basis. Of particular note in this regard is Principle 21, perhaps the most commonly referenced part of the Stockholm Declaration, which provides as follows:

> States have, in accordance with the Charter of the United Nations and the principles of international law, the sovereign right to exploit their own resources pursuant to their own environmental policies, and the responsibility to ensure that activities within their jurisdiction or control do not cause damage to the environment of other states or of areas beyond the limits of national jurisdiction.

Principle 22 actually takes the issue of responsibility to other States one step further by recognizing, in general terms, the concept of liability and compensation for damage caused outside a State's border. The development of international law to deal with liability and compensation issues is identified for future negotiation. In hindsight, it seems that these two principles were negotiated with an expectation that human activities with extraterritorial environmental impact would be the exception rather than the rule, and an assumption that specific incidents of extraterritorial impact of human activity within the borders of a State could be dealt with either on a case-by-case basis through the development of issue-specific treaties, or through the development of general principles of liability and compensation. Clearly, these expectations have proven overly optimistic, and as a result neither of the two options for dealing with environmental challenges internationally have proven to be realistic or effective to date.

At the same time, the declaration contains perhaps less well known language that suggests in hindsight a surprisingly enlightened[17] and visionary understanding of our relationship to nature. Paragraph 1 of the preamble, for example, reads as follows:

> Man is both creature and moulder of his environment, which gives him physical sustenance and affords him the opportunity for intellectual, moral, social and spiritual

---

[16] See Stockholm Declaration, *supra* note 11.

[17] Other than the lack of gender sensitivity in the wording of this provision.

> growth. In the long and tortuous evolution of the human race on this planet a stage has been reached when, through the rapid acceleration of science and technology, man has acquired the power to transform his environment in countless ways and on an unprecedented scale. Both aspects of man's environment, the natural and the man-made, are essential to his well being and the enjoyment of basic human rights – even the right to life itself.

Other provisions of the Stockholm Declaration suggest a recognition that a healthy natural environment is a precondition for human well being, and a recognition that quality of life and other goals of humanity depend upon cooperation with nature rather than a domination of nature.[18] The declaration made repeated reference to the need to address the challenges ahead in a spirit of cooperation between developed and developing countries by addressing the inequities suffered by much of the human population living in developing countries.[19] The concept of intergenerational equity was also introduced.[20]

Another principle introduced in the Stockholm Declaration of particular relevance here is the principle of integration. Principle 13 states: "In order to achieve a more rational management of resources and thus to improve the environment, States should adopt an integrated and co-ordinated approach to their development planning so as to ensure that development is compatible with the need to protect and improve the environment for the benefit of their population".[21] The principle of integration was so new and foreign to States, that it took 20 years for this principle to re-emerge in Rio and be taken seriously by the global community.

As will be discussed in Chapters 8 to the Epilogue, it is clear, in the climate change context at least, that States are still struggling to implement the concept of integration in an effective manner.[22] Domestically, one of the contributing factors to the failure to integrate effectively is arguably the creation of environment departments in many States. Ironically, the creation of these departments also goes back to the Stockholm Declaration, which urges States to institutionalize environmental protection. It was this reference to the need for domestic institutions that resulted in the establishment of departments of the environment in many States. The creation of these departments has in turn made domestic integration

[18] See Stockholm Declaration, *supra* note 11, Preamble (paragraph 6), and Principle 5.

[19] *Ibid.* Principles 11, 12.

[20] *Ibid.* Principle 2.

[21] *Ibid.* Principle 13.

[22] On a related point, the period from Stockholm to Rio saw a transition in MEAs from sectoral agreements to agreements that cut across a range of sectors.The three Rio Conventions, the Convention to Combat Desertification, the Convention on Biological Diversity, and the UNFCCC, are all examples of this trend toward broader, framework law making.This in turn, as discussed in Chapters 8 to the Epilogue, has challenged States to tackle integration both internationally and domestically.

more difficult by isolating environmental concerns from more mainstream concerns such as health, education, and economic development.

As mentioned, another development in 1972 that qualifies as a turning point in the evolution of international environmental law is the creation of the United Nations Environment Programme (UNEP).[23] The establishment of UNEP by the General Assembly[24] of the United Nations was in direct response to the recommendations contained in the action plan that accompanied the Stockholm Declaration. The initial mandate of UNEP was to oversee and guide the implementation of the action plan agreed to in Stockholm. UNEP was to do this by serving as a catalyst for international coordination of national action, but without any formal authority to implement in the absence of international consensus. This has been followed up with more specific mandates assigned to UNEP by the General Assembly from time to time.[25] Structurally, UNEP is run by a governing council and an executive director, as well as a secretariat to carry out the instructions of the executive director. UNEP is funded mainly through direct voluntary payments from member States rather than through the UN general budget.

While an overall assessment of UNEP is beyond the scope of this book, it is fair to conclude that UNEP has played a key role in raising awareness of environmental issues in the international community since 1972, and was a catalyst for many of the multilateral environmental agreements entered into since 1972. While the influence of UNEP over the substance of these agreements is very much open to debate, it seems clear that UNEP was responsible for initiating the negotiations, and was instrumental in ensuring that the principles of the Stockholm Declaration were considered in the negotiation process.

In addition, the period from 1972 to 1992 saw the emergence of the international environmental community as an influential player in the development of international environmental law. In particular, a wide range of international environmental non-governmental organizations (ENGO's) such as the IUCN, Greenpeace, the World Wildlife Fund, Friends of the Earth and others became active participants in international negotiations, both as observers and as advisors

---

[23] For a more detailed discussion of UNEP, see Annette Petsonk, "The Role of the United Nations Environment Programme (UNEP) in the Development of International Environmental Law" (1990) 5 Am. U.J. Int'l L. & Pol'y 351, Mark Allan Gray, "The United Nations Environment Programme: An Assessment" (1990), 20 Envtl. L. 291, and Dena Marshall, "An Organization for the World Environment: Three Models and Analysis" (2002) 15 Geo. Int'l Envtl. L. Rev. 79.

[24] Patricia Birnie, *supra* note 9, at 48, and 53.

[25] *Ibid.* at 54.See also Dena Marshall, *supra* note 23, at 82, on the 1997 Nairobi Declaration, which provided a renewed mandate for the UNEP.

to State negotiators. Their influence can be traced through to Rio as well as individual MEAs such as the UNFCCC.[26]

Substantively, the period from Stockholm in 1972 to Rio in 1992 saw an explosion of the number of international instruments dealing with environmental issues. Examples include the 1979 *Convention on Long-Range Transboundary Air Pollution*,[27] the *Law of the Sea Convention* of 1982[28], the *Vienna Convention on the Protection of the Ozone Layer*,[29] the *Basel Convention on the Transboundary Movements of Hazardous Wastes and their Disposal*,[30] and the *Protocol on Environmental Protection under the Antarctic Treaty*,[31] to name a few. These agreements demonstrate the complexity of the environmental issues facing the global community, the increasing number of environmental issues reaching a crisis level, and the unpreparedness of the international community to address these issues in a timely, effective, and coordinated manner.

Almost without exception, the problems tackled in these agreements are still with us today. Furthermore, in the process of trying to address some of these problems, others have been created or exacerbated. Finally, most have been addressed without making any real progress on the fundamental issues raised but not resolved in the Stockholm Declaration, such as the conflict between state sovereignty and extraterritorial responsibility,[32] the connection between environmental protection and equity, and the implication of the connections among economic development, quality of life, and protection of nature for the development and implementation of international environmental law.

At least some of these failures seem to have been recognized by UNEP and the world community in the years leading up to the 20th anniversary of Stockholm,

---

[26] For an assessment of the influence of ENGOs on the climate change negotiations, see Michele Betsill, "Environmental NGOs Meet the Sovereign State: the Kyoto Protocol Negotiations on Global Climate Change" (2002) 13 Colo. J. Int'l Envtl. L. & Pol'y 49; Richard A. Rinkema, "Environmental Agreements, Non-State Actors, and the Kyoto Protocol: A 'Third Way' for International Climate Action?" (2003) 24 U. Pa. J. Int'l Econ. L. 729; and A. Rest, "Enhanced Implementation of International Environmental Treaties by Judiciary – Access to Justice in International Environmental Law For Individuals and NGOs: Efficacious Enforcement by the Permanent Court of Arbitration" (2004) 1 Macquarie Journal of International and Comparative Environmental Law 1.

[27] *Convention on Long-Range Transboundary Air Pollution*, 13 November 1979, 18 I.L.M. 1442 (entered into force 16 March 1983).

[28] *United Nations Convention on the Law of the Sea* (UNCLOS), 10 December 1982, 21 I.L.M. 1261.

[29] *Vienna Convention on the Protection of the Ozone Layer*, 22 March 1985, 26 I.L.M. 1529 (entered into force 22 September 1988).

[30] *Basel Convention on the Transboundary Movements of Hazardous Wastes and their Disposal*, 22 March 1989, 28 I.L.M. 657.

[31] *Protocol on Environmental Protection under the Antarctic Treaty*, 21 June 1991, 30 I.L.M. 1455.

[32] Includes both global and intergenerational responsibility.

resulting in the establishment of the Brundtland Commission, the publication of *Our Common Future*,[33] and the follow-up negotiations leading up to the 1992 United Nations Conference on Environment and Development (UNCED) in Rio. In the next Section, the impact of the Rio Conference on international efforts to deal with these challenges will be briefly considered.

4. FROM RIO[34] TO THE PRESENT

The time leading up to UNCED in 1992 was marked on the one hand by another flurry of multilateral environmental agreements, particularly on a number of global environmental challenges. Ozone Layer Depletion,[35] Climate Change[36] and Biodiversity[37] are perhaps the most notable global environmental issues for which agreements were negotiated during this period. On the other hand, the time leading up to UNCED was also marked by efforts to reflect on the evolution of environmental law since Stockholm. The Brundtland Commission with its publication of *Our Common Future* in 1987[38] is generally recognized as the most influential initiative in this regard. Brundtland provided the context for negotiations leading up to Rio, particularly with respect to the negotiations leading to the Rio Declaration and Agenda 21.[39]

Given this, the similarities between the Stockholm and Rio Declarations are surprising, suggesting limited progress over the 20 years in our collective understanding of the problem and possible solutions. A number of key principles from Stockholm, such as principles 21 and 22 are repeated with minor changes in Rio.[40]

---

[33] World Commission on Environment and Development, *Our Common Future* (Brundtland Report) (Oxford & New York: Oxford University Press, 1987).

[34] *Rio Declaration on Environment and Development*, UN CEDOR, Annex, Agenda Item 21, UN Doc. A/CONF.151/26/Rev.1 (1992) [hereinafter Rio Declaration].See also *Report of the United Nations Conference on Environment and Development*, (Rio de Janeiro, 3-14 June 1992) ACONF 151/26 vol. 1, online: United Nations <http://www.un.org/documents/ga/conf151/aconf15126-1annex1.htm>.

[35] See, *supra* note 29. See also *Montreal Protocol on Substances that Deplete the Ozone Layer*, 16 September 1987, amended at London on 29 June 1990, amended at Copenhagen on 25 November 1992, amended at Vienna in 1995, amended at Montreal on 17 September 1997, and amended at Beijing on 3 December 1999, 1522 U.N.T.S. 3, Can. T.S. 1989 No. 42, 26 I.L.M. 1550 (entered into force 1 January 1989), online: United Nations Environment Programme <http://www.unep.org/ozone/pdf/Montreal-Protocol2000.pdf>.

[36] See *United Nations Framework Convention on Climate Change,* Intergovernmental Negotiating Committee for a Framework Convention on Climate Change OR, 5th Sess., Annex, UN Doc. A/AC.237/18 (PartII)/Add.1 (1992), 31 I.L.M. 849, online: UNFCCC <http://unfccc.int/resource/docs/a/18p2a01.pdf> [UNFCCC or The Framework Convention].

[37] See *Convention on Biological Diversity of the United Nations Conference on the Environment and Development*, June 5, 1992, U.N. Doc. DPI/1307, reprinted in 31 I.L.M. 818 [CBD].

[38] Brundtland Report, *supra* note 33.

[39] Rio Declaration, *supra* note 34.

[40] *Ibid.* Principles 1, 13.

The main change in the Rio Declaration is its attempt, following the approach advocated by the Brundtland Commission, to integrate environmental, social equity, and economic issues. Rio arguably advocates, at least in developing countries, development as a precondition for addressing environmental issues. In the process, the concept of sustainable development has all too often become an excuse for the environmental impact of development, rather than a measuring stick to separate sustainable from unsustainable human activity.

Two additions to the Rio Declaration with the potential for a particularly positive contribution to the evolution of environmental law are the precautionary and the polluter pays principles.[41] Both principles have been widely applied and their status under customary international law is being openly debated. Both have influenced the negotiation of specific instruments and the evolution of regimes since 1992, as has the concept of sustainable development and the concept of integrating economic, environmental and equity issues in tackling environmental challenges.

After a flurry of activity leading up to and following UNCED in 1992, developments in international environmental law have been modest in the past decade. While issues such as ozone layer depletion, for example, caught the attention of the international community during this dynamic period leading up to UNCED, issues such as biodiversity and climate change seem to be progressing at a much slower pace since the signing of the respective framework conventions in 1992. The World Summit on Sustainable Development, held in 2002 in Johannesburg, was by most accounts a minute step forward at best.[42]

What has been the impact of UNCED and the Brundtland Commission in the development of international law since 1992? At the heart of *Our Common Future* and Rio was the idea of sustainable development, a concept that seemed to have the potential to break the impasse that had developed between environmental, economic and equity interests. More specifically, following the definition of sustainable development by the Brundtland Commission, it was the concept of meeting the needs of the present without compromising the ability of future generations to meet their needs that was to break this impasse. While still anthropocentric, at least at first glance it appeared to provide a way forward, by asking developed countries to focus on needs and on assistance to developing countries.

---

[41] *Ibid.* Principles 15, 16.

[42] See, for example, Carl Bruch & John Pendergrass, "The Road from Johannesburg: Type II Partnerships, International Law, and the Commons" (2003) 15 Geo. Int'l Envtl. L. Rev. 855, George Pring, "The 2002 Johannesburg World Summit On Sustainable Development: International Environmental Law Collides With Reality, Turning Jo'Burg Into 'Joke'Burg'" (2002) 30 Denv. J. Int'l L. & Pol'y 410, and John C. Dernbach, "Making Sustainable Development Happen: From Johannesburg To Albany" (2004) 8 Alb. L. Envtl. Outlook 173.

One part of this way forward was to make the shift from societies driven by the generation and subsequent satisfaction of wants that have little to do with meaningful progress in an individual or global sense to a focus on needs and on ways to minimize the environmental impact of meeting those needs. The second part is a commitment to assist developing countries in their efforts to meet the needs of their citizens, also with minimal environmental impact. In terms of environmental impact, there were four broad categories to consider, the depletion of resources, the creation of waste, the preservation of biodiversity, and the release of pollution.[43]

Clearly, this shift to needs and to meeting the needs with minimal environmental impact has not taken place in any organized or systematic way at the international level. Instead, the effect of *Our Common Future* and Rio appears to have been a focus on economic development as the driver of environmental protection at some ill-defined point in the future when we are so well off that we will be able to afford environmental protection. Very little seems to have happened since Rio in international environmental law in terms of mandating development to achieve the dual objective of focussing on needs as opposed to wants,[44] and of finding ways to minimize the environmental impact of meeting the needs of present generations.

It is surprising, given the central role of "needs" in most commonly used definitions of sustainable development, that there has not been more of a public debate over the meaning of "needs". Given this, it is not surprising that little has been done to treat wants differently from needs by, for example, requiring development for the purpose of meeting wants to meet a higher test of demonstrating its net positive contribution to sustainability. Rather, it seems to be assumed that generating and meeting "wants" increases short-term economic activity, and that this approach will inevitably improve equity and environmental protection, when clearly, so far at least, the opposite has been true.

Finally, by way of summary, there are a number of principles that have emerged from the evolution of international environmental law to date. First in time, but still very much a factor today in the development and implementation of international environmental law is the concept of state sovereignty over natural resources modified by state responsibility for environmental impact outside its territory as set out in Principle 21 of the Stockholm Declaration.[45] More recently,

---

[43] See, for example, A. Tynberg, "The Natural Step and its Implications for a Sustainable Future" (2000) 7 Hastings W.-Nw. J. Envtl. L. & Pol'y 73, for an elaboration on these four categories of environmental impact of human activity as a basis of guidance toward sustainable development.

[44] See, for example, B. A. Harsh, "Consumerism and Environmental Policy: Moving Past Consumer Culture" (1999) 26 Ecology L.Q. 543, and A. Rochette, "Stop the Rape of the World: An Ecofeminist Critique of Sustainable Development" (2002), 51 U.N.B.L.J. 145.

[45] And more or less reaffirmed in Principle 2 of the Rio Declaration.

this principle has been supplemented with the concept of sustainable development, perhaps the driving force behind the development of international environmental law since 1992. Still emerging to varying degrees in terms of influence over the evolution of international environmental law are the precautionary principle, the polluter/pay principle, and the concept of common but differentiated responsibility.[46]

Other changes in international environmental law over the past decade or so include the changing role of non-governmental actors in the development and implementation of international law, and the emergence of environmental impact assessment, economic instruments, monitoring, reporting, and verification as favoured implementation tools.[47] As the stakes have become higher and the sheer number of obligations has increased, there has also been a renewed interest in the question of compliance. One crucial area identified in Stockholm in which progress has been slow is with respect to the development of international environmental law on liability and compensation for extra territorial environmental damage. While some specific instruments may provide opportunities to establish liability and seek compensation, there is no overall framework for the establishment of liability and compensation as contemplated in Principle 22 of the Stockholm Declaration. In fact, Principle 13 of the Rio Declaration at best mirrors the expressed need to develop international law in this area, without demonstrating any tangible progress.[48]

The results of the World Summit on Sustainable Development (WSSD) meetings in Johannesburg in August 2002 have done little to set a clear direction for international environmental law. Based on the results of the WSSD, there is currently no appetite for global governance on environmental issues. Progress on soft law development internationally appears slow at best. In the meantime, the decade between Rio and Johannesburg saw considerable developments in terms of MEAs dealing with environmental issues on biodiversity, desertification, climate change, ozone layer depletion among others.

Of these MEAs, the regime on ozone layer depletion is generally considered to be the most successful. Not surprisingly, it has served as the template for most of the others, including the climate change regime. In the final part of this Chapter, the influence of the Ozone Layer Depletion (OLD) regime on the international response to climate change is therefore considered.

---

[46] Rio Declaration, *supra* note 34, principles 15, 16, and 7 respectively. See also Alhaji B. M. Marong, "From Rio to Johannesburg: Reflection on the Role of International Legal Norms in Sustainable Development" (2003) 16 Geo. Int'l Envtl. L. Rev. 21.

[47] For a discussion of these issues in the climate change context, see Rinkema and Betsill, *supra* note 26.

[48] For an overview of the work of the International Law Commission on transboundary environmental harm, see Patricia Birnie, *supra* note 9, at 105.See also discussion in Section 4 of the Epilogue.

5. OZONE LAYER DEPLETION: A TEMPLATE FOR CLIMATE CHANGE?

By many accounts the ozone layer regime, developed through the Vienna framework convention and a series of protocols starting with the Montreal Protocol,[49] is considered to be the greatest success of the regime building approach to international environmental law. In fact, it is perhaps the greatest success story for international environmental law to date, in that it appears to have responded to a serious global threat in an effective, coordinated and timely manner. The framework convention was signed in 1985 in Vienna, followed by the Montreal Protocol in 1987.

While the framework convention did not include binding obligations to reduce the use or release of ozone-depleting substances, it provided the substantive and administrative framework for the negotiation of those targets. What followed was a series of agreements in the form of protocols and amendments and adjustments that resulted in the complete phase out of some of the most harmful substances by 1996 in developed countries. In addition, Parties have over time negotiated a growing list of restricted substances to be phased out in developed countries first, with the developing world to follow.

There are still significant challenges ahead for the ozone regime, most notably the implementation of the phase out in developing countries, and the black market in ozone-depleting substances that has developed in some countries as a result of the restrictions imposed through the regime.[50] Nevertheless, the ozone regime has been more successful in addressing the specific threat of ozone layer depletion than could have been expected in 1985, when the framework convention was signed. What is less clear is what impact the implementation of the ozone regime has had or will have on the broader issues of sustainability and equity.

Because of the success of the ozone regime, and the fact that the climate change regime was just a few years behind in gaining international recognition as a global environmental threat,[51] it is not surprising that the ozone regime has become a

---

[49] *Supra* note 35.

[50] For a more detailed discussion of some of the challenges with the OLD regime, see Frederick Pool Landers Jr., "The Black Market Trade in Chlorofluorocarbons: The Montreal Protocol Makes Banned Refrigerants a Hot Commodity" (1997) 26 Ga. J. Int'l & Comp. L. 457, Elizabeth DeSombre & Joanne Kauffman, "The Montreal Protocol Multilateral Fund: Partial Success Story" in Robert O. Keohane & Marc A. Levy, eds., *Institutions for Environmental Aid: Pitfalls and Promise* (Cambridge: The MIT Press, 1996) 89, Jennifer Clapp, "The Illegal CFC Trade: An Unexpected Wrinkle in the Ozone Protection Regime" (1997) 9 International Environmental Affairs 259, and Joel A. Mintz, "Keeping Pandora's Box Shut: A Critical Assessment of the Montreal Protocol on Substances That Deplete the Ozone Layer" (1989) 20 U. Miami Inter-Am. L. Rev. 565.

[51] The UNFCCC was signed in 1992 compared to 1985 for the ozone layer framework convention.

template for climate change. The climate change regime has followed the general approach of setting an administrative and substantive framework for the development of the regime first. This framework provides a platform for the negotiation of substantive obligations that will lead to the implementation of a solution to the environmental challenge in question. While there are a few notable differences, the general approach for the two issues has been remarkably similar.[52] There are also striking similarities between the two issues, at least on the surface. Both issues involve the release of a number of different gases, initially thought to be harmless, into the atmosphere. Both involve a large number of non-point sources. Both involve an uneven distribution of the burden and benefits among States, whether considered from the perspective of addressing or not addressing the problem. At the same time, there are also significant differences between the two issues.

The effect of ozone layer depletion is felt disproportionately in developed countries, it involves a tangible threat to humans in the form of skin cancer, and the hole in the ozone layer is measurable immediately. In addition, in terms of the evolution of international action on ozone layer depletion, the economic cost of addressing this issue turned out to be relatively modest, and the most powerful State, the US, stood to gain economically from the implementation of the solutions implemented through the ozone regime. Not surprisingly, the US was the main champion for the ozone regime, and proved to be an effective champion on this issue.

Climate change, on the other hand, is expected to be felt disproportionably in developing countries. The effects of climate change are not tangible, not certain, and not immediate. The economic cost of addressing climate change is on a very different scale than is the case for ozone layer depletion, and the European Union, who stands to gain economically from the implementation of many of the solutions proposed for climate change, has to date been a less effective champion on this issue. Finally, it is not clear whether and how climate change can be addressed without coming to grips with the broader equity and sustainability issues. Because the solution to the ozone layer problem essentially involved substituting chemicals, the broader issues could easily be left aside. Not surprisingly, given these differences, the climate change regime has evolved much slower than the ozone regime.

---

[52] Two differences in the regimes worth noting are the adjustment clause in the Montreal Protocol, which allowed for an acceleration of the targets without consensus, and the fact that the targets in the Montreal Protocol were more principle based, whereas the targets in the Kyoto Protocol are pledge based. As will be discussed, further on, these two factors are considered to have contributed to the slower progress on climate change.However, it is difficult to separate cause and effect, in terms of whether the differences are the result of a difference in the international commitment to address these issues, or whether other factors led to the differences in the regimes, which, in turn, slowed progress on climate change.

6. CONCLUSION

Within this context, the potential importance of the climate change regime for the future of international environmental law begins to emerge. In many respects, it is much easier to predict the regime's influence if it turns out to be successful in meeting the challenge. In that case, the regime may place international environmental law firmly on the road of developing individual regimes to deal with major environmental challenges as they emerge. This still leaves the question of the potential influence of soft law and customary law in the development of these regimes. Other remaining questions include how these regimes will interact and how to integrate the efforts under the various regimes into an overall move toward sustainability. These issues are considered again in Section 3 of Chapters 8 and in the Epilogue.

On the other hand, if the climate change regime fails, or fails to meet the expectations created by the OLD regime, what does that mean for international environmental law? Would the failure of the climate change regime suggest that the OLD regime is an aberration, or can it still be a template for an effective international response to global environmental challenges? If regimes are not an effective way for international cooperation on environmental issues, what are the options? The success or failure of the climate change regime will contribute significantly to answering these questions. The Epilogue to this book therefore returns to these broader questions and considers the likely influence of the climate change regime on the future of international environmental law.

2

From Negotiation to Implementation

1. INTRODUCTION

In this Chapter, the current state of the climate change regime is explored, with an emphasis on the Kyoto Protocol[1] and subsequent agreements by the Conference of the Parties to the UNFCCC[2] in Marrakech, New Delhi and Milan to operationalize and implement its key provisions. The focus here is on the various commitments addressed and obligations established for the period from its entry into force in February 2005 to the end of the first commitment period in 2012. Negotiations for the further evolution of the regime post-2012 are addressed in Chapter 9. Given the central role science has played in the evolution of the climate change regime to date, a brief overview of the state of the science is warranted as context for the assessment of the regime.

2. THE IPCC & THE SCIENCE OF CLIMATE CHANGE

The state of the science is central to any discussion and evaluation of international efforts to address climate change. The main source of information on the science for purposes of this book is the third assessment report (TAR)[3] of the Intergovernmental Panel on Climate Change (IPCC).[4] The IPCC is the closest the scientific community has come to developing a comprehensive consensus on the

---

1 *Conference of the Parties to the Framework Convention on Climate Change: Kyoto Protocol*, 10 December 1997, U.N. Doc. FCCC/CP/1997/L.7/add. 1, 37 I.L.M. 22 (1998), [hereinafter the Kyoto Protocol].

2 *United Nations Framework Convention on Climate Change*, Intergovernmental Negotiating Committee for a Framework Convention on Climate Change OR, 5th Sess., Annex, UN Doc. A/AC.237/18 (PartII)/Add.1 (1992), 31 I.L.M. 849, online: UNFCCC <http://unfccc.int/resource/docs/a/18p2a01.pdf> [UNFCCC or The Framework Convention].

3 Robert T. Watson et al, eds., Climate Change 2001: Synthesis Report: A Contribution of Working Groups I, II and III to the Third Assessment Report of the Intergovernmental Panel on Climate Change (Cambridge, England: Cambridge University Press, 2002); John T. Houghton et al, eds., Climate Change 2001: The Scientific Basis: Contribution of Working Group I to the Third Assessment Report of the Intergovernmental Panel on Climate Change (IPCC) (Cambridge, England: Cambridge University Press, 2002); James J. McCarthy et al, eds., Climate Change 2001: Impacts, Adaptation & Vulnerability: Contribution of Working Group II to the Third Assessment Report of the Intergovernmental Panel on Climate Change (IPCC) (Cambridge, England: Cambridge University Press, 2002); Bert Metz et al, eds., Climate Change 2001: Mitigation: Contribution of Working Group III to the Third Assessment Report of the Intergovernmental Panel on Climate Change (IPCC) (Cambridge, England: Cambridge University Press, 2002).

4 For brief history and terms of reference, see Ibid. TAR Synthesis Report, Foreword, page vii.

state of scientific knowledge on the causes of climate change and options for responding to the challenge. At the same time, it is fully recognized that the process has its limitations. It is in fact neither comprehensive nor is there a consensus on the conclusions reached by the IPCC. For example, there have been and continue to be scientists who criticize the IPCC for over or understating the problem and the level of certainty of the science of climate change.[5] There also continue to be scientific discoveries that shed new light on our understanding of the nature of the problem and possible responses alike. Finally, there are uncertainties in our understanding of past and future climates that are not likely to ever be fully resolved.[6] Nevertheless, the IPCC is clearly the most credible source of scientific information on climate change, and therefore an appropriate basis for law and policy consideration. Unless otherwise indicated, this book therefore accepts the third assessment report of the IPCC as the most credible source on the science; it is the source on which the Parties to the UNFCCC agreed to rely in their efforts. This is done in the full recognition of the ongoing scientific debate on whether the IPCC has been too cautious or too aggressive in its predictions.

This first assessment report was released in 1990. It was very cautious in drawing conclusions about any human impact on the global climate. The second assessment report, released in 1995,[7] was the first to declare a discernable human impact on the global climate system. The most recent update available at the time of publication was the third assessment report (TAR) released in 2001. The TAR is therefore used as the primary source of the scientific context for this book. It is supplemented with other sources that shed some light on new discoveries since 2001, such as the *Arctic Climate Impacts Assessment*.

---

[5] See, for example, P. Chylek, et al, eds., 1st International Conference on Global Warming and the Next Ice Age, (Conference Proceedings, Halifax, Nova Scotia, Canada, 19-24 August 2001) (Halifax: Dalhousie University, 2001).

[6] Most importantly, perhaps, historical data on the earth climate are limited due to the need to rely on secondary information such as ice core samples. Similarly, future predictions are limited by our still limited understanding of the exact nature of the interactions between numerous interactions within the climate systems, limiting the ability of computer models to make accurate predictions about the exact impact human activity is likely to have on the global climate in the future.

[7] *IPCC Second Assessment Climate Change 1995: A Report of the Intergovernmental Panel on Climate Change* (Cambridge, England: Cambridge University Press, 1996); John T. Houghton *et al*, eds., *Climate Change 1995: The Science of Climate Change: Contribution of Working Group I to the Second Assessment of the Intergovernmental Panel on Climate Change* (Cambridge, England: Cambridge University Press, 1996); Robert T. Watson, Marufu C. Zinyowera & Richard H. Moss, eds., *Climate Change 1995: Impacts, Adaptations and Mitigation of Climate Change: Scientific-Technical Analyses: Contribution of Working Group II to the Second Assessment of the Intergovernmental Panel on Climate Change* (Cambridge, England: Cambridge University Press, 1996); James P. Bruce, Hoesung Lee & Erik F. Haites, eds., *Climate Change 1995: Economic and Social Dimensions of Climate Change: Contribution of Working Group III to the Second Assessment of the Intergovernmental Panel on Climate Change* (Cambridge, England: Cambridge University Press, 1996).

The basic theory behind human-induced climate change is that a number of gases released in significant quantities into the atmosphere as a result of various human activities trap energy from the sun in the earth's atmosphere, thereby causing an increase in the amount of energy stored in the atmosphere, on the surface, and in the oceans. This increase in energy stored in the earth's biosphere results in an increase in global average air and water temperatures. It also has a number of secondary effects on the climate, such as regional changes in temperature, changes in wind patterns and ocean currents, changes in extreme weather events, sea-level rise, etc. These global, regional and local changes in the climate are predicted to impact on biological systems, availability of renewable resources, food production, and other human activities that are dependent on access to natural resources and on climate stability.[8]

There are a number of gases that have the physical properties to contribute to the trapping of energy. In fact, this so-called greenhouse effect is generally recognized to be a natural effect essential to life on earth. The main, naturally occurring, energy trapping, gases in the atmosphere are water vapour, carbon dioxide, methane and nitrous oxide. Of these, humans have added significantly to the concentration of carbon dioxide, methane, and nitrous oxide since the start of industrialization. Between the years 1750 and 2000, concentrations of carbon dioxide increased 31% from 280 to 368 ppm, methane increased 151% from 700 to 1,750 ppb, and nitrous oxide increased 17% from 270 to 316 ppb.[9] These trends are continuing to date.

These increases in greenhouse gas (GHG) concentrations have set in place a chain of events that, according to the IPCC, has led to an increase in global average temperature of about 0.6 degrees Celsius over the past 50 years, and has also already resulted in some sea-level rise, an increase in extreme weather events, changes in precipitation, changes in ocean temperatures, and numerous other climate changes. While these trends are expected to continue and intensify, future prognoses vary depending on predictions about human population growth and economic development paths, particularly in developing countries. Most importantly, they depend upon the success of efforts to reduce GHG emissions from human activity.

To provide some perspective on the potential impact of mitigation measures in controlling climate change, the TAR poses nine questions about the state of understanding of the climate change challenge from a scientific, technical, economic and social perspective. The conclusions provided in the Synthesis Report are summarized below under the following five key categories.

---

[8] TAR Synthesis Report, *supra* note 3 at 3, Figure SPM-1.

[9] TAR Synthesis Report, *supra* note 3 at 5, Table SPM-1.

(a) The State of Knowledge in the Context of Article 2;

(b) Human-Induced Climate Change to Date;

(c) Predicted Future Climate Change without Mitigation;

(d) Opportunities, Costs and Benefits of Mitigation; and

(e) Interaction Between Climate Change and Other Environmental Issues.

(a) The State of Knowledge in the Context of Article 2

Article 2 of the UNFCCC establishes the overall objective of the climate change regime as the prevention of dangerous human interference with the climate system. The question posed in the Synthesis Report is how to determine what constitutes dangerous interference. Put differently, by how much do GHG emissions or concentrations need to be curbed to meet the objective set out in Article 2? The report concludes that there is no clear answer to this question, but rather varies regionally and will change over time.[10]

Essentially, the report concludes that we will continue to develop better information about the risks globally and regionally of increasing GHG concentrations in the atmosphere, and based on that evolving knowledge, we will have to continuously make decisions about the level of effort justified to reduce the risk. In other words, there really are no guarantees that keeping GHG concentrations below a certain level will ensure that the Article 2 objective can be met. We do know that effects in some regions of the world are already significant. The TAR provides us with parameters, but no firm numbers. In the context of early discussions about emission reduction targets post-2012, numbers such as 2 degrees Celsius and around 450 ppm CO_2 concentration have been discussed as numbers to put the Article 2 objective into concrete, measurable terms.[11] No such connection, however, was made in the Synthesis Report.

(b) Human-Induced Climate Change to Date

This part of the work of the IPCC is essentially about the separation of natural variation from human-induced climate change in the changes observed to date.[12]

---

[10] Synthesis Report, *supra* note 3, at 2 – 4.

[11] See, for example, Bill Hare & Malte Meinshausen, "How Much Warming Are We Committed to and How Much Can be Avoided?" (2004) PIK Report No. 93, Potsdam Institute for Climate Impact Research, online: http://www.pik-potsdam.de/publications/pik_reports, and "Winning the Battle Against Global Climate Change", Commission of the European Communities, Brussels, 9 February 2005 {SEC(2005) 180} online: <http://72.14.207.104/search?q=cache:crDpXng SfK8J:europa.eu.int/comm/environment/climat/pdf/comm_en_050209.pdf/>.

[12] Synthesis Report, *supra* note 3, at 4 – 8.

To explore this, the report first identifies the impact human activities have had on GHG concentrations. As a second step, the changes in the climate that have been observed are assessed. The report then considers whether the observed changes are more reasonably explained based on natural climate variation or the change in GHG concentration in the atmosphere from human activity since the industrial revolution.

The report first notes the very significant increases in GHG concentrations from pre-industrial levels.[13] It then considers observed changes in the climate over the same time period. The report notes that global average temperature has increased by 0.6 degrees Celsius. This change has been distributed unequally, in that Polar Regions have felt higher increases, and the increases have varied by season. The report also identifies changes in precipitation patterns, and extreme weather events. Sea level rise and melting of ice are also noted. As a next step, the report indicates significant changes in growing seasons for agricultural products and plant and animal habitat due to changes in the climate.

Finally, the report concludes that "most of the warming observed over the last 50 years is attributable to human activities".[14] This is based on historical data on natural variation as well as the current state of knowledge about the various natural influences on the climate system. One of the main differences between the second and third assessment reports is that the increases in global average temperature in 1995 was still generally within the range of natural variation, whereas in 2001, the TAR concludes that natural variation can no longer explain the changes to the climate system.

(c) Predicted Future Climate Change Without Mitigation

The Synthesis Report considers the state of scientific understanding of the future climate change to be expected as a result of GHG emissions and other human activities.[15] The predictions are made using six scenarios, each based on different assumptions about the future evolution of human development and our interaction with nature. The scenarios vary in a number of key areas, including the general socio-economic development path chosen, population growth, available energy sources, and the influence of environmental concerns generally on future development choices. None of these scenarios assume the implementation of any global climate change mitigation commitments.

For the year 2100, the projections are as follows. The predictions for CO_2 concentrations range from 540 to 970 ppm for the various scenarios compared to

[13] *Ibid.* at Table SPM-1.
[14] *Ibid.* at 5.
[15] *Ibid.* at 8 – 16.

the baseline of 280 ppm. Global average increases in temperature are between 1.4 and 5.8 degrees Celsius. Increases in precipitation are projected to be between 5 and 20%. Glacial retreat and melting of the ice caps are expected to continue, as is thermal expansion of the oceans, resulting in predictions for sea-level rise of 0.09 to 0.88 meters by 2100.[16] Other impacts include impacts on ecosystems, human health, food production, and water supply.

By the year 2050, the TAR predicts a temperature rise of 0.8 to 2.6 degrees Celsius, and sea-level rise between 0.03 and 0.14 meters. Other impacts are listed qualitatively. They include risks to unique and threatened ecosystems, increased risk of extreme weather events, and impacts on productive capacity of agricultural lands. For most of the categories of impacts considered, the TAR concludes that as the change in global average temperature increases, so does the risk of negative impacts. Some impacts, such as the impact of climate change on food production, may be positive for minor increases in global average temperature in some regions of the world. As the change in temperature increases, so does the likelihood that the overall impact will be negative in most regions.[17]

(d) Opportunities, Costs and Benefits of Mitigation

In this Section, the Synthesis Report considers the range of mitigation measures available and their costs and benefits.[18] The starting point is a discussion of the inertia or time delay between the reduction of GHG emissions and the re-establishment of equilibrium. The time frames indicated range from centuries for temperature to millennia for sea-level rise. In other words, the TAR makes the point that GHG emission reductions will not be able to stop climate change immediately, but mitigation is critical because it will slow the rate of change and affect what the new equilibrium will look like.[19] As an example, the average projected temperature change by 2100 is just over two degrees, whereas the projected average change at equilibrium is almost six degrees.

Based on this, the report reaches the overall conclusion that the greater and earlier the reductions in GHG emissions, the lower the impact on the climate. As the various impacts are all interconnected, as the rate of temperature increase is slowed, so are other changes to the climate system, such as precipitation, extreme weather events, etc. It is important to distinguish, however between slowing the rate of change and reversing the change. Reversal can only take place, if at all, after the climate system reaches equilibrium. Furthermore, many impacts, such

---

[16] *Ibid.* at 9.
[17] *Ibid.* at Figure SPM-3.
[18] *Ibid.* at 16 to 28.
[19] *Ibid.* Figure SPM-5.

as extinction of species, and the impact of desertification from changes in precipitation patterns, are likely to be irreversible.

(e) Interaction Between Climate Change and Other Environmental Issues

Finally, the Synthesis Report considers a number of links between climate change and other environmental issues, at a regional and global level. Issues such as biodiversity, air pollution, resource depletion, food production, and the capacity of renewable resources to meet human needs are all explored. The TAR recognizes that mitigating climate change is essential for many of these issues, and that mitigation measures themselves can have positive and negative impacts on these areas. For example, mitigating climate change through the reduction of energy consumption will improve air quality, and reduce the depletion of resources.[20]

Having explored the state of knowledge on the science of climate change, and the clear need for an effective global response, the remainder of the Chapter will assess the current state of the international climate change regime in the form of the UNFCCC, the Kyoto Protocol, and various agreements to effect its implementation.

3. THE CLIMATE CHANGE REGIME

The climate change regime was initiated in 1992 with the negotiation of the UNFCCC. Since 1997, The Kyoto Protocol has in many respects become the heart of the regime. It set the framework for the acceptance of legally binding targets, and it included specific emission reduction targets for each country included in Annex 1 of the United Nations Framework Convention on Climate Change.[21] Much of the detail remained unresolved in the 1997 Protocol. It became clear soon after Kyoto that the detail was needed not only to implement the Protocol, but that ratification by Annex I[22] Parties would not be forthcoming until these outstanding issues were resolved. Negotiations on most of these issues took years. They were concluded in November 2001 in Marrakech, a full year after a self-imposed deadline. A few issues not considered essential for ratification decisions, such as the rules for the use of sinks in the CDM, were left to be resolved at the eighth and ninth Conferences of the Parties to the UNFCCC (COPs 8 and 9) in New Delhi and Milan respectively.

---

[20] See Sections 4 and 5 of Chapter 8 for a more detailed discussion of the links between climate change and other environmental challenges.

[21] See Kyoto Protocol, *supra* note 1, Annex B.

[22] *Ibid.* Note that the targets agreed to by Annex I countries are included as Annex B of the Kyoto Protocol. The reference to the Parties, however, continues to be to Annex I of the Framework Convention, given that the Kyoto Protocol consistently makes this reference back to Annex I of the Convention, other than for purposes of reference to the specific targets adopted.

What was planned in the Buenos Aires Plan of Action[23] to be a three year process leading to a final agreement and entry into force in 2000, turned into a seven year process after the failure of negotiations in The Hague. The main step in the negotiation process was agreement in the form of the Marrakech Accords at COP 7 on November 10, 2001. The next three years were taken up with various efforts to convince key developed States to ratify Kyoto to bring it into force. A key question raised in this Chapter is whether or not the Marrakech Accords provide a meaningful direction forward on climate change for the international community. This Chapter provides an analysis of the agreement reached in Marrakech, what it means for the Kyoto Protocol and for international efforts to address climate change.

The UNFCCC provides the context for this analysis of Kyoto and Marrakech. It set the tone for the negotiations, and has set the framework both in terms of process and substance. The role of Kyoto and Marrakech in developing an international response to climate change is therefore considered in the context of the goals set out in the UNFCCC. The focus in this analysis is on the following key goals and objectives set out in the UNFCCC. Article 2 sets as the overall goal of the UNFCCC to stabilize greenhouse gas concentrations at levels that prevent dangerous human interference with the climate system, ensure that the rate of change allow nature to adapt, do not threaten food production and allows sustainable development to take place.

This overall goal is then refined through principles set out in Article 3, including equity for present and future generations, common but differentiated responsibilities, and the precautionary approach. The merits of these goals and principles are assumed for purposes of this Chapter.

(a) History of Negotiations

In December 1997, Parties to the UNFCCC agreed to the text of the Kyoto Protocol, the first international agreement on binding emission reduction targets for greenhouse gas emissions. The Protocol included maximum emissions on a country-by-country basis for developed countries for the time period of 2008 to 2012. The emissions targets were based on 1990 as a base year. This meant that countries had to limit their emissions during the five years from 2008 to 2012 so that the average emissions would be no more than the target set for that country in the Kyoto Protocol. While the target is generally seen as an annual target, it is actually a five-year target with average annual emissions related to emissions in

---

[23] *Report of the Conference of the Parties on its Fourth Session, held at Buenos Aires from 2 to 14 November 1998. Addendum. Part Two: Action Taken by the Conference of the Parties at its Fourth Session*, UNFCCCOR, UN Doc. FCCC/CP/1998/16/Add.1 (1999) [*Buenos Aires Plan of Action*].

1990. Cumulatively, these country specific targets are expected to result in reductions of emissions in developed countries of about 5% below 1990 levels. The U.S., for example, accepted a target of -7%. This means that if it ratified the Kyoto Protocol, it would have to reduce its average annual emissions during this five-year period to 7% below its 1990 emissions.

The Kyoto Protocol then provides some guidance on what countries can and cannot do to meet their targets. For example, the Protocol allows countries to offset emissions by actually removing greenhouse gases from the atmosphere. Processes that achieve this are generally referred to as 'sinks'. The Protocol also allows some flexibility to trade emission reductions. This means that if one country can achieve reductions beyond those required to comply with its own target, it can sell those reductions to another country, which will then be able to use those reductions to meet its own target. There are three mechanisms in the Kyoto Protocol to accommodate this. They are the clean development mechanism, joint implementation, and emissions trading. These Kyoto mechanisms are addressed in more detail below. They function to allow Parties to separate the question of where the reductions are achieved, from which Party can count the reductions in achieving its emission reduction target.

Finally, the Kyoto Protocol provides for assistance to developing countries to facilitate their efforts to reduce emissions. This is particularly important given that developing countries were not allocated reduction targets in the Protocol. These are the basic building blocks of the Kyoto Protocol. It does little more than provide these basic building blocks: the three credit trading mechanisms, the use of sinks to offset emissions, and general provisions for assistance to developing countries, recognizing the common but differentiated responsibilities of Parties to contribute to the solution.

Four years passed from the signing of the Kyoto Protocol to the agreement in Marrakech. While it was clear from the outset that ratification and implementation of the Protocol would take time, no one could have predicted the events of the twelve months leading up to Marrakech. The first couple of years post-Kyoto had gone more or less according to plan, with Parties slowly negotiating toward agreement on the rules that would operationalize the Kyoto Protocol, and would enable UNFCCC Parties to make an informed decision about ratification. The time frame for this was identified in the Buenos Aires Plan of Action, agreed to at the Conference of the Parties (COP) 4 in 1998.[24] The plan was developed in recognition that in order to allow sufficient time for Parties to meet their commitments, the Protocol should enter into force by 2003. This would leave two years following the coming into force of the Protocol for countries to demonstrate progress toward their targets, and five years until the start of the first commitment

---

[24] *Ibid.*

period. According to the Buenos Aires Plan of Action, therefore, agreement on the rules necessary to operationalize the Kyoto Protocol was to be reached by COP 6 in November 2000 in The Hague.

Cracks in this plan began to appear in the last meeting of the subsidiary bodies prior to COP 6,[25] when it became clear that, over time, Parties' interpretations of the Protocol were becoming more and more divergent. The sheer number of unresolved issues heading into COP 6 made a comprehensive agreement exceedingly unlikely. Uncertainty over the outcome of the U.S. presidential election also seemed to haunt the COP 6 negotiations. Deep and fundamental divisions among the main negotiating blocks crystallized. In very broad terms each of the three major negotiating blocks had, by then, carved out an overall negotiating basis that proved to be fundamentally at odds with those of the other two. The Umbrella Group (UG)[26] had focussed its position on economic efficiency of the mechanisms. The European Union (EU) was presenting its position in the context of the environmental integrity of the Protocol. The G-77 and China[27] focussed on equity for developing countries. With the EU and G-77 finding considerable common ground in the negotiations leading up to and in The Hague, the UG seemed to be more and more isolated. Each side seemed convinced that it would prevail if it just continued to hold out on all issues, and that the other negotiation blocks would give in eventually. By the time Parties realized this would not happen, and real negotiations started at COP 6 in The Hague, it was too late and no agreement was reached.[28]

While the timetable for the Buenos Aires Plan of Action could not be met, Parties did agree to pick up the pieces at a special negotiation session referred to as COP 6 bis in Bonn in 2001. Failure of the negotiations in The Hague was not

---

[25] Part 1 of the Thirteenth Sessions of the UNFCCC Subsidiary Bodies was conducted from 11 to 15 September 2000 in Lyon, France. For an overview of the negotiations in Lyon and The Hague, see Lavanya Rajamani, "Air and Atmosphere, Re-negotiating Kyoto: A Review of the Sixth Conference of Parties to the Framework Convention on Climate Change" (2000) 11 Colo. J. Int'l Envtl. L. & Pol'y Y.B. 201.

[26] The Umbrella Group (UG) generally consists of The United States, Russia, Japan, Canada, Australia, Iceland, Norway, New Zealand, and at times the Ukraine. The United States remained a participant in the UG in 2001 in Bonn and Marrakech even after the Bush Administration decided not to ratify the Protocol. New Zealand, Norway and Iceland voted against the UG position in the last days of negotiations in Marrakech, and the Ukraine seemed more aligned with the CG 11 negotiating block (consisting of eastern European countries) in Marrakech than with the UG. This left Russia, Japan, Canada and Australia as the core of the UG for purposes of negotiations in Marrakech, four countries with a very different view on how the Protocol should be operationalized, and due to the ratification formula, with a bargaining position that enabled it to dominate the negotiations.

[27] The G-77 consists of most developing countries, including small island States, oil producing countries, and the developing countries of Africa, Asia, and South America.

[28] See Sean S. Clark, Mark C. Trexler & Laura H. Kosloff, "Installment Six of the Climate Treaty Debate: A Report on COP-6" (2001) 15 Nat. Resources & Env't 180.

initially seen as a fatal blow to the Protocol. By the spring of 2001, however, the Bush administration in the U.S. confirmed publicly what many had suspected: that it had no intention of initiating the ratification process in the U.S., and would not participate in any further negotiations on the Protocol. The administration feared harm to the U.S. economy and refused to enter into an international agreement on climate change that would bind the U.S. without also binding developing countries.[29] With the single largest emitter and most powerful nation out of the process, the Kyoto Protocol seemed doomed.

The initial reaction from close trading partners of the U.S., such as Japan, as well as Australia and Russia, seemed to be that without the U.S. they would also not ratify the Protocol. Efforts by the EU, supported by Canada and other UG countries, prior to the resumptions of COP 6 negotiations in July 2001 in Bonn, however, prevented any formal decision by any Party other than the U.S., and allowed negotiations to continue under the Kyoto Protocol. This was in part due to the efforts of the EU and others, and in part due to the absence of any concrete alternative proposal by the U.S. administration leading up to Bonn. Nevertheless, expectations for Bonn were low. Surprisingly, the EU came into the negotiations with a very different approach than it had in The Hague eight months earlier. It essentially offered the UG most of what it had demanded in The Hague in return for reaching an agreement on the major outstanding issues. This was accomplished in Bonn, and the detail was worked out at COP 7 in Marrakech. The G-77 Parties were offered some concessions on development assistance, but developing countries were otherwise not a major factor in influencing the outcome of negotiations after The Hague.

The fact that the Kyoto Protocol came into force in 2005 itself is a remarkable turn of events, not only because it succeeded without participation from the most powerful nation, but because, without the U.S., ratification by almost every Annex I country was required to bring the Protocol into force.[30] In an effort to bring countries such as Japan, Russia, and to a lesser extent Australia and Canada, on board, many concessions were made, particularly by the EU and the G-77. What is left of the Kyoto Protocol? Will it be effective? Will it be fair? Does it set a framework for effective long-term action to reduce human impact on the climate?

---

[29] For information on the U.S. position on climate change, see the U.S. EPA website <http://www.epa.gov>. The initial positions of the Bush Administration are summarized in a position paper entitled *Climate Change Review* (2001), online: U.S. EPA <http://yosemite.epa.gov/oar/globalwarming.nsf/UniqueKeyLookup/SHSU5BNM7H/$File/bush_ccpol_061101.pdf>.

[30] See Kyoto Protocol, *supra* note 1, art. 25. The U.S. represents about 35% of Annex I emissions, Russia about 17%, Japan about 9%, Australia 2%, Canada 3%, and most of the rest split between the EU and eastern European countries. This data is based on information in the first national communications of Parties under the UNFCCC, submitted on or before 11 December 1997, and used in the negotiations on Article 25.

Can it survive without the U.S.? Will other countries ratify the Protocol without U.S. participation? These questions are explored ahead.

(b) The Marrakech Accords

With the conclusion of negotiations on the issues raised in the Buenos Aires Plan of Action, and the entry into force of the Protocol, the focus of the international community has now shifted to implementation. With this shift into a new phase, it is timely to consider how much was accomplished in the Kyoto negotiation phase; a phase that took us from a framework convention with general statements of objective in 1992 to binding targets and detailed rules on how to meet them in 2001.

How much progress toward the goals and objectives of the UNFCCC are Kyoto and Marrakech likely to provide to the international community? What is the role of Kyoto and Marrakech, collectively, in meeting the goals and objectives of the framework convention? What is the role of the Kyoto Protocol in achieving the environmental and equity goals set in the UNFCCC? In answering this question, the focus will be on the targets taken on by developed country Parties as well as the various mechanisms available to these Parties to meet their targets.

(c) Implications for the Kyoto Protocol

(i) *The Kyoto Mechanisms*[31]

The Kyoto Protocol includes as alternatives to domestic action a number of so-called flexibility mechanisms available to Annex I Parties to supplement domestic action with reductions outside their own jurisdictions. The three mechanisms established for this purpose are the Clean Development Mechanism, Joint Implementation, and Emissions Trading. The analysis of these mechanisms is directly relevant to the first question posed, the role of the Protocol in achieving the environmental and equity objectives of the framework convention. In this context, it will be important to consider whether these mechanisms undermine the environmental objectives of the UNFCCC, or whether they assist in bringing countries into the process in an equitable, constructive and meaningful way.

The Kyoto Mechanisms were included in the Protocol, at least in part, in recognition that the country-specific targets agreed to for Annex I countries provided only a crude tool of balancing the relative economic cost of Parties to

---

[31] For a general discussion of the issues related to the mechanisms, see Sophia Tsai, "UNFCCC Technical Workshop on Mechanisms of the Kyoto Protocol" (1999) 10 Colo. J. Int'l Envtl. L. & Pol'y 220.

reduce emissions. Only after the conclusion of the first commitment period will a thorough analysis of the relative economic costs and benefits to the respective Parties to the Protocol be possible. In the meantime, the mechanisms provide a release valve to ensure that if one Party's target is disproportionately expensive to achieve, it can delay reductions in its own State and instead support reductions in another State. This objective is partly met through all three mechanisms, but primarily through emissions trading.

Another objective for at least two of the mechanisms was to address capacity-building issues in developing States and economies in transition. These groups of States are expected to make major capital investments in energy-producing technologies in the near future. Inherent in the inclusion of these mechanisms was a recognition that influencing the choices made at this stage might have significant long-term benefits. This applies particularly to developing States such as China, India and Brazil that are at a stage in their development where their energy needs are likely to increase significantly in the years to come. The mechanism established for this purpose is the Clean Development Mechanism. This concept also applies to economies in transition, which refers to eastern European economies that fell apart after the collapse of the Soviet Union, and which are in the process of rebuilding their economies. The mechanism established to facilitate this process for economies in transition is Joint Implementation.

(ii) *The Clean Development Mechanism (CDM)*[32]

The Clean Development Mechanism was a last-minute addition to the Kyoto Protocol.[33] The basic concept was to give Annex I States a release valve in case domestic action became too expensive, and to provide developing States with much needed development assistance in the form of technology transfer and economic activity. Parties agreed that if reductions could be achieved more cost-effectively in a developing State that has no reduction target,[34] that State should

---

32 For a general discussion on CDM, see Richard Stewart *et al*, *The Clean Development Mechanism: Building International Public-Private Partnerships Under the Kyoto Protocol*, UN Doc. UNCTAD/GD5/GF5B/Misc.7 (2000) at 9; William L. Thomas, Daniel Basurto & Gray Taylor, "Creating a Favorable Climate for CDM Investment in North America" (2001) 15 Nt. Resources & Env't 172; Sean Michael Neal, "Bringing Developing Nations on Board the Climate Change Protocol: Using Debt-for-Nature Swaps to Implement the Clean Development Mechanism" (1998) 10 Geo. Int'l Envtl. L. Rev. 163.

33 See Jacob Werksman, "The Clean Development Mechanism: Unwrapping the 'Kyoto Surprise'" (1998) 7 Rev. Eur. Community & Int'l Envtl. L. 147.

34 It is assumed, therefore, that the developing country has no direct, short-term incentive to make the reductions. Certainly, no such direct incentive exists under the Protocol itself. The question of what a developing country would do, however, in the absence of assistance under this mechanism is a very difficult question to answer, and would have to consider other motivations for taking action to reduce emissions, including, but not limited to, the possibility of a future target,

be able to join forces with an Annex I State[35] to achieve those reductions. Thus, the reductions would count toward the target of the Annex I State. In turn, the Annex I State provides assistance to the developing country to achieve reductions that would otherwise not have been realized. It became quite clear through the years of subsequent negotiations that putting this fairly simple concept into practice would be a challenge. The main question for purposes of this Chapter is what balance the CDM strikes between retaining the environmental integrity of the developed State targets and starting the process of bringing developing States into the process of emissions reductions.

(A) Baselines

One of the main challenges with CDM projects in the negotiations was to determine how emissions in a host country compare with and without the CDM project. The challenge was how to predict, at the time the CDM project is certified, what the host country would have done without the project.[36] This requires the establishment of a baseline for the CDM project against which the emissions from the project are to be measured to determine whether (and how much) emissions have been reduced by the project.

Provisions regarding baselines for CDM are contained in the Annex to the Marrakech decision on modalities and procedures for a clean development mechanism at paragraphs 45 to 52.[37] The options for methodologies are set out in paragraph 48. Under this provision, project participants, which include the host and funding Parties, may choose a methodology for establishing a baseline for a project from the following options:

1. Existing actual or historical emissions;
2. Emissions from a technology that represents an economically attractive course of action, taking into account barriers to investment; and

---

or an effort to demonstrate to Annex I countries that their efforts are inadequate, and not based on best efforts.

35 Annex I Parties have emission reduction targets under the Protocol, and thereby an incentive to support efforts to reduce emissions, if they can get credit for the reductions realized.

36 Either for the life of the proposed CDM project or even just until the end of the first commitment period.

37 See *Report of the Conference of the Parties on its Seventh Session*, Conference of the Parties, United Nations Framework Convention on Climate Change (UN FCCC), 29 October – 10 November 2001, FCCC/CP/2001/13/Add.1(Decisions 1/CP.7 - 14/CP.7), FCCC/CP/2001/13/Add.2(Decisions 15/CP.7 - 19/CP.7), FCCC/CP/2001/13/Add.3(Decisions 20/CP.7 - 24/CP.7), FCCC/CP/2001/13/Add.4(Decisions 25/CP.7 - 39/CP.7 & Resolution 1/CP.7 – 2/CP.7), online: UN FCCC <http://unfccc.int/2860.php> [hereinafter COP 7 or Marrakesh Accords], Annex, paras. 45-52 at page 36-37.

3. The average emissions of similar project emissions of comparable project activities undertaken in the previous five years in like circumstances and whose performance is among the top 20 percent of their category.

The first option is most likely to be applied for energy efficiency or conservation projects, or for straight energy source replacement projects. Given that CDM projects are likely to combine economic development and GHG emissions reduction, it is more than likely that CDM project activities will be associated with an increase in overall energy consumption, which makes the first option unattractive to project participants as a baseline. For most CDM projects, Parties are therefore more likely to rely on options two and three. The net effect of this approach to establishing baselines is an overall overestimation of the emissions reductions achieved from CDM projects. Negotiators clearly decided to err on the side of encouraging CDM projects over protecting the environmental integrity of the Annex I targets. Based on the early experience of the Executive Board for the CDM, it would appear that in spite of this, CDM activity may be limited.[38]

(B) CDM Projects and Sustainability

A second, very controversial, issue has been what types of projects should qualify for the CDM. What factors other than the project's impact on greenhouse gas emissions should be considered? Does it matter whether the project is otherwise environmentally sound? Does it matter whether the project assists the host country in its effort to become sustainable, or to remain sustainable? Who gets to decide whether a project meets any of these criteria? What is the role of the public, either internationally or locally, in making this determination? Should there be standards set for environmental assessment process requirements to consider these issues and ensure public access to the decision-making process?

The Marrakech Accords deal with these issues in the following manner. The host Party is given the final decision-making authority to determine whether the project activity assists it in achieving sustainable development.[39] One exception to this is that Annex I Parties agreed to refrain from using reductions achieved from nuclear projects.[40] While this approach may seem reasonable on the surface, and was argued on the basis of the sovereignty of developing countries to choose their own path to sustainability, there are a number of serious flaws in this reasoning.

---

[38] See CDM Executive Board Meeting Reports, available at http://cdm.unfccc.int/EB/Meetings.

[39] See *supra* note 37, Decision 17/CP.7 (Modalities And procedures for a clean development mechanism as defined in Article 12 of the Kyoto Protocol), at page 20.

[40] *Ibid.* at page 20. The same principle applies to Joint Implementation projects. Nuclear energy projects are therefore also excluded from JI.

The practical implication of this part of the Marrakech Accords will be that a developing country will be approached by an Annex I country or a private company with an offer that is linked to a specific technology or project. The "choice" of the host country at that point will be to decide whether to look a gift horse in the mouth and turn down the offer of assistance, or to take what is being offered. The suggestion that the host developing country is in a position to make a sound objective decision about what is consistent with sustainable development is naive at best. Secondly, this approach makes no distinction between countries where individuals affected by a proposed project have a voice and an opportunity to be heard and influence the decision made by the host country, and the many countries around the world where that is not likely to happen.

Furthermore, the suggestion that international involvement to ensure that CDM projects are consistent with sustainable development poses a threat to the host country's sovereignty is also not defensible on closer examination. The effect of international oversight into the question of sustainability of the project, is not a question of imposing anything on the host countries, but is rather about whether the international community is prepared to recognize the greenhouse gas emission reductions achieved from the project as an alternative to domestic action in an Annex I country. In the name of host country sovereignty, this approach would force the international community to accept credits for a project that creates other environmental problems while reducing greenhouse gas emissions. How this is less of a sovereignty issue for members of the international community than conditions placed on assistance to a host country is difficult to follow.[41]

The one alleviating factor to reduce the pressure on host countries to accept undesirable technologies is that under the Marrakech Accords unilateral CDM projects appear to be permitted. This would allow developing countries to seek out private companies who are in possession of technology that the host country chooses to adopt, and try to develop CDM projects on their own. The battle throughout the negotiations on this issue was really a battle over market access between the EU and the UG for technologies that these countries are able to offer. Not surprisingly, given the lack of past efforts on greenhouse gas emission reductions domestically, UG countries have relatively few technologies to offer in this regard. Furthermore, most of the technologies developed in UG countries with the potential to reduce greenhouse gas emissions have raised serious concerns from a sustainability perspective. Examples include nuclear power plants, large

---

[41] This was the basis on which the UG was arguing against the development of any criteria for the types of projects to be accepted under the CDM. It is interesting to note that the technologies supported by many UG countries are technologies that likely would have had difficulty surviving close scrutiny under sustainability criteria.

scale hydro projects, so-called clean coal technology, and forest plantations as sinks projects.[42]

From the perspective of consistency, fairness, and environmental integrity, it would have been clearly preferable if Parties had made a collective decision about what technologies either do or do not qualify as environmentally sound. This could have streamlined the environmental assessment process, by focussing on site-specific concerns. It could also have started the much needed process of getting out of compartmental thinking and reaction to one symptom at a time.

It could have signalled an effort to confronting the fundamental problem of too many human beings causing too much environmental harm in meeting their needs and wants.[43] Agreement here could have reduced the risk that action on climate change through the CDM at the international level will come at the expense of biodiversity and sustainability.[44] In other words, this was a missed opportunity for a more integrated approach to climate change mitigation.

One of the alternative solutions proposed by some countries for addressing these concerns about the effects of CDM projects in the host countries was to include requirements for meaningful environmental assessments and public participation, both in terms of the local and the international public. These proposals lost out in the name of reducing the cost of going through the CDM verification process, which was based on a legitimate concern that the cost of complying with the process requirements may more than offset the incentive provided by the credits, particularly with the expected lower price of carbon as the result of a combination of Russian hot air, and the loss of the major buyer, the U.S.

What was lost, unfortunately, in the negotiations, was the opportunity to streamline the process without giving up on the idea of environmentally sound projects, as discussed above. In addition, the problem can also be solved through an international commitment to require meaningful environmental assessments for all internationally funded or supported projects, thereby eliminating the con-

---

[42] The EU and the G-77 put forward proposals on positive or negative lists at COP 6 in The Hague in 2000. The positive list was proposed as a list of projects that would be eligible for CDM credits. The proposed list contained mostly projects which were likely to involve EU companies. The alternative propose, a negative list of projects that would not be eligible, contained projects such as nuclear plants, large scale hydro, coal plants, etc., mostly projects UG companies were likely to be interested in. As a result, the list approach was dropped, and an opportunity to streamline the eligibility and verification process without compromising the environmental objective of the CDM mechanism, was lost.

[43] For a more detailed discussion of this topic, see Bradley A. Harsch, "Consumerism and Environmental Policy: Moving Past Consumer Culture" (1999) 26 Ecology L.Q. 543.

[44] The opportunities for integration of climate change and other environmental challenges such as biodiversity is discussed in some detail in Sections 4 and 5 of Chapter 8.

cern that CDM projects would be unable to compete.[45] Such an approach would also have provided opportunities to further the objectives of the UNFCCC through any development assistance offered to economies in transition, by ensuring that all projects are sustainable and climate friendly. Another solution to the procedural cost concern would have been to require each Annex I Party to hold a certain minimum number of CDM credits.

In conclusion, in the context of the first question posed at the outset, it is clear that equity and the environment lost out to reducing the cost of generating CDM credits. This, in combination with the difficulties in establishing baselines, raises some serious questions as to whether the CDM process is likely to live up to its "hype" of serving the dual role of making compliance cheaper for Annex I Parties while assisting developing countries without undermining environmental objectives.

(C) Process Issues

The fourth major challenge was who would control the process, and who would be making decisions on how to apply the modalities for baseline and verification to a particular project? Would it be a political process or one based on expertise? Would there be regional representation based on general UN practice, based on whose interests were at stake, or would the selection process for the body established to do this work be based on proven expertise and impartiality?

The body set up under the Marrakech Accords to implement the CDM is the CDM Executive Board.[46] The Executive Board is made up of ten members to be elected by the conference of the Parties serving as the meeting of the Parties to the Protocol (COP/MOP). Non-Annex I Parties will have a majority on the Executive Board. Members are to be selected based on their technical and policy expertise, and be elected in their personal capacity, not as representatives of the State. Meetings of the Executive Board are open to the public, and documents are to be made available to the public.

The validation of individual CDM projects is actually not carried out by the Executive Board, but rather by operational entities accredited by the Executive Board.[47] These operational entities have the responsibility to validate CDM pro-

---

[45] In the Canadian context, this has been a controversial debate in the context of the application of the *Canadian Environmental Assessment Act*, S.C. 1999, c. 33, to the Canadian International Development Agency (CIDA), and the Export Development Corporation (EDC). Efforts to ensure proper environmental assessments of CIDA projects have been partly successful, but there has been virtually no progress with EDC.

[46] See Marrakech Accords, *FCCC/CP/2001/13/Add.2*, Draft decision -/CMP.1 (Article 12), Annex, Part C at 27.

[47] *Ibid.* Draft decision -/CMP.1 (Article 12), Annex, Part D at 30.

ject activities and verify actual emissions against the established baseline.[48] The Executive Board oversees this process by means of annual activity reports submitted by each operational entity. Operational entities will be retained by project participants to carry out the validation and verification and report the results to the Executive Board. The Executive Board formally registers the CDM project after its review of the report of the operational entity. There are limits imposed on the time period during which credits can be issued for a given project.[49]

Participation in a CDM project is voluntary. If a Party chooses to participate, however, Part F[50] of the Annex requires it to have a designated national authority as a body responsible for CDM project involvement of that country. Credits from CDM projects can be earned effective from the year 2000. Baseline and credit approval can take place as soon as the Executive Board has completed the accreditation of operational entities to carry out that process. Sinks projects under CDM are limited to afforestation and reforestation projects, and no Annex I country can use more than the equivalent of 1% of its assigned amount from sinks credit under the CDM to meet its first commitment period target.[51] The rules for afforestation and reforestation projects under the CDM were developed by the 9th Conference of the Parties in the fall of 2003.[52] Finally, 2% of the certified emission reductions are to be set aside under an adaptation fund to assist developing countries that are particularly vulnerable to the adverse effects of climate change.[53]

(iii) *Joint Implementation (JI)*[54]

Joint implementation has only limited relevance for the issues raised in this Chapter. Essentially it is a hybrid between Emissions Trading and the Clean Development Mechanism directed at economies in transition. It actually consists of two tracks, one resembles emissions trading, and the other is project-based and resembles the Clean Development Mechanism. The reason for the two tracks is

---

48 *Ibid.* Draft decision -/CMP.1 (Article 12), Annex, Part E at 31.

49 *Ibid.* Draft decision -/CMP.1 (Article 12), Annex, Part G, para. 49 at 37.

50 *Ibid.* Draft decision -/CMP.1 (Article 12), Annex, Part F at 32.

51 *Ibid.* Decision 17/CP.7 (Article 12), para. 7 at 22.

52 See COP 9, Milan, Decision 19/CP.9, FCCC/CP/2003/6/Add.2.

53 See Marrakech Accords, Decision 17/CP.7 (Article 12), para. 15 at 23.

54 See Marrakech Accords, Decision 16/CP.7(Article 6) at 5; *Ibid.* Draft decision -/CMP.1 (Article 6) at 6. For a more detailed discussion on joint implementation, see *e.g.* Glenn Wiser, "Joint Implementation: Incentives for Private Sector Mitigation of Global Climate Change" (1997) 9 Geo. Int'l Envtl. L. Rev. 747; Alex G. Hanafi, "Joint Implementation: Legal and Institutional Issues for an Effective International Program to Combat Climate Change" (1998) 22 Harv. Envtl. L. Rev. 441; and Chester Brown, "Facilitating Joint Implementation under the Framework Convention on Climate Change: Toward a Greenhouse Gas Emission Reduction Protocol" (1997) 14 Envtl. and Planning L. J. 356.

that there was concern among Annex I countries who were interested in joint implementation activities that some of the countries with economies in transition would have capacity problems with respect to some of the eligibility requirements for emissions trading.[55]

The main difference between the two tracks relates to who gets to verify the additionality[56] of emission reduction units to be issued to the funding Party. For track one, this is done by the host country, based on the principle that because it has its own target to meet, there is no incentive for the host country to overestimate the reductions achieved. Track two is based on the assumption that the host country does not know where it stands domestically; it has not met an eligibility requirement (such as annual reporting on emissions based on accepted standards), and is therefore essentially treated like a Non-Annex I country without an emissions reduction target. This justifies the involvement of a JI Executive Board similar to the CDM Executive Board, and a similar role for operational entities to validate the project and verify emission reductions. While there are some minor differences between each of the tracks and their corresponding mechanisms, due to the similarity a detailed review of JI is not warranted as part of this study.

One point of note is that JI projects cannot earn credits until 2008, whereas CDM projects can earn credits effective 2000. A JI project can, however, similarly to CDM projects, be approved as soon as the administrative structure, including the Executive Board, is established.

(iv) *Emissions Trading (ET)*[57]

On the one hand, emission trading is the most straightforward of the three mechanisms, in that it deals with the exchange of credits between Annex I countries only. It does not involve the creation or validation of new emission reduction or sequestration credits, nor is it concerned with baselines. It simply allows Annex I countries to exchange the right to emit greenhouse gases during a given com-

---

55 The issue of mechanisms' eligibility is discussed in the general section on the Kyoto Mechanisms, Section 3(i), above. The specific requirements for reporting on emissions and sinks is discussed in the context of Articles 5, 7, and 8 of the Kyoto Protocol, Section 3(vi), below.

56 Additionality refers to the need to verify that the reduction credits claimed are additional to any reductions that would have taken place without the joint implementation project. There are different ways to define additionality, one is financial, others may look at technology or other ways of determining what would have happened without the joint implementation of the project. The project may not have taken place, the project may have gone ahead with higher emissions, or the project may have gone ahead without change.

57 For a discussion of emissions trading in the Canadian context, see Andrew Bachelder, "Using Credit Trading to Reduce Greenhouse Gas Emissions" (2000) 9 Envtl. L. & Prac. 281; and Richard B. Stewart, James L. Connaughton & Lesley C. Foxhall, "Designing an International Greenhouse Gas Emissions Trading System" (2001) 15 Nat. Resources & Env't 160.

mitment period.[58] On the other hand, this is where the whole system of flexibility mechanisms created under the Kyoto Protocol comes together. This created some unique challenges for the negotiation of the Marrakech Accords. It is here that the issues raised at the start of the Chapter come to a head. It is with respect to the emission trading rules that crucial decisions were made about the environmental integrity and equity of the Kyoto Protocol. It is here that monitoring, reporting, verification and compliance all come together to determine the effectiveness of the overall process being established through Kyoto and Marrakech.

(A) Fungibility

The term fungibility is one of a number of new terms introduced in the course of negotiations leading up to Marrakech. Fungibility refers to the interchangeability of emissions or sequestration credits. For UG Parties, the objective of the negotiations was to ensure that a carbon credit could be traded on the open market, and that there would be no difference in value depending on whether that credit was generated through a Clean Development Project, Joint Implementation, a sink, or whether it was part of the assigned amount a Party was allocated based on its Kyoto target. The difficulty with this is that many Parties did place different relative values on credits depending on how they were generated. In addition, there were policy objectives other than the creation of an open low transaction cost market in carbon credits that were at conflict with this objective.

As a starting point, there was general agreement that if a country was able to reduce its emissions to below the assigned amount,[59] any portion of the assigned amount not required for compliance could be traded under Article 17 of the Kyoto Protocol. A key question was to what extent other credits generated would be treated differently for trading purposes. To answer this seemingly straightforward question, Parties had to address a number of specific issues that proved complicated and difficult to overcome. How much of the action taken by Annex I

---

[58] It is important to note that the Marrakech Accords indicate that the Kyoto Protocol does not grant any title or entitlement to emissions (see Marrakech Accords, *FCCC/CP/2001/13/Add.2*, Decision 15/CP.7 (Mechanisms), at 2). Nevertheless, the assigned amount, any removal units, and emission reduction credits generated under the Kyoto Protocol are recognized to indicate that a Party holding those units has emitted greenhouse gases in accordance the rules set out in the Kyoto Protocol and the Marrakech Accords.

[59] The assigned amount means the emissions a country is permitted to release during a commitment period solely based on the emission reduction target it accepted, without considering additional credits generated through the use of sinks, the Clean Development Mechanism, or Joint Implementation.

countries to meet their emission reduction targets would be taken domestically?[60] Should the same conditions apply to AAUs, CERs, ERUs and RMUs[61] with respect to the right to carry over unused credits to future commitment periods?[62] Who should be held liable in case a Party sells credits it actually needs for its own compliance? Should trading take place between Parties only, or should private entities be permitted to trade any or all of the credits generated?

These issues were resolved in the Marrakech Accords generally in favour of unrestricted trading, with a few largely symbolic measures to address competing concerns about the environmental integrity of the trading system developed. Parties are free to trade CERs and RMUs in addition to being able to trade part of their assigned amount. Trading can be between Parties and can involve private entities. There is a limit imposed on banking of CERs,[63] and banking of RMUs is prohibited. This is, however, merely symbolic given that the Marrakech Accords do not prevent recycling, whereby a Party would keep CERs and RMUs for compliance purposes and bank its assigned amount units instead.[64]

(B) Commitment Period Reserve (CPR)[65]

The commitment period reserve is a compromise solution to a problem that was a major stumbling block in the negotiations, and a particularly obvious illustration of the different visions Parties brought to the negotiations. In the context of the questions posed at the beginning, it was also a choice between

---

[60] This issue is generally referred to as supplementarity. It arises out of the Kyoto Protocol Articles 6(d) and 17, which essentially state that use of the mechanisms shall be supplemental to domestic action, but do not quantify what is supplemental or even whether this is a mandatory, quantitative requirement.

[61] AAUs refer to Assigned Amount Units, CERs are Certified Emission Reductions, ERUs refer to Emission Reduction Units, and RMUs are Removal Units. These units refer to various types of carbon credits generated under the Protocol. See Section 3(B), for a more detailed discussion.

[62] This is generally referred to as banking, and is generally permitted under Article 3.13. Banking allows a Party to retain credits generated during the first commitment period and use them to meet its second commitment period target, rather than forcing it to trade all credits it does not require for compliance to another Party. See Section 3(iv) below for a discussion of the various terms used in the context of emissions trading, such as banking, retiring, issuing, canceling, and recycling credits.

[63] The limit for CERs is 2.5% of a Party's assigned amount.

[64] Recycling can circumvent all these measures given that the amount of credits generated through the Clean Development Mechanism and the use of sinks will be considerably less on average than the amount of emission credits held by a Party based on its target, the assigned amount. The limitations placed on banking only become an issue for a Party that holds CERs and RMUs that are in total close to or higher than the actual emissions of that Party, without counting any of the assigned amount units held by that Party. This is why, in practice, the limits imposed on banking of RMUs and CERs are merely symbolic.

[65] See Marrakech Accords, *FCCC/CP/2001/13/Add.2*, Draft Decision 1-/CMP. 1 (Article 17) Annex, para. 6-10, at 54.

effective compliance and an effective trading system. The concern expressed by the G77, the EU and a number of other countries was that a country might decide to oversell, keep the revenues generated from the overselling, fail to comply, and then refuse to accept the consequences of noncompliance, or even leave the Protocol altogether. While the worst case scenarios must be considered unlikely, given that the Protocol will not operate in an international vacuum, there is a real risk that a Party, whether intentionally or as a result of an error in judgement or an unforeseen event, will sell more than it can afford to. This in turn could create a challenge for the compliance system that may be difficult to resolve. Given the limited enforcement mechanisms available in international law, preventing such problems was considered crucial by many.

During the course of the negotiations, a whole range of solutions were proposed to this problem, ranging from absolute seller liability to absolute buyer liability.[66] The problem with buyer liability from a market perspective is that carbon credits will inevitably have different values attached to them depending on the source. A buyer will consider the likelihood of the originating country failing to meet its target in deciding what price it will be willing to pay, in recognition of the risk that, in case the originating country oversold, the credits are worthless. On the one hand, this means that buyer liability can be a very effective means of promoting compliance, because ability to demonstrate compliance affects the price a country will be able to get for credits it wishes to sell, and buyer liability imposes a responsibility on the buyer of the credit to become part of the compliance effort to protect its investment. On the other hand, buyer liability would have destroyed the concept of an open market by creating different values in credits depending on the source. It does this because under buyer liability, the value of a ton of carbon depends on the seller's ability to meet its obligations.

What was agreed to in the end was a compromise. As long as a Party keeps a reserve of 90% of its assigned amount, or 100% of its most current reviewed emissions inventory (whichever is less), the principle of seller liability applies. Parties are discouraged from selling below those levels. If a Party sells below those levels, buyer liability applies, which means the buyer will not be able to use those credits unless the seller restocks its reserve to above 90% of the assigned amount or 100% of emissions. The reference to 100% of emissions allows countries whose actual emissions have dropped by more than 10% below their target to sell more of their assigned amounts.[67] The buyer liability approach on the CPR

---

[66] Seller liability essentially means that the country that oversells is out of compliance. This means relying on the compliance system for incentive and motivation not to oversell. Buyer liability would mean the buyer of the credits could not use the credits to meet its commitments. This would place an incentive on the buyer of credits to ensure that the seller has not oversold.

[67] It is expected that this will apply mainly to economies in transition, such as Russia, whose emissions have dropped to 70%-80% of 1990 levels as a result of the economic collapse following the disintegration of the Soviet Union.

has been operationalized through the use of a transaction log which tracks the trading of any AAU, ERU, CER, and RMU. Part of the tracking process will include calculation of the status of the seller's CPR status at the time of transaction. This means that the transaction log will notify the potential buyer if the seller's CPR reserve has dropped below the required minimum. In that case, the buyer can terminate the sale or proceed with the understanding that the credits cannot be used by the buyer until the seller has replenished its CPR.

One final note on the commitment period reserve. The EU and the G-77 had taken the position that compliance with the reserve limits should be an eligibility requirement for access to the mechanisms generally. This would mean that in case of a country overselling, not only would the buyer not be able to use the credits, but the seller would be prohibited from any further selling until the commitment period reserve was replenished. This proposal was rejected by the UG, and was not included in the Marrakech Accords.[68] The compromise struck has the effect of limiting the extent of the problem, but otherwise the agreement reached errs on the side of reducing the cost of compliance and in the process weakening the compliance system.[69]

(C) Mechanisms Eligibility

One cross-cutting issue with respect to the use of the mechanisms was the question of eligibility. Eligibility refers to the question of when and under what circumstances Parties should be permitted to use the three mechanisms established in the Kyoto Protocol. This issue is most relevant to the question of reporting and compliance. Use of the mechanisms provides significant reporting and verification challenges. At the same time, the desire to use the mechanisms is likely to be a strong motivator to meet any eligibility requirements imposed, as long as they are proportional to the benefit of being able to use the mechanisms. It is not surprising, for example, that mechanisms eligibility was considered by some to be an effective incentive to ensure that an amendment under Article 18 with legally binding consequences would be ratified by all Parties to the Protocol.

Having said this, there were a number of additional motivations for limiting mechanisms eligibility. Some countries were simply against the use of the mechanisms, and were arguing for strict eligibility requirements as a way to minimize their use. For others, some of the reporting, review and verification procedures were seen as crucial to ensuring that credits generated or traded under the mechanisms were legitimate and deserved to be treated on par with domestic action. Overall the negotiations were over access to the mechanisms on the one hand,

[68] See Sections 3(vi) and (vii), below.

[69] See also Chapter 4.

and use of eligibility as a motivator to comply with other requirements under the Kyoto Protocol and the Marrakech Accords on the other.

The European Union was generally in favour of strict eligibility requirements. This is consistent with its approach to implementation and compliance, which seems to focus on domestic action and the use of the so-called "Bubble" under Article 4 of the Kyoto Protocol. Article 4 will allow the EU to trade within the EU without having to resort to use of the Kyoto mechanisms.[70] The G-77 and China were generally in favour of strict eligibility criteria, although for varying reasons. The UG was generally considered to be most interested in making use of the mechanisms, and it was clear that access to "hot air"[71] from Russia was part of the compliance strategy for a number of UG countries.

The list of UG countries interested in liberal trading rules included the U.S. when it was still actively engaged, but also Japan, Australia, and Canada. For these countries a focus of the mechanism eligibility negotiations was whether hot air from Russia and other economies in transition would be freely available. In this context, UG countries were focussed on keeping out of the Marrakech Accords any conditions for eligibility that Russia might have difficulty meeting or might be unwilling to meet. This included qualitative reporting requirements on sinks inventories,[72] any link to acceptance of binding consequences,[73] and the failure of a Party to comply with Commitment Period Reserve requirements.[74]

On the other hand, mechanisms eligibility was seen by many as a very effective incentive to comply with reporting and procedural requirements of the Kyoto Protocol. This is in part because it provides an internal compliance incentive, which means that it does not require an action or sanction outside the Kyoto process. This avoids difficult enforcement issues and, depending on the definition

---

[70] Note, however, that during the course of the negotiations the EU includes only one economy in transition, former East Germany. The Bubble otherwise does not give the EU a compliance advantage over countries like the U.S., given that both are subject to the same rules for purposes of supplementing action in developed countries with action in economies in transition and developing countries, which are limited to Joint Implementation and the Clean Development Mechanism. "Bubble" refers to the process of replacing individual country targets with a collective target for all members of the "bubble".

[71] Hot air refers to the difference between actual emissions and the assigned amount for economies in transition, which has resulted in excess credits in those countries generated by economic collapse rather than emission reduction efforts. For a discussion of Hot Air, see Christine Batruch, "'Hot Air' as Precedent for Developing Countries? Equity Considerations" (1999) 17 UCLA J. Envtl. L. & Pol'y 45.

[72] See discussion in Section 3(vi).

[73] See discussion on compliance in Sections vi and vii.

[74] For the CPR requirements, see Marrakech Accords, *FCCC/CP/2001/13/Add.2*, Draft decision -/CMP.1 (Article 17), Annex, para. 6 – 10, at 54. See also discussion on CPR in Section 3(vi).

of binding consequences, provides an effective compliance tool without the requirement of an amendment under Article 18.[75]

In the end, only limited use was made of mechanisms eligibility as a motivator for Parties to comply with requirements under the Kyoto Protocol and the Marrakech Accords. In Bonn, Parties had agreed that in order to be eligible to use the mechanisms, a State has to meet the following requirements:[76]

It has to be a Party to the Kyoto Protocol;

a) It has to have satisfactorily established its assigned amount in accordance with Articles 3.7, 3.8 and 7.4;

b) It has to have its national system in place for emissions inventories in accordance with Article 5.1; and

c) It has to have its national registry in place in accordance with Article 7.4.[77]

The Marrakech Accords incorporate these decisions made in Bonn and include the following additional decisions on eligibility:

a) Maintenance of the Commitment Period Reserve was decided not to be an eligibility requirement, meaning that a Party could sell beyond the limits set under the reserve provisions and still have full access to the mechanisms;[78]

b) There is a requirement to report on sinks inventories under Article 3.4 as part of a Party's annual reporting requirements, but this is not a qualitative requirement. This treats sinks reporting differently than emissions reporting, for which a Party has to meet qualitative[79] requirements under Articles 5, 7, and 8 in order to retain its eligibility to use the mechanisms;[80]

---

[75] For a more detailed discussion of this, see Chapter 4.

[76] These are reflected in the Marrakech accords, *Ibid.* Draft decision -/CMP.1 (Article 17), Annex, para. 2, at 52.

[77] For a copy of the political agreement in Bonn, see UNFCCC, *Review of the Implementation of Commitments and of other provisions of the Convention: Preparations for the First Session of the Conference of the Parties Serving as the Meeting of the Parties to the Kyoto Protocol (Decision 8/CP.4)*, UNFCCCOR, 6th Sess., pt. 2, UN Doc. FCCC/CP/2001/L.7 (2001), online: FCCC <http://www.unfccc.int/resource/docs/cop6secpart/l07.pdf>.

[78] See Marrakech accords, *FCCC/CP/2001/13/Add.2*, Draft decision -/CMP.1 (Article 17), Annex, para. 2 at 52.

[79] *Ibid.*, Draft decision -/CMP.1 (Article 17), Annex, para. 2(e) at 53.

[80] See also discussion under Section 3(vi) and Sinks, Section 3(v), this Chapter. In the last moments of negotiations, Canada was able to include a further amendment on sinks' eligibility under Article 3.4 of The Kyoto Protocol, which results in each category of 3.4 sequestration being treated separately. This means that failure to meet requirements under Articles 5, 7, and 8 for forest management disqualifies a Party from issuing sinks credits for forest management until the problem has been resolved, but allows it to issue soil sequestration credits under 3.4, as long as the 5, 7, and 8 requirements are met with respect to soil sequestration. This illustrates the approach

c) The opportunity to link mechanisms eligibility to acceptance of an amendment under Article 18 on compliance was lost as a result of a failure to agree on binding consequences. Russia, Australia, and Japan opposed binding consequences and, as a result, opposed the link because it could have put them in a position of having to choose between ratifying an amendment containing binding consequences or losing access to the mechanisms. Solutions proposed for this in the last few days of negotiations were rejected by the four remaining UG members, including Canada. As a result, only acceptance of the compliance procedures, not an amendment on binding consequences has become an eligibility requirement;[81]

d) For Joint Implementation, there is an exemption for the requirement to meet reporting criteria in case of project-based Joint Implementation projects. In other words, if a host country can meet the reporting requirements for its emissions' inventories, assigned amounts, etc., it will be eligible under Track One of Joint Implementation. If it cannot meet those eligibility requirements, it can still take part in Joint Implementation, but only under Track Two. The basic difference between the two tracks is that the First Track is more like Emissions Trading, whereas Track Two is more like the Clean Development Mechanism.[82]

In summary, some use was made of eligibility to the mechanisms to improve compliance with monitoring, reporting, and verification procedures. Overall, however, this must be considered an opportunity lost for internal compliance. Eligibility was the main cross-cutting issue on the mechanisms addressed in Marrakech. A few others are covered under Articles 5, 7, and 8 Section (vi) below.

(v) *Land Use, Land Use Change and Forestry (Sinks)*[83]

Given the temporary nature of sink storage, and the challenges involved in accurately estimating the contribution of sinks to reducing GHG concentrations in the atmosphere, the rules for the use of sinks have a critical role to play in understanding the environmental impact of the Protocol. Monitoring, reporting, and verification all present major challenges for sinks, as does the influence of nature, such as forest fires just before the end of a commitment period. Sinks will increase the chance of a Party being out of compliance as a result of an unforeseen event, in spite of careful planning and implementation.

---

in the Marrakech Accords of maximizing access to mechanisms and sinks, and to limit them only when there is a direct concern about the quality of the credits in question. The opportunity to use eligibility as a motivator to bring countries into compliance was lost. This is particularly noteworthy and worrisome in light of the failure to agree on the binding nature of the consequences.

81 See Marrakech Accords, *FCCC/CP/2001/13/Add.2*, Decision 15/CP.7 (Mechanisms) at 2; *Ibid.* Draft decision -/CMP.1 (Mechanisms) at 3.

82 See also Joint Implementation, Section 3(iii). This again demonstrates the focus on ensuring the quality of the specific credits generated as the objective of eligibility criteria, not the use of eligibility as a motivator for compliance.

83 See Marrakech Accords, *FCCC/CP/2001/13/Add.1*, Decision 11/CP. 7 (Land use, land use change and forestry) at 54; *Ibid.* Draft decision -/CMP.1 (Land use, land use change and forestry) at 56.

Sinks are an issue under the Kyoto Protocol in three contexts:

- Article 3.3 with respect to land use changes in the context of forests;
- Article 3.4 with respect to additional human-induced activities related to land use, land use change and forestry; and
- The use of sinks under the Clean Development Mechanism.

The issue of sinks has been a controversial one ever since their inclusion in the Kyoto Protocol.[84] It is worthwhile at the outset to establish exactly what the relevant provisions of the Protocol provide.

Article 3.3, as a starting point, is relatively limited in scope. It requires the inclusion of net changes in carbon stock in Annex I countries compared to 1990 as a result of afforestation, reforestation and deforestation to be included in the accounting of a Party's emissions for purposes of meeting its emissions reduction target. Article 3.3 is limited to land use changes for forests, and is further limited to direct human-induced changes, excluding natural afforestation, deforestation and reforestation. Furthermore, what is accounted for under Article 3.3 is the change in carbon stock during the commitment period as a result of the land use change, not the total change in carbon stock since 1990 (the base year for identifying what land use changes have occurred). Essentially, Article 3.3 creates an incentive for Annex I countries to maximize the forest cover within their jurisdictions from 2008 on.

The most controversial issue with respect to Article 3.3 has been where to draw the line between reforestation on the one hand, and deforestation and afforestation on the other. In the end, the Parties agreed to a definition of afforestation as the forestation of any land that has not been forested for at least 50 years, reforestation as the forestation of land that was not forested on 31 December 1989, and deforestation as the conversion of forests to non-forested land.[85] Other difficult issues included the definition of a forest,[86] and the role of harvesting in the accounting of carbon stock.[87] This is relevant because of the reference in the definition of deforestation to the conversion of land to non-forested land. Harvesting with either natural or human-induced regrowth would not meet that definition. It is left to Parties to report on how harvesting is distinguished from deforestation. In the end, while there were fundamental disagreements over whether or not sinks should be used by Parties to meet their emission reduction targets, there was general

---

[84] *Supra* note 1.

[85] See Marrakech Accords, *FCCC/CP/2001/13/Add.1*, Draft decision -/CMP.1 (Land use, land use change and forestry), Annex, Part A at 58.

[86] *Ibid.*

[87] See *Ibid.* Draft decision -/CMP.1 (Land use, land use change and forestry), Annex, Part B at 59.

recognition that this issue had been included under Article 3.3 in Kyoto. As a result, the remaining issues were resolved relatively easily.

Article 3.4 has been much more problematic and controversial. It essentially provides an opportunity for Annex I Parties to claim credits for removing greenhouse gases from the atmosphere through means other than land use change in the context of forests. In practice this can include forest management activities, and the sequestration of carbon through other means such as in soils, among others.

There were a number of fundamental issues which made negotiations on this provision difficult. First, given that there was no clear expectation by the Parties as to whether and to what extent they would be able to make use of this provision, it is reasonable to assume that Article 3.4 was not a major factor in the negotiation of countries' targets under the Protocol. If this was the case, it raises the possibility of a country essentially getting an economic windfall relative to other Parties, if it turns out that the rules agreed to for Article 3.4 provide it with a cheap alternative to meet a major portion of its commitment. In other words, use of Article 3.4 could have been and was seen by countries who could not take advantage of this provision as an attempt by countries with large land mass to renegotiate their targets. For some countries, Article 3.4 had the potential to cover more than half the gap between the country's Kyoto target and business-as-usual projections.[88]

Secondly, one of the objectives of the Protocol was that it should achieve an overall reduction of emissions of at least 5% below 1990 levels in Annex I countries. If a major effort were to be put into activities under Article 3.4, this collective obligation would no longer be met.

Thirdly, sinks carry with them the inherent difficulty of their lack of permanence. Should sequestration be treated on par with emission reduction, knowing that any credit claimed by a Party now will turn into a debit at some point in the future, and most likely become an issue in the negotiation of future commitment period targets? Should Parties be permitted to essentially borrow from future commitment periods in the interest of minimizing the cost of meeting the first commitment period target, thereby enticing those States to ratify?

The response to all these questions from two of the three major negotiating blocks was "no" for most of the negotiations. The EU and the G-77 were opposed to allowing any additional sequestration activities under Article 3.4 for the first commitment period. This was one of the reasons, perhaps the main reason, why negotiations failed in The Hague. The change in position by the EU on this issue

---

[88] For a good overview of the issues related to sinks, see Chris Rolfe, *Sink Solution* (2001), online: West Coast Environmental Law Association <http://www.wcel.org/wcelpub/2001/13458.pdf>.

between The Hague and Bonn was also the main reason why the political agreement was reached in Bonn, and why it was possible to finalize the implementation of the Buenos Aires Plan of Action in Marrakech. As a result, there are generous provisions in the Marrakech Accords allowing for the use of Article 3.4 that clearly go beyond what was contemplated, or at least agreed to, in Kyoto.

As a starting point, Parties are permitted to use the following activities to generate credits under Article 3.4: revegetation, forest management, cropland management, and grazing land management.[89] Of these, forest management has been the most controversial, in part because it is generally recognized to have the greatest sequestration potential, and because of difficulty in separating natural sequestration from sequestration induced by human forest management activity. In the end, Parties agreed to country by country caps on the amount of credits a Party can generate from forest management under Article 3.4.[90] The numbers are loosely based on an 85% discount factor,[91] and a 3% cap on forest management activities, meaning a country may be able to increase its assigned amount by up to 3% based on forest management alone.

Other sinks activities under Article 3.4 are not affected by this cap, and can be used to the extent that a Party is able to demonstrate anthropogenic sequestration. Any land accounted for under Article 3.4 becomes part of the permanent inventory of the Party, which means that all future anthropogenic emissions by source and removals by sinks are accounted for in future commitment periods. This presumably means that any carbon stored in these forests as a result of human activity will become an emission in case of fire or any other activity resulting in their release back into the atmosphere, whether or not the release itself is caused by human interference. A country, of course, has the option not to designate any or all of its forested land for forest management, and thereby avoid both the obligation to account for changes in carbon stock for existing forests, and the risk of future debts in case of forest fires or other natural disasters leading to release of the carbon stored.

Another major concession to UG countries most interested in using Article 3.4 was the full separation of the various activities permitted. If a Party cannot meet

---

[89] See Marrakech Accords, *FCCC/CP/2001/13/Add.1*, Draft decision -/CMP.1 (Land use, and use change and forestry), Annex, Part C at 59.

[90] See Marrakech Accords, *Ibid.* Draft decision -/CMP.1 (Land use, land-use change and forestry) Annex, Appendix at 63; and *Ibid.* Decision 12/CP.7 (Forest management activities under Article 3.4 of the Kyoto Protocol: the Russian Federation) at 64, which increases the Russian cap from 17 to 33 megatons.

[91] The discount factor is intended to ensure that removals other than from anthropogenic forest management are not counted. Such natural removals include removals resulting from elevated carbon dioxide levels, indirect nitrogen deposition, and the dynamic effects of the age structure resulting from activities prior to the reference year.

the inventory requirements to issue forest management credits it can, for example, still issue cropland management credits, as long as it meets the inventory requirements for cropland management. At the heart of this issue was whether the reporting requirements and standards for sinks reporting are only put in place to ensure that the particular credits claimed can be and are verified, or whether these requirements for eligibility to claim sinks credits also operate as an enforcement tool to encourage Parties to comply with reporting and other obligations generally in the Protocol. The issue of qualitative annual sinks reporting as an eligibility requirement for the use of the mechanisms is another example of this. UG countries took the position that as long as a country had to meet sinks reporting requirements before issuing sinks credits, the integrity of the Protocol was preserved. Others argued that extending the effect of failing to report properly on sinks inventories to mechanisms eligibility was an effective means to encouraging compliance with sinks reporting requirements, and would avoid difficulties at the end of the commitment period, when the stakes are likely to be much higher because the options for making up discrepancies will be limited and are not likely to be attractive.[92] As discussed under Articles 5, 7, and 8 ahead, there are no qualitative eligibility requirements on sinks reporting for the first commitment period in the Marrakech Accords.[93]

Finally, in the context of the Clean Development Mechanisms, the issue of whether sinks projects would qualify for credits had to be determined. On the face, it might have seemed that sinks projects were excluded under Article 12, given its reference to emission reductions only. Article 6, for comparison, made specific reference to removal by sinks as eligible for credits under Joint Implementation. Nevertheless, Parties agreed that there could be limited use of sinks under the Clean Development Mechanism. Pressure to allow for this came from Annex I countries that saw this as a low-cost opportunity to meet their emission reduction targets, and a number of Latin-American countries who saw this as an opportunity to receive funding for reforestation projects in their countries.

The main challenge with sinks projects under the CDM relates to impermanence, which refers to the fact that all natural sinks eventually re-release the carbon stored back into the atmosphere. In case of sinks in countries that have an emissions reduction target, this problem is addressed by requiring the country to add those emissions into its emissions inventory when the carbon credited is re-released. In the context of the CDM, however, this may not be possible, given that the host countries currently do not have emission reduction targets or assigned

---

[92] It can generally be expected that the price of carbon credits will increase toward the end of the commitment period, as some Parties struggle to come into compliance. Even if that is not the case, the ability to bank credits will ensure an increase in the price, as long as the second commitment period target is considered to be a challenge for Parties.

[93] See Section 2 of Chapter 4 for a more detailed discussion of the compliance issue raised above.

amounts. These and other outstanding issues on sinks in the CDM could not be resolved in Marrakech, and were left to be addressed at the 9th Conference of the Parties, with technical assistance from the IPCC.[94] It was decided in Marrakech that sinks projects would be limited to afforestation and reforestation during the first commitment period, and that no Annex I country could claim credits from such projects in excess of 1% of its assigned amount.

In conclusion, it is difficult to see how sinks will play a constructive part in any long-term regime to address climate change, given the challenges involved. It is difficult to see significant net benefits, either from an environmental or an equity perspective.[95] At the same time, sinks are likely to generate incredible monitoring, reporting, verification and compliance challenges. Matters such as the temporary nature of sinks storage, the loss of other benefits of reductions, the difficulties in separating natural from anthropogenic carbon storage, and the long-term cost of accounting for the carbon stored all serve to make this a diversion from the overall objectives of the UNFCCC. Moreover, whatever the value of sinks in achieving the objectives of Kyoto and the UNFCCC, the way they were negotiated, as an alternative, rather than an addition to emission reductions, means sinks have watered down the ability of Kyoto to meet the objectives of the UNFCCC.

(vi) *Reporting and Review (Articles 5, 7, and 8)*

Articles 5, 7, and 8 of the Kyoto Protocol are designed to ensure that decisions about compliance and the use of the mechanisms are based on accurate, reliable, and consistent information from all Parties. To this end, Article 5 requires Annex 1 Parties to put in place a system for national emissions estimations on an annual basis in accordance with agreed upon methodologies. Article 5 allows for adjustments to be made to the emissions estimation if the methodologies are not followed. Article 7 then proceeds to require Parties to use those national systems to report annually on emissions by source and removal by sink, again, in accordance with agreed upon methodologies. Article 8, finally, provides for review, verification and adjustment of the information provided by expert review teams to ensure that Parties' annual reporting on emissions and sinks is accurate, consistent, and complies with the agreed upon methodologies.

---

[94] See *Report of the Conference of the Parties on its Ninth Session*, Conference of the Parties, United Nations Framework Convention on Climate Change (UN FCCC), 1-12 December 2003, FCCC/CP/2003/6/Add.1 (Decisions 1/CP.9 - 16/CP.9), FCCC/CP/2003/6/Add.2 (Decisions 17/CP.9 - 22/CP.9 & Resolution 1/CP.9), online: UN FCCC <http://unfccc.int/2860.php> [hereinafter COP 9].

[95] See also Sections 4 and 5 of Chapter 8 on the relationship between climate change, biodiversity and other environmental issues.

The negotiations under Articles 5, 7, and 8 introduced a number of new terms to the vocabulary of the participants. Other key terms were taken from provisions of the Kyoto Protocol. The following are some key terms used in the Marrakech Accords in this context. In addition to CERs,[96] ERUs[97] and AAUs,[98] which were introduced in the Kyoto Protocol, the Marrakech Accords introduced the concept of a Removal Unit (RMU).[99] This refers to a credit generated through sinks activity under Articles 3.3 and 3.4.

Once credits are generated, a number of different things can happen to them. Terminology to describe these different options has evolved to describe processes other than the obvious one of trading or transferring credits. Issuing of credits refers to the first step, the process of verifying and certifying credits. In other words, issuing credits refers to the process of bringing credits into existence. The process of cancelling credits refers to the opposite process, whereby credits are essentially taken out of circulation without being used to meet a country's commitments. This could be done, for example, if a Party wished to pay the price of emission credits not for investment purposes or for purposes of compliance, but to prevent emissions beyond the reductions required under the Kyoto Protocol. There are other isolated incidents where Parties may be required to cancel credits.

Most credits are expected to be retired by one Party or another at the end of the commitment period. Retirement of credits refers to the process of using credits to offset emissions during the commitment period. One other option for Parties is to hold certain credits to meet subsequent commitment period targets rather than use up or trade the credit in the commitment period in which it was generated. This option is referred to as banking credits or carry over of credits. One final term of note is the recycling or laundering of credits. These terms are used interchangeably depending on whether the person using the term approves or disapproves of the process. Recycling and laundering refers to the process of purchasing one type of credit with restrictions attached to it, such as an RMU that cannot be banked, and switching the credit purchased for an AAU or other unrestricted credit held by the Party, so that the AAU is banked and the RMU is retired for compliance purposes, rather than the AAU originally held for that purpose.

---

[96] Certified Emission Reductions, generated under Article 12 of the Kyoto Protocol. See Marrakech accords, *FCCC/CP/2001/13/Add.2*, Draft decision -/CMP.1 (Article 12), Annex, para. 1, at 26.

[97] Emission Reduction Units, generated under Article 6 of the Kyoto Protocol, see Marrakech accords, *Ibid.* Draft decision -/CMP.1 (Article 6), Annex, para. 1, at 8.

[98] Assigned Amount Units, assigned to Parties based on their assigned targets, see Marrakech accords, *Ibid.* Draft decision -/CMP.1 (Article 17), Annex, para. 1, at 52.

[99] See Marrakech accords, *Ibid.* Draft decision -/CMP.1 Modalities for the accounting of assigned amounts under Article 7, para. 4, of the Kyoto Protocol, Annex, paras. 4, 11 at 57, 59.

Much of the work done by the negotiating group on these three Articles was technical and not controversial. There were in the end, however, a number of key issues that had major implications for the use of the mechanisms and sinks as well as for the operation of the compliance system. Key issues in the negotiations included:

(1) Accounting of assigned amount under Article 7.4;

(2) Annual reporting on sinks;

(3) Reporting and review of Article 3.14 information;

(4) Composition of Expert Review Teams; and

(5) Relationship between the power of the Expert Review Teams to make conservative adjustments and compliance issues.

(A) Modalities for the Accounting of Assigned Amount, Article 7.4

There is a direct link between the challenges in the negotiations over the modalities for the assigned amount under Article 7.4, and the different views of Parties over the role of the mechanisms. It became clear in the course of the negotiations that the G-77 on the one hand and the Umbrella Group on the other hand, developed very different visions of how the mechanisms would be and should be used by Parties.

For the G-77, whatever credits were generated under any of the mechanisms would be held either by the host or the funding country until the end of the commitment period, and would be traded at that time between Parties depending on which Annex I countries had been able to meet their targets through domestic action and the use of project-based mechanisms. In other words, trading would take place among Parties at the very end to financially reward those who exceeded their target and "punish" those who were unable to meet their target without trading. In the meantime, the purpose of the mechanisms was to generate credits as between the host and funding country, not to trade them.

The Umbrella Group, on the other hand, had developed a market-based vision of how the mechanisms would operate, with assigned amounts and credits generated under the mechanisms traded on the stock market, with credits generated mainly by private entities, and with as few restrictions on trading as possible.[100] The objective would be for countries to be able to determine based on the price of carbon on the stock market, what the limit of expenditure per ton of carbon should be for domestic action. In other words, the UG wanted investors to be able

[100] For a discussion of the role of private entities, see Jean Acquatella, *Private Finance and Investment Issues in GHG Offset Projects* (Geneva: International Academy of the Environment, 1998).

to look to the stock market for guidance on what level of domestic effort would be cost effective in a country's effort to meet its emission reduction target.

These two very different visions on how the mechanisms would operate remained until the last few days of negotiations in Marrakech, and created ongoing challenges in resolving much of the detail on how to operationalize the Kyoto mechanisms. The way this impacted on the negotiations on Articles 5, 7, and 8 and, in particular, the modalities for accounting of assigned amount under Article 7.4 was as follows. The G-77 saw the assigned amount as fixed, and wanted to keep CER's ERU's and RMU's separate from each other and from the assigned amount for the duration of the commitment period. The UG, on the other hand, saw CER's, ERU's and RMU's as essentially freely added to and subtracted from the assigned amount, making credits generated indistinguishable from credits assigned to Parties at the outset as part of their assigned amounts. In the end, as discussed above under mechanisms, this issue was resolved in favour of the UG, with some symbolic safeguards put in place.[101]

(B) Annual Reporting on Sinks Activity under Articles 3.3 and 3.4

There is a clear requirement under Article 7 for Annex I Parties to include in their annual inventories and their annual reporting information on removal of greenhouse gases by sinks. The issues left to be determined were whether there would be a qualitative requirement similar to emissions inventories, and whether there would be a link to eligibility to use the mechanisms. It is important to note that the process of reporting annually on removal by sinks is separate from the process of verifying credits for removal of greenhouse gases through the use of sinks in accordance with Articles 3.3 and 3.4 and the resulting issuance of RMU's.

The purpose of requiring the reporting on sinks activities annually under Article 7 is to be able to better identify problems either with a country's ability to verify sinks credits or more generally, whether a country's overall compliance strategy, including its reliance on sinks, is likely to result in overall compliance with its target. This is particularly important in the context of uncertainty over legally binding consequences, based on the assumption that the less ability you have to motivate compliance through the imposition of consequences, the more important it is for the overall compliance system to be able to identify problems early and work with Parties to address problem areas.

In the end, another opportunity to develop an effective internal compliance mechanism that would improve the likelihood of compliance, either in the absence of or without having to resort to binding consequences, was lost. While there is

[101] See discussion on fungibility, under Kyoto Mechanisms, Emissions Trading, Section 3(iv), above.

a link between mechanisms eligibility and the requirement to report annually on sinks, there is no link to any qualitative requirement for reporting on sinks. This means an incentive to report something, but not to report in a manner consistent with what will be required when a Party seeks to have its sinks credits certified to enable it to issue RMU's for trading or compliance. That means there is reduced ability of the compliance system to catch sinks inventory problems early, and to prevent compliance problems at the end of the compliance period, when choices for addressing non-compliance are limited and generally not likely to be attractive.

While there was no link between the quality of reporting on sinks and eligibility, Parties did agree on some qualitative requirements for reporting. They include a requirement to report on the geographic location of the boundaries for each unit of land subject to Articles 3.3 and 3.4. The purpose of this requirement is at least twofold, to allow land that a country chooses to include in its sinks inventory to be tracked over time, and to allow for the monitoring of the biodiversity impacts of measures to improve the sink capacity of these lands. Reporting to demonstrate that the activities under Article 3.4 are human-induced was also included in the qualitative requirements, as was reporting on Parties' legislative efforts to address biodiversity concerns.[102]

(C) Reporting and Review of Article 3.14 Information

Article 3.14 requires Parties to strive to implement their commitments in such a way as to minimize adverse social, environmental and economic impacts on developing country Parties. The nature of this obligation was a contentious issue during the course of the negotiations, mainly due to pressure from OPEC countries, who consider this to require Annex I countries to take measures to minimize impacts on their economies. Given that the economies of OPEC countries are heavily reliant on the production and export of fossil fuels, OPEC countries took the position that Article 3.14 requires Annex I countries to take measures to ensure that reductions in consumption of fossil fuels are achieved in line with reductions in production within Annex I countries, so as to minimize any impact on the economies of OPEC countries. OPEC countries were looking for a number of ways to turn Article 3.14 into an enforceable obligation. One such attempt was to include qualitative reporting requirements on efforts to comply with this Article, and to make compliance with those reporting requirements an eligibility requirement for the use of the mechanisms.

---

[102] See Marrakech accords, *FCCC/CP/2001/13/Add.1*, Draft decision -/CMP.1 Land use, land-use change and forestry, Annex, Part E at 61; and *Ibid. FCCC/CP/2001/13/Add.3,* Draft decision -/ CMP.1 Guidelines for the Preparation of the Information Required under Article 7 of the Kyoto Protocol, Annex, Part D, at 22.

There was no overall agreement in Marrakech on what Article 3.14 substantively requires. With respect to reporting, Parties do have to report annually, but there are few qualitative criteria,[103] and reporting is not an eligibility requirement. Parties are required to report on efforts to phase out domestic subsidies for fossil fuel production, to assist in the development of non-energy uses of fossil fuels, and to assist developing countries in making the transition away from production of fossil fuels for energy purposes. Article 3.14 compliance issues can only be referred to the facilitative branch. There are, therefore, no consequences attached to non-compliance with Article 3.14.

(D) Composition of Expert Review Teams[104]

Composition of the Expert Review Teams was a long-standing issue in the negotiations, similar in nature to other disputes over composition of bodies to be established for the implementation of the Kyoto Protocol. What made negotiations on the Expert Review Teams somewhat different was the level of expertise required to carry out the function on the one hand, and the serious implication of the work of the Expert Review Teams on the other. This process of reviewing the information provided by Parties on their compliance with various obligations under the Kyoto Protocol is at the core of the compliance regime, and at the core of the implementation of the Protocol. It is not surprising, therefore, that the composition of the teams was particularly controversial. Positions on the selection of experts for these review teams ranged from geographical representation and essentially political appointments to representation from Annex I countries only, with appointment based on expertise.

In the end, the Marrakech Accords include some provision to limit the potential for the process to be politicized, and place the onus on the secretariat to select experts to a specific review team in a manner that ensures fairness and geographical balance. The roster from which the secretariat will select team members can be nominated by Parties. The selection process is left to be decided by the Parties collectively through the COP/MOP.

(E) Inventories and Conservative Adjustments[105]

The issue here relates to the power of Expert Review Teams to make conservative adjustments to national inventories submitted if those inventories do not

[103] See *Ibid. FCCC/CP/2001/13/Add.3,* Draft decision -/CMP.1 Guidelines for the Preparation of the Information Required under Article 7 of the Kyoto Protocol, Annex, Part H, at 25.

[104] See *Ibid. FCCC/CP/2001/13/Add.3,* Draft decision -/CMP.1 Guidelines for Review Under Article 8 of the Kyoto Protocol, Annex, Part E, at 42.

[105] For the decision that adjustments to emissions inventories shall be conservative, see *Ibid. FCCC/*

comply with the methodologies developed, or in case of gaps in reporting. Conservative adjustment in this context refers to the process of estimating a country's emissions, if the information provided is incorrect or incomplete. This measure was put in place due to the vital importance of access to a consistent and accurate estimation of emissions. The annual emissions of Parties during the commitment period provide the foundation for determining compliance, or what amount of a combination of assigned amount and credits a Party has to have in its possession at the end of the commitment period in order to be found to be in compliance with its target.

The conservative adjustment power serves two purposes, to encourage Parties to provide complete information in accordance with the methodologies developed so that all Parties are evaluated equally, and to resolve problems with reporting in a manner that ensures gaps in information are resolved in favour of protecting the integrity of the targets set. This is crucial because an accurate assessment of a country's emissions is only possible with the full cooperation of the Party and, even then, emissions can only be estimated. It is practically impossible to determine accurately actual emissions given their diverse sources.

One controversial issue in the negotiations was what to do when the conservative adjustment in itself was not a sufficient response to a Party's failure to comply with its reporting requirements on its annual emissions. This was resolved by providing as an additional enforcement measure the loss of eligibility to use the mechanisms, when the difference between what Parties were reporting and the adjustments made, exceeded certain limits. The limits imposed were a 7% difference in a single year, or a cumulative difference of more than 20%.[106]

The reporting and verification procedures are on the whole, quite strong. Consistent with negotiations dealing with trading generally, however, the process errs on the side of producing an efficient trading system over one that maximizes the environmental and equity benefits of the actions taken to comply with the Kyoto obligations. Significant future adjustments must be considered to protect the environmental and equity objectives of the UNFCCC if the cost of carbon on the international market becomes a significant driver for a Party's policy decisions on how to meet their Kyoto obligations.

CP/2001/13/Add.3, Draft decision -/CMP.1 Good practice guidance and adjustments under Article 5, paragraph 2, of the Kyoto Protocol, para. 5, at 12.

[106] See *Ibid. FCCC/CP/2001/13/Add.3,* Draft decision -/CMP.1 Guidelines for the Preparation of the Information Required under Article 7 of the Kyoto Protocol, para. 3 at 19.

(vii) *Compliance*[107]

It has been clear since the early days of negotiations under the UNFCCC that the stakes are high under the Kyoto Protocol, and efforts by individual Parties can be expected to depend on their assessment of the likelihood that other Parties will comply. In fact, many countries are perceived to consider the Kyoto Protocol as much an economic agreement as an environmental one. Given the economic implications, and the importance of equitable burden sharing, a strong compliance system was therefore considered crucial to this agreement.[108]

The heavy reliance on flexible market mechanisms places the Protocol in a unique position on compliance, at least as compared to other environmental agreements.[109] Compliance systems with effective enforcement tools are rare in international law, and almost nonexistent in the context of Multilateral Environmental Agreements (MEAs). To make matters more difficult, the Kyoto Protocol includes a provision, in Article 18, which requires any binding consequences to be adopted by way of amendment. This essentially prevents any Party from being subject to binding consequences as a precondition for ratifying the Protocol, given that any amendment to Kyoto can only be put forward once the Protocol has come into force, and that Parties will be able to decide whether to ratify any amendment independent of the decision to join the Protocol. At the same time, the economic stakes are considered to be high,[110] and much of the disagreement over interpretation and implementation of the Kyoto Protocol has been about the effort to be made by one country relative to another.

In this context, an effective compliance regime is likely to be essential as it will give each Party the assurance that other countries will do their part or face

---

[107] This Section provides a general description of the main features of the compliance regime designed for the Kyoto Protocol. A more detailed assessment of the compliance regime is provided in Chapter 4. For a general discussion of compliance issues in the context of climate change, see Peggy Rodgers Kalas & Alexia Herwig, "Dispute Resolution under the Kyoto Protocol" (2000) 27 Ecology L.Q. 53; David G. Victor, "Enforcing International Law: Implications for an Effective Global Warming Regime" (1999) 10 Duke Envtl. L. & Pol'y F. 147; Jutta Brunnée, "A Fine Balance: Facilitation and Enforcement in the Design of a Compliance for the Kyoto Protocol" (2000) 13 Tul. Envtl. L.J. 223.

[108] See Section 1 of Chapter 4 for a more detailed discussion of the importance of compliance with the Kyoto obligations for the future of the climate change regime, and for international law more generally.

[109] See Donald M. Goldberg *et al*, *Responsibility for Non-Compliance Under the Kyoto Protocol's Mechanisms for Cooperative Implementation* (Washington, D.C.: The Centre for International Environmental Law and Euronatura-Centre for Environmental Law and Sustainable Development, 1998), online: CIEL <http://www.ciel.org/Publications/pubccp.html>.

[110] While this assessment may still be accurate for domestic action, the price of international credits to supplement domestic action is generally considered to be much lower as a result of the U.S. pull-out. For a post-Marrakech analysis of the international price of carbon, see the Point Carbon Price Forecasting website: <http://www.pointcarbon.com/article.php?articleID=1648>.

consequences that will retain or restore the economic balance the Parties are so concerned about. Some of the Parties most concerned about the economic impact of compliance with their obligations under the Kyoto Protocol were also most strongly opposed to binding consequences, suggesting that non-compliance may be an option contemplated by those countries in case the economic burden of meeting their obligations was considered to be too high.[111] In spite of its limitations, and probably because of the high economic stakes, the compliance system developed under the Marrakech Accords promises to be by far the most effective compliance regime under any Multilateral Environmental Agreement.[112] The negotiation process itself, the ultimate enforcement tool of restoration of tons in the next commitment period with a penalty rate and a compliance action plan, as well as the fact that the next commitment period targets have not yet been negotiated, are all factors that can enhance the integrity of the compliance system by providing mechanisms for compliance on the one hand and affording opportunities to address any apparent inequities in the allocation of the burden to date on the other.

The focal point of the compliance system is the obligation of Annex I countries under Article 3.1 to meet their emissions reduction target in accordance with the rules and procedures agreed to in Marrakech. There are, however, numerous other obligations in the Kyoto Protocol. These commitments fall into two categories: obligations necessary to verify that a country has met its emissions reduction target, and obligations not essential to that process. Reporting obligations under Articles 5, 7, and 8 of the Kyoto Protocol clearly fall into the first category. Without the information required under those provisions, it is simply not possible to determine whether a country has complied with its national emissions reduction obligation under Article 3.1.

Obligations under Articles 10 and 11 of the Kyoto Protocol with respect to assistance to developing countries in the form of financial aid, capacity-building, and technology transfer clearly fall into the second category. In addition, there are a number of obligations that fall into a grey area, in that they have some link to the emissions reduction targets, but there were very divergent views among the Parties on how relevant that link was. This, in turn, had implications for how

---

[111] Australia, Japan, and Russia remained opposed to binding consequences throughout the negotiations. Canada formally accepted the concept of binding consequences late in the process, but supported those opposed in their efforts to prevent a decision on this issue before the first meeting of the Parties to the Kyoto Protocol. Of interest is that before it pulled out of the negotiations in the winter of 2001, the U.S. was a strong supporter of binding consequences.

[112] See Jeff Trask, "Montreal Protocol Noncompliance Procedure: The Best Approach to Resolving International Environmental Disputes?" (1992) 80 Geo. L. J. 1973; Elizabeth R. DeSombre, "The Experience of the Montreal Protocol: Particularly Remarkable, and Remarkably Particular" (2001) 19 UCLA J. Envtl. L. & Pol'y 49, and Elizabeth P. Barratt-Brown, "Building a Monitoring and Compliance Regime under the Montreal Protocol" (1991) 16 Yale J. Int'l L. 519. See also Brunnée, *supra* note 107.

these obligations were treated in the compliance system. These grey area obligations include the following:

- Article 3.1, reference to the goal of reducing overall emissions of Annex B countries by at least 5% below 1990 levels;
- Article 3.2, obligation to make demonstrable progress by 2005;
- Article 3.14, obligation to implement commitments in such a way as to minimize adverse social, environmental, and economic impacts on developing countries;
- Articles 6(d), 17, obligation that use of mechanisms are to be supplemental to domestic action; and
- Article 2, obligation to cooperate in the development and implementation of policies and measures to enhance their individual and combined effect.

Against this backdrop, we can now consider the actual compliance system established in Marrakech.[113] The Marrakech Accords set up a compliance committee that will function in the form of a plenary, a bureau, and two branches, one for the purpose of facilitating countries efforts to comply with obligations under the Kyoto Protocol, and the other to enforce compliance with specific obligations. The bureau, as well as the two branches, will select its members from the overall compliance committee. The composition of the overall compliance committee is determined by the composition of the facilitative and enforcement branches. Each branch is composed of one member from each of the five regional groups of the United Nations, one member representing small island States, and two members each from Annex I countries and Non-Annex I countries. While this was relatively uncontroversial for the facilitative branch, composition of the enforcement branch was one of the last issues resolved, due to a reluctance of Annex I countries to be judged by a majority of Non-Annex I members of the branch. Decisions are to be made by consensus whenever possible. In case consensus is not possible, a majority of three-quarters is required for any decision of the committee or one of its branches.

The plenary is responsible for the reporting to the COP, and for the overall administration of the compliance process. The bureau serves the function of receiving and reviewing questions of implementation brought to the compliance committee and determines which branch of the compliance committee is responsible for responding to the issue raised. The facilitative branch is generally responsible for assisting Parties in their efforts to meet their obligations under the Kyoto Protocol. This includes providing advice, and otherwise facilitating compliance with respect to commitments under Articles 3.1, 5.1, 5.2, 7.1, and 7.4.

[113] See Marrakech accords, *FCCC/CP/2001/13/Add.3*, Procedures and Mechanisms Relating to Compliance under the Kyoto Protocol, Decision 24/CP.7 (including the Annex), at 64.

With respect to these provisions, the mandate of the facilitative branch overlaps with that of the enforcement branch, which has a mandate to determine compliance and impose consequences of non-compliance with these provisions. In addition to providing advice on Articles 3.1, 5, and 7, the facilitative branch has the exclusive mandate to address questions of implementation with respect to supplementarity under Articles 6, 12, and 17, Article 3.14 dealing with effects of mitigation measures on developing countries, and reporting on demonstrable progress under Article 3.2.

Overall, therefore, the jurisdiction of the enforcement branch was limited to provisions that had a clearly accepted link to the emissions reduction target under Article 3.1. All other obligations were in the end agreed to be subject to facilitation only, not subject to enforcement. Decisions of the enforcement branch regarding compliance with Article 3.1 will generally follow the review of the final reports submitted by a Party under Article 8 at the end of the commitment period, which is expected to be concluded about 15 months after the end of the commitment period. Before a determination of noncompliance is made at this point, Parties are provided with an opportunity to come into compliance by purchasing the necessary credits from another Party. Under Part XIII of the compliance annex, a Party may buy credits for compliance purposes up to 100 days after the expert review process for the commitment period under Article 8 is declared by the conference of the Parties to be concluded.[114]

The importance of the distinction between the two branches becomes clear when one compares the consequences applied by each of the branches. The facilitative branch, under part XIV of the Annex on compliance, can take the following actions:

- Provision of advice and facilitation of assistance;
- Facilitation of financial and technical assistance, including technology transfer and capacity-building; and
- Formulation of recommendations to a Party on what could be done to address concerns about a Party's ability to comply with its obligations.[115]

The enforcement branch has the power to apply the following consequences:

- Make a declaration of noncompliance;
- Require a Party to submit a compliance action plan, which would include an analysis of the causes of non-compliance, measures to be taken to return to compliance, and a timetable for implementing the measures;

---

[114] *Ibid.*, Part XIII, Additional Period for Fulfilling Commitments at 74.

[115] *Ibid.*, Part XIV, Consequences Applied by the Facilitative Branch at 75.

- Suspend eligibility to use the mechanisms, if a Party is found not to meet one of the eligibility requirements; and
- In case of failure to meet its emissions reduction target under Article 3.1, the branch shall deduct from the Party's assigned amount for the second commitment period 1.3 times the amount of excess emissions from the first commitment period.[116]

The substantive penalty for not meeting the first commitment period target therefore is a reduction of the assigned amount in the second commitment period with a multiplier or penalty rate.[117] The question of what to do at this stage of the process was the subject of considerable disagreement during the negotiations. The challenge from a compliance perspective is that by 2012, it will be too late to actually reduce emissions during the commitment period, either through domestic or international action. A Party that has not been able to gather enough credits to offset its emissions has few options left at this point, and no control over whether they are available. The general options left at this stage are to require the Party to restore the reductions in the second commitment period, to impose a monetary penalty, or both.

The compliance challenge at the end of a commitment period, on the one hand, highlights the important role of the facilitative branch to identify these concerns early enough be able to assist a country to avoid noncompliance at the end of the commitment period. On the other hand, recognizing that the facilitative process alone is not likely to prevent non-compliance in all cases,[118] it raises the question of how to effectively respond at this stage. If other Parties have exceeded their targets, the noncomplying Party will be able to use emissions trading as a means to come into compliance. Without the U.S., and given the amount of "hot air" and sinks credits expected to be available from Russia and Ukraine, it is likely that this will be sufficient to allow countries to come into compliance. Given the possibility at the time that the U.S. might still join Kyoto, and in case of other unforeseen events that cause international credits to be unavailable,[119] some other means of restoring the balance and motivating compliance had to be found.

---

[116] *Ibid.*, Part XV, Consequences Applied by the Enforcement Branch at 75.

[117] Also referred to as borrowing or restoration.

[118] Especially considering the uncertainty introduced through the expanded use of sinks. Parties are not required to do qualitative reporting until they try to issue sinks credits. This could mean considerable differences between a Party's expectation and credits issued. In addition, any area included under articles 3.3 or 3.4 could lose its sinks credits in the case of a forest fire. This means considerable uncertainty over the amount of sinks credits a Party will have at the end of a commitment period. As a result, a Party could be out of compliance in spite of reasonable efforts to meet its target. Similar uncertainties would be introduced if a Party decided to purchase credits from a Party that was not complying with its commitment period reserve requirement, as it might again not know early enough whether those credits will be available for compliance purposes.

[119] Such as little or no effort domestically by a large number of Annex I countries.

There were two options considered in the early stages of negotiations, the restoration rate included in the Marrakech Accords, and a compliance fund. The compliance fund would have required payment rather than cancellation of assigned amounts in the next commitment period.[120] The compliance fund concept was eventually dropped over disagreement on how to ensure that the funds collected would be spent in such a way as to ensure the environmental harm of non-compliance would be undone. This debate centred on who would administer the fund, whether the funds would be spent domestically or internationally, whether payment into the fund would be voluntary or mandatory, and whether there would be a limit to the price to be paid per ton, or whether it would be linked to the price of carbon at the end of the commitment period with a multiplier. These issues could not be resolved in The Hague. The compliance fund was eliminated as an option at that stage, and did not resurface in Bonn or Marrakech.

The general process in the compliance system is that a question is brought to the bureau for a determination of which branch has jurisdiction. There are three ways issues can come before the compliance committee: as a result of a review of a country's submissions by an expert review team under Articles 5 and 7, at the initiative of a Party that realizes it requires assistance in meeting one of its obligations, or at the request of another Party that questions compliance of a Party with one of its obligations.[121] In cases of issues over which both branches have jurisdiction, such as reporting obligations under Articles 5 and 7, the two branches are expected to work in parallel, one assisting the Party in its efforts to correct the problem, the other to apply consequences associated with the breach.

The process described is generally open. However, there are provisions in the compliance agreement[122] that can reduce or eliminate the transparency of the process to a point where it risks losing its credibility. There are broad powers, for example, to prevent information from being made public until after the conclusion of the process. Similarly, there is provision for the hearings of the enforcement branch to take place in private. These powers could undermine the progressive provisions in the compliance agreement, which allow for submissions from intergovernmental and non-governmental organizations. Depending on the exercise of discretion, particularly by the enforcement branch, the process could be either open and transparent, or cloaked in secrecy. Most Parties to the negotiations seemed to value transparency and participation by non-Parties, and it was mainly

---

[120] See Glenn Wiser & Donald M. Goldberg, *The Compliance Fund: A New Tool for Achieving Compliance under the Kyoto Protocol* (Washington, D.C.: The Center for International Environmental Law, 1999), online: CIEL <http://www.ciel.org/Publications/ComplianceFund.pdf>.

[121] Marrakech accords, Procedures and Mechanisms Relating to Compliance under the Kyoto Protocol, Decision -/CP.7, Part VIII, General Procedures, para. 3 at 70.

[122] *Ibid.*, Part VIII, General Procedures, paras. 4-6 at 70; *Ibid.*, Part IX, Procedures for the Enforcement Branch, para. 2 at 71; *Ibid.*, Part X, Expedited Procedures for the Enforcement Branch, para. 1 at 72.

due to the strong bargaining position of Russia, as a key Party for the coming into force of the Protocol without ratification by the U.S., that these powers to restrict access were included. It would be reasonable to expect, particularly given the composition of the two branches, that these powers will not be exercised.

The agreement provides for an appeal process, but it is limited to due process issues. The Conference of the Parties (COP) will serve as the appeal body, and the decision being appealed will stand pending the appeal. This will ensure that the appeal process, which can take some time given that the COP generally only meets once a year, is not used as a way to delay application of consequences of non-compliance.

The most difficult issue throughout the negotiations on compliance, and the one compliance issue that remained unresolved until the last moments in Marrakech was the issue of legally binding consequences. Leading up to The Hague, there was a growing momentum toward legally binding consequences, with only three countries remaining opposed by the end of negotiations in The Hague.[123] As a result, Parties started to turn their attention to the Article 18 problem of how to make such consequences applicable to everyone who ratified the Protocol. One proposed solution was to make acceptance of the amendment a condition of eligibility to use the mechanisms. Other proposals would have tied acceptance of legally binding consequences to the acceptance of the second commitment period targets.

In the course of negotiations in Bonn, momentum continued to be in favour of legally binding consequences, however, the negotiating dynamics had changed as a result of the U.S. decision to pull out of the negotiations. This gave Japan and Russia a much stronger position to oppose legally binding consequences, and these two countries were successful in achieving a last-minute change to the Bonn agreement to delay the decision on this issue until the first meeting of the Parties to the Protocol, after it came into force. This had the effect of making it even more difficult to establish a link between acceptance of an amendment under Article 18 and some other step or process that countries would want to participate in, thereby assuring that the same compliance rules would apply to everyone.

The difficulty with retaining the link between mechanisms eligibility and an amendment under Article 18 post Bonn was that the content of the Article 18 amendment has not been determined. This could result in Parties agreeing to a link without knowing what would be included under the Article 18 amendment. This problem arises out of the amending formula under Article 20, which requires only 3/4 majority for agreement on the text of an amendment. This means Parties could be forced through the eligibility link to accept something, without having

[123] Japan, Russia, and Australia.

control over its content. One solution proposed was to make it a condition of the link between the Article 18 amendment and mechanisms eligibility that the text of the Article 18 amendment be adopted unanimously by the meeting of the Parties to the Protocol. This solution was rejected. As a result, even if there is agreement to amend the Protocol to provide for legally binding consequences, there may not be any way to ensure ratification of that amendment until the second commitment period. This is unfortunate given the importance for each Party to know that all Parties are committed to meeting their emission reduction obligations.

If Parties had been serious about preserving the possibility of legally binding consequences, as agreed to in Bonn, the problem could have been overcome. By not choosing to agree to the link between eligibility and the amendment on legally binding consequences in case of an amendment reached by consensus, an opportunity to strengthen the compliance regime was lost. This approach would, on the one hand, have protected States who are not yet convinced that legally binding consequences are the way to go. On the other hand, it would have preserved a way to ensure that in case of consensus on this point at COP/MOP 1, countries could ratify the amendment with some assurance that everyone would. It is unfortunate that this opportunity was lost in Marrakech. The direction of the negotiations on this issue would suggest that some countries were still looking to the compliance system as a release valve in case they had underestimated the effort required for compliance.

(viii) *The Role of Developing Countries*[124]

There is a tendency to focus on Annex I countries in assessing the Kyoto Protocol, because for the first commitment period, only Annex I countries have accepted binding targets for emission reductions. Furthermore, as discussed above, developing countries were not a significant factor in the crucial stages of the negotiations, from the failed meeting in The Hague in November 2000 to the agreement in Marrakech a year later. The relationship that develops between Annex I and developing countries is, however, crucial for a number of reasons, even at this stage. First of all, developing countries are likely to be the first to

---

[124] For a general discussion of equity issues under the Kyoto Protocol, see Batruch, *supra* note 71; V. Bhaskar, "Distributive Justice and the Control of Global Warming" in V. Bhaskar & Andrew Glyn, eds., *The North the South and the Environment* (London, England: United Nations University Press, 1995) 102; Michael Grubb, "Seeking fair weather: ethics and the international debate on climate change" (1995) 71 Int'l Aff. 463; and Eileen Claussen & Lisa McNeilly, *The Complex Elements of Global Fairness* (Washington, D.C.: Pew Center on Global Climate Change, 1998), online: Pew Center on Global Climate Change <http://www.pewclimate.org/projects/pol_equity_bak/pol_equity.pdf>.

suffer the consequences of climate change, and are least likely to have the capacity to respond. They also have contributed the least to the problem to date.

The Framework Convention recognizes this, as do the Kyoto Protocol and Marrakech Accords by requiring Annex I countries to go first in reducing emissions, by providing for adaptation funding,[125] by requiring Annex I countries to strive to minimize effects of mitigation measures on developing countries,[126] and by providing for capacity-building, technology transfer and financial assistance to developing countries.[127] Developing countries did make some progress in pushing their issues, but progress was limited, especially after the failure of negotiations in The Hague and the subsequent U.S. withdrawal from the Kyoto negotiations.

A critical question for the next few years will be whether Annex I countries have agreed to do enough under Kyoto to have the basis under the Framework Convention to convince developing countries to take on emissions targets. Without developing country targets of some form, it is unlikely that UG countries will take on the kinds of targets required to make meaningful progress toward the long-term goal of stabilizing concentrations at safe levels. It is expected that this debate will begin in earnest at COP 11. Depending on developments in the U.S., the negotiations over developing country targets for the second commitment period are likely to be further complicated by the question of U.S. participation in future commitment periods, especially given that the current U.S. opposition to Kyoto is in large part due to the absence of targets for developing countries. These issues and the role of developing countries in the future evolution of the climate change regime are considered in some detail in Chapter 9.

4. CONCLUSION

For anyone looking for Parties to make commitments that will significantly reduce the threat of climate change in a single step, the Kyoto Protocol and the Marrakech Accords are clearly a disappointment. The impact of Annex I countries meeting the first commitment period targets on the world's climate is likely to be

---

125 Marrakech accords, *FCCC/CP/2001/13/Add.2*, Decision 17/CP.7 (Article 12), para. 15, at 23. 2 % of CDM credits are designated for an adaptation fund.

126 See Kyoto Protocol, art. 3.14; and Marrakech accords, *Ibid. FCCC/CP/2001/13/Add.1,* Decision 9/CP.7, Matters Relating to Article 3, paragraph 14, of the Kyoto Protocol at 48.

127 See Marrakech accords, *Ibid. FCCC/CP/2001/13/Add.1,* Decision 2/CP.7, Capacity Building in Developing Countries (non-Annex I Parties) at 15; Decision 4/CP.7, Development and Transfer of Technologies (decisions 4/CP.4 and 9/CP.5) at 22; Decision 5/CP.7, Implementation of Article 4, Paragraph 8 and 9, of the Convention (decision 3/CP.3 and Article 2, Paragraph 3, and Article 3, Paragraph 14, of the Kyoto Protocol) at 32; Decision 6/CP.7, Additional Guidance to an Operating Entity of the Financial Mechanism at 40; Decision 7/CP.7, Funding under the Convention at 43, and Decision 10/CP.7, Funding under the Kyoto Protocol at 52.

negligible. Even before additional concessions were made in the negotiations leading to the Marrakech Accords, the impact of the first commitment period targets was considered to be relatively small.

Due to its heavy reliance on sinks and flexibility mechanisms, the Kyoto Protocol will also be a disappointment to those who are looking for these targets to put Annex I countries on the road to making the fundamental changes necessary to achieve more significant reductions in the longer term. Examples of such long-term measures would include changes in urban planning, changes to transportation infrastructures, shifts to renewable sources of energy, as well as investment in conservation. While the Kyoto Protocol does not preclude any of these measures, it is based on the principle of identifying the lowest cost measures to meet the short term (i.e., first commitment period) target without, in itself, encouraging Parties to consider the longer term, either from a climate change perspective or with a few minor exceptions, to consider other environmental, social, or equity issues in deciding how to meet the Kyoto target.

The immediate challenge will be to learn through implementation. Parties need to confirm through experience how various countries will be impacted, and be willing to make the necessary adjustments to keep and bring all countries into the Kyoto framework, including the U.S. and developing countries. Only at that point can meaningful international targets to reduce greenhouse gas emissions be expected. In the meantime, a key benefit of continuing these negotiations in the context of the UNFCCC and Kyoto is that there seems to be a commitment to the baseline of 1990, which provides some incentive for voluntary early action by all Parties, particularly Annex I Parties, whether or not they are Parties to the Protocol, or intend to become Parties for purposes of the first commitment period.

The long-term challenge, however may be to prevent the economic measures established in Kyoto and Marrakech from becoming the end in itself as opposed to one limited means for reaching it. If the Kyoto Protocol results in a long-term course of measuring progress on climate change in terms of short term lowest cost measures to reduce emissions,[128] then Marrakech will prove to be a mirage.[129] To be fair to the process set up through the Kyoto Protocol and the Marrakech Accords, they do not prevent Parties from considering those collateral costs and benefits, but they do very little to encourage consideration of these factors. The political reality in many of the Annex I countries is that a long-term vision of a sustainable future may be difficult to implement.

---

128 Regardless of what the collateral costs and benefits without some form of full cost accounting to ensure that the measures chosen are actually in our long-term best interest.

129 For a more detailed discussion of the limitations of the market to promote good public policy on social and environmental issues, see Lily N. Chinn, "Can the Market be Fair and Efficient? An environmental justice critique of emissions trading" (1999) 26 Ecology L.Q. 80.

A real opportunity to assist countries in that process has been lost in not allowing those countries to point to the international regime under the Kyoto Protocol as the reason for taking such a course. Instead, whether or not countries look to the price of carbon in the stock market to decide what to do to achieve their Article 3.1 targets regardless of lost opportunities to achieve long-term environmental, social, equity, and economic objectives, will be left to the political climate in each country to decide. It is that choice more than anything else that will answer the ultimate question posed in this Chapter, will Marrakech lead Parties to sustainability? Marrakech itself clearly is not the final solution to climate change, and time will tell whether it will help or hinder Parties' search for it.

It is important to keep in mind that many European countries have been able to achieve emissions on a *per capita* basis around half of those in North America without any international framework to preserve an economic level playing field. Properly motivated, similar results are possible in North America and Australia, and the real motivation will not come from the Kyoto Protocol or any other international agreement, but out of a realization that renewable energy, energy conservation and efficiency, reduction in consumption and production, public transportation, and the protection of our forests, oceans and biodiversity are in the best interests of humankind. The current state of developments in international law on climate change is simply a reflection of how far from a sustainable path Australia, the U.S. and Canada currently are.

One of the difficulties for the negotiations has been the lack of commitment to the long-term objective. Instead, the focus has been on the short term economic fairness or unfairness that has dominated the economic balance part of the negotiations. Major adjustments to the allocation are inevitable, considering the significant inequity in terms of *per capita* emissions. One solution to this has been the concept of "contract and converge".[130] It refers to an approach of reducing global emissions while bringing *per capita* emissions closer together. Such an approach will put increasing pressure on three countries, Australia, U.S. and Canada, with *per capita* emissions almost double the average emissions in other Annex I countries.

The process of working out the relative burden of the various participants is complicated by the fact that the political will to act is the lowest in countries that have the highest *per capita* emissions. At the same time, this is where most of the reductions, at least on a *per capita* basis, will have to be made in the long-term. So while this experiment with the Kyoto targets will provide Parties with useful

[130] See Global Commons Institute, online: GCI <http://www.gci.org.uk/>, for technical support and information concerning "Contraction and Convergence." A planning model, "Contraction and convergence Options," is also available for download.

information on which to base negotiations for long-term targets, the battle over the relative effort of various Parties has just begun.

It would be naive to assume, that, with some minor adjustments to the Kyoto targets based on the experience with the first commitment periods, countries can simply agree to reduce everyone's assigned amount proportionally for future commitment periods, and at the same time add targets for developing countries. Some countries, such as Sweden, have already reduced their *per capita* emissions to about a third of those in Canada and the U.S. Can they really be expected to accept further reductions until emissions in the U.S. and Canada have come down to somewhere close to their *per capita* emissions, accounting for national circumstances? Similarly, with the relatively modest targets accepted by countries like the U.S. and Canada, can the international community really expect developing countries to accept targets that are anything more than symbolic until the emissions from Annex I countries on a *per capita* basis are somewhere in the same range as *per capita* emissions from developing countries?

What can we expect out of all this? Will international efforts be limited to the lowest common denominator in the long-term? There are some signs of hope in this regard. The Kyoto Protocol has come into force against significant odds. There are signs that individual nations are prepared to lead the way, or follow a developing international consensus and thereby raise the bar. Sweden, for example, has announced a national commitment to not only exceed its share of the EU obligation under the Protocol, but to do so without reliance on sinks or the mechanisms. This means a significant step forward, given that *per capita* emissions in Sweden are already the lowest of any Annex I country.[131] Sweden is taking on the role of continuing to make the case that reductions do not have to come at the cost of quality of life. There are also signs that the U.K. and other European States are preparing to lead the way in making drastic long-term reductions, with long-term emission reduction goals of 60% or more.[132] The need for such reductions has also been considered in North America, even if so far mainly at the provincial and State levels.[133] Finally, there are significant efforts under

---

131 See Sweden's CO_2 Equivalent Per Person Emissions in 1994 compared with the figures for the U.S. and other Annex I countries. Sweden reported 7.7 tonnes while the U.S. had 23.7 for the same reporting period. See generally the UNFCCC's online Greenhouse Gas Inventory Database <http://ghg.unfccc.int/>.

132 See the Department of Trade and Industry (UK), *Energy White Paper: Our Energy Future–Creating a Low Carbon Economy* (Norwich: The Stationary Office, 2003), online: DTI Energy Group <http://www.dti.gov.uk/energy/whitepaper/ourenergyfuture.pdf>.

133 For a copy of the resolutions issued during the New England Governors/Eastern Canadian Premiers Annual Conference held in August 2001 at Westbrook, Connecticut, see the Government of Nova Scotia website: <http://www.gov.ns.ca/news/details.asp?id=20010827004>. The detail with respect to the agreement on climate change is contained in: The Committee on the Environment and the Northeast International Committee on Energy of the Conference of New England Governors and Eastern Canadian Premiers, *Climate Change Action Plan 2001* (New

way in the U.S. to reduce emissions, at least at the local and State levels. This gives some hope that in the absence of a national commitment to this issue in the U.S., States such as California may lead the way on this issue in North America in the way they have on vehicle emission standards.[134] To conclude, let's return to one of the questions posed in Section 3 of this Chapter. Assuming ratification and compliance, what is the role of the Kyoto Protocol in achieving the environmental and equity goals set in the UNFCCC?

We have seen as a starting point that Kyoto as negotiated in 1997 was a modest step forward at best in terms of the environmental and equity goals of the UNFCCC. The greenhouse gas emission reductions are minor, and they are limited to developed countries (most likely minus the U.S.). In terms of equity, Kyoto requires little of developing countries for now and takes some modest steps in terms of capacity-building and other assistance to developing countries. The combination of very modest obligations for developed countries and little in terms of obligations or assistance for developing countries points to the fundamental issue to be resolved before meaningful progress can be made: what is the relative obligation of the developed versus the developing world to address climate change?

These issues are explored in more detail in Chapter 9. In the meantime, having assessed the effectiveness of the Kyoto Protocol and the Marrrakech Accords in offering a global response to climate change, it is time to turn to one of the central questions for the future of the regime, will Parties to the Kyoto Protocol comply with their obligations? Will the compliance system as negotiated in Marrakech hold up to the pressures of the first commitment period? These questions are explored in the following Chapters.

England Governors/Eastern Canadian Premiers, 2001), online: The Council of Atlantic Premiers <http://www.cmp.ca/negecp/en-ccap.pdf>.

134 See, for example, Corporate Average Fuel Economy regulation (CAFE), briefly discussed in Section 4 of Chapter 5.

3

From Compliance Theory to Compliance System Design

1. INTRODUCTION

> Too many people assume, generally without having given any serious thought to its character or its history, that international law is and always has been a sham. Others seem to think that it is a force with inherent strength of its own....Whether the cynic or sciolist is less helpful is hard to say, but both of them make the same mistake. They both assume that international law is a subject on which anyone can form his opinion intuitively, without taking the trouble, as one has to do with other subjects to inquire into the relevant facts.[1]

This statement in many ways defined perceptions and debates about the role and importance of international law for much of the past 50 years. More recently, however, the debate has shifted from whether international law is relevant to how it influences State behaviour. Understanding when and why States cooperate and when and why they comply with commitments made through international treaties in particular has become a preoccupation of international lawyers and political scientists alike for at least the last decade.

The answer to when and why States comply is considered by some to hold the key to understanding how to improve international cooperation at a time when such cooperation appears crucial in such diverse areas as arms control, security, human rights, development and environmental protection. Andrew Guzman, for example, makes the obvious point that compliance has become one of the central questions of international law.[2]

Much of the discussion and public perception of compliance with international law is based on expectations drawn from and assumptions made based on domestic law experiences, as well as perceptions of international law based on high-profile, but not necessarily representative examples of success or failure of international

---

[1] J.L. Brierly, *The Outlook for International Law* (Oxford: Clarendon Press, 1944) 1-2, cited in H. J. Morgenthau, Politics Among Nations: the Struggle for Power and Peace (Brief ed.), revised by K. W. Thompson (New York, NY: McGraw-Hill Inc., 1993) 253.

[2] See, for example, Andrew Guzman, "A Compliance Based Theory of International Law" (2002) 90 Cal. L. Rev. 1823, at 1826.

law to achieve cooperation on global or regional problems.[3] These assumptions and perceptions include the view that deterrence through enforcement action is the main and most effective way of ensuring compliance. It is interesting to note that while the tendency based on this assumption may be to compare international compliance with national criminal or regulatory enforcement and other regulatory compliance strategies, contract law has at least historically provided a much closer domestic analogy to treaty law in the international context.[4] Furthermore, while these perceptions may influence our understanding of compliance, it is important to recognize that they may have little actual relevance in the international law context.[5]

On the other end of the spectrum, the assumption is that nations will only agree to take on obligations that they are relatively certain they will meet and that are in the best interest of the State, and that this is the overriding factor in the high level of State compliance with international obligations observed to date. The issue of compliance raises questions about the role of state sovereignty in international law,[6] the absence of global government with enforcement powers, and the roles and responsibilities of nations, government, individuals, and non-state actors in bringing about compliance. A consideration of compliance in the international law context quickly leads to a very complex set of possible motivations and factors. The following Section will identify some of these issues in the context of the current state of thinking on compliance with MEAs, the evolution of that thinking over time, and our current understanding of available compliance tools and their effectiveness.

Due to the existence of institutions to impose and implement strict enforcement measures domestically, enforcement has generally played a much more significant

---

[3] For a good discussion of domestic compliance and enforcement issues, see M. D. Zinn, "Policing Environmental Regulatory Enforcement: Cooperation, Capture and Citizen's Suits" (2002) 21 Stan. Envtl. L. J. 81. For a discussion of some high-profile failures of international cooperation, see Abram Chayes & Antonia Handler Chayes, The New Sovereignty: Compliance with International Regulatory Agreements (Harvard: Harvard University Press, 1995), at 34 to 67.

[4] More recently, with the focus on multilateral treaties and framework conventions followed by protocols, at least some international treaties resemble legislation and regulations more than contracts.

[5] See, for example, Benedict Kingsbury, "The Concept of Compliance as a Function of Competing Conceptions of International Law" (1997-1998) 19 Mich. J. Int'l L. 345. Kingsbury argues that our understanding of what compliance means will depend on our understanding of the relationship between law, behaviour, objectives, and justice (see that work at page 346). This means that the concept of compliance is influenced by the theoretical foundation and the context in which it is discussed. As we move from individuals to States, for example, the concept of behaviour and justice may change. As we move from domestic to international law, certainly the relationship among these factors can be expected to change.

[6] See A. C. Dowling, "Un-Locke-ing a Just Right Environmental Regime: Overcoming the Three Bears of International Environmentalism - Sovereignty, Locke, and Compensation" (2002) 26 Wm. & Mary Envtl. L. & Pol'y Rev. 891.

role in ensuring compliance with rules, norms and standards domestically than in international law. Even in the domestic context, however, the debate about the relative effectiveness of enforcement versus alternative ways of achieving compliance is an ongoing debate.[7] The limited range of enforcement tools practically available in international law to date has perhaps contributed to a more serious consideration of other factors and tools that influence compliance internationally. One key challenge identified is the implementation of enforcement measures in the absence of authoritative international law enforcement institutions. As a result of considerable theoretical work suggesting there may be more effective ways to achieving compliance internationally than through the threat of penalty or other enforcement action, more and more alternative ways of achieving compliance have begun to be implemented internationally in recent years.

At this stage, it is perhaps useful to clarify the terminology used. Compliance here refers to the end product of a Party meeting its obligation in international law. As defined by Raustiala, it means "a state of conformity or identity between an actor's behaviour and specified rule", and in the international context more specifically: "an actor's behaviour that conforms to a treaty's explicit rules".[8] Enforcement, on the other hand, refers to the process of penalizing a Party for not complying with its obligations. In other words, enforcement is one of a range of compliance tools that can be and are used to enhance or ensure compliance with a given obligation, norm, or commitment. Compliance on the other hand is outcome-oriented, regardless of whether it is achieved through effective enforcement, other compliance tools, external influences, or through chance.[9] In addition to the distinction between compliance and enforcement, there are a number of terms used in the discussion of compliance that are perhaps also best clarified at the outset. A common understanding of these terms is essential for a meaningful discussion on this issue.[10] The first group of terms relate to the nature of the duty on a State to act. There are a number of terms commonly used to describe such

---

[7] See, for example, David R. Boyd, Unnatural Law: Rethinking Canadian Environmental Law and Policy (Vancouver: UBC Press, 2003) at 228, and Andrew J. Roman et al, "The Regulatory Framework" in Geoffrey Thompson et al, eds., Environmental Law and Business in Canada (Aurora, Ont.: Canada Law Book Inc., 1993).

[8] Kal Raustiala, "Compliance & Effectiveness in International Regulatory Cooperation" (2000) 32 Case W. Res. J. Int'l L. 387, at 391. See also R. Fisher, Improving Compliance with International Law (Charlottesville, VA: University Press of Virginia, 1981).

[9] The role of enforcement is at the heart of the debate between rational and managerial schools of compliance theory, and the debate on whether the high rate of compliance is an indication that enforcement is not necessary or that States tend to accept obligations in international treaties only if they are easy to implement and comply with. See, for example, Chayes and Chayes, *supra* note 3.

[10] See Markus Ehrmann, "Procedures of Compliance Control in International Environmental Treaties" 13 Colo J. Int'l Envtl L & Pol'y 377. See also Ronald B. Mitchell, "Compliance Theory: an Overview" in James Cameron, Jacob Werksman, and Peter Roderick, eds., Improving Compliance with International Environmental Law (London: Earthscan Publications Ltd., 1996) 3.

duties; norms, commitments and obligations. Each suggests a different nature and extent of a duty to act.

Norms are generally considered to include a broad range of substantive and procedural concepts, including principles, standards, and rules. Norms can be explicitly expressed in writing within a treaty, or evolve out of custom, conduct or expectations among States. The term principle is sometime used to similarly suggest a general agreement to a framework for determining appropriate action in the future, without any specific duty to act in a specific manner in specific circumstances. This would suggest that there can be compliance with norms and principles, but perhaps that norms and principles, so defined are not subject to enforcement.

Commitment generally suggests an assumption of a specific duty, perhaps without direct or binding consequences in case the commitment is not fulfilled. In other words, commitment suggests a duty on States to act without a corresponding right or expectation on the part of other States to expect the results or benefits of that action. While there is no overall consistency in the use of these terms, a commitment is considered here to be more than a norm in that it involves a formal acceptance of a duty to implement. In other contexts, the term commitment has been used to describe duties that are subject to enforcement measures as well.[11]

Obligation, finally, is considered here to mean something more than a commitment, in that it implies an additional expectation of and provision for a direct consequence in case the obligation is not fulfilled. The term obligation can also suggest that other States are negatively impacted by a failure to comply whereas commitment suggests a benefit to others if the commitment is met, but not necessarily a right to that benefit.

This leads to the terminology to define the extent to which the obligation, commitment or norm is met or achieved by a State, group of States, or non-state actor. Implementation generally refers to the process of reflecting international norms, commitments or obligations in domestic legal systems, but not necessarily to turn the duty into domestic action.[12] In other words, in the context of a treaty obligation, implementation refers to the steps taken by a Party to fulfill its obligation, not to whether those actions achieve compliance or whether those actions

---

[11] It is important to note that these terms are not used consistently either in the literature or in international instruments such as the Kyoto Protocol. In the Kyoto Protocol, for example, the term commitment is used to refer to something that has become an obligation through the design of the compliance system and the consequences negotiated for non-compliance. See Chapter 4 generally on the Kyoto Compliance System.

[12] Edith Brown Weiss & Harold K. Jacobson, Engaging Countries' Strengthening Compliance with International Environmental Accords (Cambridge: The MIT Press, 1998), at 4.

are effective in meeting the objectives of the agreement. This is what distinguishes implementation from compliance and effectiveness.

Implementation by the government of a Party State will only inevitably lead to compliance where the obligation is completely in the hands of the national government of the treaty Party, such as an obligation to put in place domestic legal systems to reflect international norms or principles. Whenever the obligation goes beyond setting up a process and includes an obligation to achieve a substantive outcome, implementation and compliance diverge in that a Party may implement a treaty without complying with its substantive obligations.

For example, in case of a commitment to provide for a constitutional right to a clean environment for present and future generations, implementation and compliance with that commitment may be the same. In case of an obligation to ensure a clean environment for present and future generations, however, implementation may still take the form (at least in part) of providing for a constitutional right to a clean environment. Compliance, however, requires an assessment of whether the right to a clean environment is actually ensured for present and future generations. This would, as a starting point, require an assessment of case law applying the constitutional right to a clean environment in States that have made this commitment or taken on this obligation. Effectiveness, furthermore, starts with a consideration of the objective of the set of norms, principles, commitments and obligations that make up the treaty or regime in question.[13]

Considering the effectiveness of the treaty or regime then requires an assessment of the actions taken in response to the duty assumed in the context of how far it has advanced the objective. This means implementation and compliance are important factors in determining the effectiveness of an international treaty, but determining effectiveness requires further analysis.[14] In the environmental treaty context, effectiveness as distinct from compliance would ask the following questions: 1. Did the Parties achieve the treaty goals? 2. Did decisions made in the context of the treaty (such as emission reduction targets set) correspond to expert advice? 3. Did the environmental conditions, which lead to the treaty, actually improve compared to what would have happened without the treaty?[15]

In the context of compliance, for purposes of determining effectiveness of certain compliance measures, it is important to be able to identify treaty induced

---

[13] *Supra* note 10, Mitchell. For a good general discussion of effectiveness of international environmental law, see also: J. F. C. DiMento, "Lessons Learned" (2000/2001), 19 UCLA J. Envtl. L. & Pol'y 281.

[14] Brown Weiss, *supra* note 12, at 5. See also Ronald B. Mitchell, "Institutional Aspects of Implementation, Compliance, and Effectiveness" in Urs Luterbacher and Detlef Sprinz, eds., International Relations and Global Climate Change, (Cambridge, Mass.: MIT Press, 2001), 221-244.

[15] Mitchell, *supra* note 10, at 24.

compliance, which refers to State behaviour that is demonstrably influenced by the existence of the treaty, including the treaty's obligations and its compliance measures. Finally, the relationship between compliance and effectiveness needs to be considered. This involves consideration of whether the act of compliance or non-compliance furthers or hinders the effort to achieve the objective of the regime. In the context of climate change, it will be important to distinguish, if possible, between compliance that furthers and hinders the pursuit of the objective of preventing irreversible harm to the climate system. Similarly, it will be important to distinguish non-compliance based on whether it hinders or furthers the effectiveness of the climate change regime. A further consideration in this regard is a possible distinction between the purpose and effect of the act of compliance or non-compliance. For example, it is conceivable that empirical evidence shows that an act of compliance delayed the achievement of the objective of the UNFCCC. On the other hand, it is possible for an act of non-compliance by a State or group of States to actually have the effect of facilitating the achievement of the objective of the UNFCCC.

Compliance therefore refers to the end result of the process of taking a Party who has taken on an obligation or commitment in international law to a point where it has fulfilled the commitment or obligation. This process may seem straight forward on the surface, but it raises a number of issues. They include whether compliance is a yes or no proposition, who determines the nature of the obligation or commitment, and what measures are available to whom to promote, enhance or ensure compliance by a given Party. What is the role of international institutions, other Parties, non-Parties, and non-State actors in promoting compliance? What is the relationship between the specific obligation or commitment under consideration and the broader framework of international environmental law specifically or international law more generally? What is the relationship between compliance measures and state sovereignty?

In the next Section, the various theoretical foundations for State behaviour in response to international obligations are briefly introduced. Included are realist, rationalist, norm-driven and liberal theories of compliance with international law. These theories are considered from the perspective of what they contribute to the understanding of compliance and non-compliance, and what range of compliance tools might be effective in securing the highest possible level of compliance with international obligations.

2. THEORIES OF STATE BEHAVIOUR IN THE CONTEXT OF COMPLIANCE

Over time, a number of different theories have evolved to try to explain State behaviour with respect to each other and more specifically in response to international norms, commitments and obligations taken on in the context of international treaties. The following is a general overview of the most influential theories in the context of compliance. This overview is offered mainly as context for what is to follow, an assessment of the most prominent compliance debate over the past decade, the debate between managerialism and deterrence, as to which is the dominant influence on compliance with international law.

(a) Realist Theory

For realists, the absence of international enforcement institutions undermines the relevance of international rules, standards, norms, or laws. At the heart of realist theory of international law is the view that international society is in a state of anarchy. Anarchy is seen as the absence of an authority to enforce rules, or the decentralization of power.[16] This state of international anarchy, according to realists, results in individual States seeking to maximize power.[17] In other words, decisions about whether to comply with international rules, laws, or norms are based on decisions made by States on how to maximize power with respect to other States. Realism agrees with rational theories in viewing State action in the context of self-interest, but it differs in that it considers self-interest solely in terms of State power. In this light, compliance behaviour for realists is driven by a combination of the existing power distribution among States and decisions made by individual States on how to shift the power balance in their favour.[18] Realists reject the suggestion that international law does or even can guide State behaviour other than to the extent that international law is already a reflection of existing power relations among States.

(b) Rational Choice Theory

Rational choice theory has been a dominant theory of State behaviour and compliance for some time.[19] Rationalist theories of State behaviour generally consider States to act as rational entities making decisions that are or are at least perceived by the State to be in its best interest. It builds on realist theories of international relations by looking beyond State power in considering States' self-interests. In the process of going beyond State power as the goal, rational choice focuses on the process rather than the goal. In other words, it does not replace or supplement State power with other goals, but rather leaves the question of what

---

[16] Kenneth W. Abbott, "Modern International Relations Theory: A Prospectus for International Lawyers" (1989) 14 Yale J. Int. L. 335.

[17] See, for example, Claire R. Kelly, "Realist Theory and Real Constraints" (2004) 44 Va. J. Int'l L. 545.

[18] See Kingsbury *supra* note 5 at 350 – 351.

[19] See, for example, K. Abbott, *supra*, note 16. See also Raustiala, *supra* note 8, at 400.

the States' goals and priorities are in making choices about international cooperation and compliance, and instead focuses on the process of making rational choices about the interests of the State.

According to rational choice theory, factors considered by States in making choices about compliance with an international obligation can range from specific benefits expected from compliance, specific sanctions expected in case of non-compliance to less direct consequences such as reputational concerns. The focus of rationalist theories is on the structure of interests, actors, power, and incentives at play.[20] Raustalia distinguishes between coordination games, collaboration games, and non-cooperative game theory in explaining the basic choices a State can face, and the respective basic influences on compliance with a given rule.

Coordination games refer to situations where all Parties benefit from the existence of a given rule, standard or agreement compared to a baseline scenario of no rule. The example given is a rule that everyone drive on one side of the road or the other. Once a decision is made and the rule is established that everyone drive either on the right or the left, there is no benefit to anyone to break the rule intentionally, and everyone actually benefits from following the rule.

Collaboration games describe situations where there is a benefit to all Parties in complying with the rule or standard, but there are individual Parties who may be able to achieve greater benefit from non-compliance. This means that motivation to comply is mixed, in that there is an overall collective incentive to comply, but there are individual incentives not to comply. This will usually apply in situations where one Party's non-compliance will not significantly affect the collective benefit, but as the number of Parties who fail to comply increases, the collective benefit starts to erode.

An example might be an agreement to reduce or eliminate fishing in an area to allow a stock to recover. All Parties may have a significant interest in ensuring the recovery of the stock. If only one Party continues to fish, the stock may still recover. However, as the number of Parties that continue to fish increases, the loss of the collective benefit obviously outweighs the individual benefit of breaking the agreement.

Based on this comparison, Raustiala concludes that agreements that reflect the collaboration scenario should exhibit lower levels of compliance than those involving a coordination scenario. He offers international trade rules as an example of the former, and shipping navigational rules as an example of the latter, and concludes that at least in those examples, the relative rate of compliance is as expected.

---

[20] Raustiala, *supra* note 8, at 400.

The scenario perhaps most commonly referenced to illustrate the challenge of creating a scenario where all States are rationally motivated to comply with a given rule is the prisoners' dilemma.[21] In the basic scenario, there are two players who are not permitted to communicate with each other. They are each told that they have a choice to cooperate or defect. If they both cooperate, they each benefit equally. If one cooperates and the other defects, the player who defects benefits more than if they both cooperate, the other significantly less. If both defect, they both benefit less than if they both cooperated, but more than if they cooperate and the other does not.

The game illustrates very well the difficulty with many global environmental challenges. While there is an overall incentive to cooperate, in that the net combined benefit to the players is highest if they do cooperate, there are two factors that work against cooperation. One is the concern over being taken advantage of by cooperating when the other does not. The other is the reverse of the first, that one can get the benefit of the other player's cooperation without having to pay the price.

Rational choice theory itself has developed at least two distinct streams, institutionalism and political economy theory, reflecting the legal and social science perspective of rational choice theory respectively. They are briefly described below.

(i) *Institutionalism*

"Treaties marry a problem to a solution".[22] Institutionalism looks at the role of institutions in facilitating the process of enhancing the motivation for cooperation. Institutions[23] can do this by facilitating the distribution of obligations and benefits in solving a problem to ensure there is an incentive to join the institution and to comply with its rules. Institutionalists, as rationalists, assume that States will make rational decisions about their self-interest in deciding whether to join and whether to comply with the rules of the institution.

---

[21] See for example, Abbott, *supra* note 16, at 358.

[22] Raustiala, *supra* note 8, at 403.

[23] For a discussion of international institutions in an environmental context, see O. Yoshida, "Soft Enforcement of Treaties: the Montreal Protocol's Non-compliance Procedure and the Functions of Internal International Institutions" (1999) 10 Colo. J. Int'l Envtl. L. & Pol'y 95. For a good discussion of the difference between institutions and organizations, see also A. Najam, "Neither Necessary, Nor Sufficient: Why Organizational Tinkering Will Not Improve Environmental Governance" in Biermann, F., *et al*, *A World Environment Organization: Solution or Threat for Effective International Environmental Governance?* (Aldershot, Ashgate Publishing Company, 2005), at 237.

In other words, it is the institutions and their design in terms of the package of obligations and benefits that ensures opportunities for all to benefit from mutual compliance and an allocation of benefits and responsibilities that ensures everyone has the necessary incentive to comply. There are two interconnected but separate phases to consider; the decision to join the institution, and the ongoing process of making decisions about the level of effort to comply with the rules of the institution created once a decision is made to join.

Institutionalism is the international lawyers version of rationalist theories of compliance, in other words it focuses on the role of institutions in setting incentives for compliance and disincentives for non-compliance. Institutionalism looks at the benefits built into the design of an international institution combined with the deterrence included in the form of enforcement measures and/or penalties or consequences for non-compliance. Institutionalists tend to focus on the role of formal provisions in the agreement in considering the incentives and disincentives to comply with a given set of obligations.

In short, institutions are seen as a way to overcome the prisoner's dilemma essentially by changing the rules of the game. Institutions can facilitate communications between the players, overcome the fear that the other player will defect, and establish rules that ensure all players benefit more from cooperation than from defection, regardless of the choice made by the other players.[24]

(ii) *Political Economy Theory*

Political economy theory of compliance[25] is the social science version of rational compliance theory, but with a leaning toward realism in its focus on economics and power relationships. Not surprisingly, it takes a somewhat broader view than the institutional theory in that it goes beyond institutional incentives and disincentives to take a broader look at the benefits and penalties that influence States' rational choices about whether they comply with international obligations that go beyond the specific benefits and penalties incorporated into the institution itself. In other words, political economists would argue that there are many factors that influence a State's rational choice about compliance that are outside the agreement entered into,[26] and that both incentives within and outside the agreement have to be considered to gain a complete understanding of the rational choice made by individual States about their level of commitment to an interna-

[24] See Brett Frischmann, "A Dynamic Institutional Theory of International Law" (2003) 51 Buffalo L. Rev. 679, at 719.

[25] See, for example, Kingsbury, *supra* note 5, at 351.

[26] For a discussion of the difference between the institutionalist view and the political economy view, see George W. Downs, "Enforcement and the Evolution of Cooperation" (1997 - 1998) 19 Mich. J. Int'l L. 319, at 320 to 321.

tional obligation. Examples of factors outside the agreement might include the threat of economic sanction or other retaliation by a more powerful State against a less powerful State who, in failing to comply with obligations under an international agreement, harms the interests of the more powerful State.

(c) Norm-Driven Theories

Norm-driven theories focus on the role and influence of norms and ideas on State behaviour. These theories look at influences that cannot be explained simply by pointing to tangible material costs or benefits to a State. Rather, the influences are argued to result from the sharing of ideas and from the interaction between States that result in common views on a problem and the solutions. This process is thought to contribute to States complying with agreements reached through the process of internalizing the international norms that have been developed.[27]

According to Kingsbury, "socio-psychological studies suggest that individuals are much more likely to conform their behaviour to norms to which they have an internal volitional commitment, and that such commitments are correlated with perceptions that the relevant rule is fair".[28] He then relates this to State behaviour to suggest that being involved in developing the norm and more generally accepting the process for developing the norm as well as the legitimacy of the norm itself will contribute to a State's willingness to comply with the norm in spite of relatively high cost and low direct benefit.

The implication appears to be that the influence of norms on State behaviour results in compliant behaviour where a rational decision about the costs and benefits of compliance to the State alone might have resulted in non-compliant behaviour. A few questions arise from this suggestion. First, do norms replace rational choice as the primary driving force behind State behaviour, or is it a secondary influence? A related question is whether the influence of norms depends on the extent of the gap between costs and benefits to a State. Finally, how is it possible to separate compliance with norms as a rational choice, based on the calculated benefit of cooperating, from compliance, based on a desire to cooperate that goes beyond rational choice? It is not clear that the empirical work has been done or even how empirical work would be done to answer these questions. For example, is it possible to clearly separate normative from rational choice influences on State behaviour? If so, is it possible to identify one as the dominant or exclusive influence over State behaviour in a particular context? In the absence of this empirical work, it may be hard to determine the relative weight of these various influences on State behaviour in response to international obligations. It

[27] Raustiala, *supra* note 8 at 405.

[28] Kingsbury, *supra* note 5, at 355.

would appear reasonable to assume that the answer will depend on the State involved, the issue involved, and the particular circumstances surrounding the international obligation in question.

Under the general categories of norm-driven theories, constructivism, and managerialism are of particular interest here. They are therefore briefly described here.

(i) *Constructivism*

Constructivism is the dominant norm-driven theory of compliance.[29] The focus is on the "power of shared ideas" and the concept that State interests and States are "socially constructed, in part by norms, rather than as pre-theoretical givens".[30] Constructivists look at the influences of norms, ideas and social interactions not from a perspective of the costs and benefits a State determines can be gained from cooperating or following a norm, but from a less tangible desire to conform and cooperate that will lead States to follow either pre-existing norms or norms jointly developed even if to do so is not in the best interest of the individual State in question. An example of a pre-existing norm could be the norm that laws and rules developed through international agreements are to be followed.

It is important to consider exactly how institutionalists and constructivists see the role of interaction and discourse among actors differently, given that they have similar starting points. Both focus on the role of the institutions developed through the treaty in question as a major influence on State behaviour when it comes to compliance with the obligations agreed upon. The key difference between the two is that institutionalists see these interactions leading to rational choices made based on self-interest and constructivists see them as mechanisms for developing a joint commitment to a common objective that is removed from or independent of rational decisions about the best interest of participating States.

What is not clear is whether and why these two are mutually exclusive, why decisions made by States cannot at times be influenced more by rational choice that cooperation is in the best interest of a State, and at other times, when there is no event that forces such a rational choice, the constructivist influence of norms is more dominant in explaining State behaviour. The other open question is to what extent the impact of either change depending on the nature of the obligation, especially in terms of how fundamental a change is required for compliance, or how deep the required cooperation is.

---

[29] See generally, Raustiala, *supra* note 8.

[30] Raustiala, *supra* note 8, at 405.

(ii) *Managerialism*[31]

Managerialism is largely norm-driven, but it also incorporates some aspects of rationalists theories of compliance,[32] particularly institutional perspectives. While managerialists focus on the influence of norms and ideas in shaping compliance, there is really nothing in the managerialist approach to contradict the notion that the end result of compliance is achieved as a result of rational decisions made by States. The rational choice would be to cooperate and comply due to the overall benefit of international cooperation. This would presume a conclusion that compliance is generally in the best interest of each participating State, whether or not the direct costs and benefits arising out of the treaty in question in isolation would lead to a decision to comply. In other words, managerialism may be at odds with institutionalism, but it certainly seems to inherently accept that political economy theories of compliance provide at least a partial explanation for why States comply.

In the final analysis, managerialism is really an approach that is built upon the observed outcome that States generally meet their obligations unless there are capacity problems, uncertainties about the obligations or unforeseen changes in circumstances. It really does not focus very much on how States reach the point of engaging in compliant behaviour in the first place, in the sense that it does not consider in any detail the level of effort needed to meet the obligations studied. In other words, it has not to date responded to the criticism that the reason States comply with obligations most of the time is that the obligations they take do not require deep cooperation. Deep cooperation is used in this context to refer to cooperation that requires a State to do significantly more that what is in its independent self-interest.[33]

Managerialism assumes that the "majority of treaty violations are neither premeditated or [sic] deliberate, but are caused instead by three factors: (1) ambiguity and indeterminacy of language, (2) limitations in the capacity of Parties to carry out their undertakings, and (3) the temporal dimensions of the social, economic, and political changes contemplated by regulatory treaties".[34] Implicit in this position is that it is unlikely that coercion will lead to high levels of

---

[31] For the leading work on a managerial approach to compliance in international law, see New Sovereignty, *supra* note 3.

[32] For a detailed discussion of the role of norms from a managerial perspective, see New Sovereignty, *supra* note 3, at 112. See also Raustiala, *supra* note 8 at 407, and Friedrich v. Kratochwil, "How Do Norms Matter?" in Michael Byers ed., The Role of Law in International Politics: Essays in International Relations and International Law (Oxford and New York: Oxford University Press, 2000) 35.

[33] This point will be explored more fully in Section 3 in the context of the managerial versus deterrence-based compliance.

[34] New Sovereignty, *supra* note 3, at 10 to 14. See also Downs (1997/98), *supra* note 26, at 328.

compliance, and that coercion is more likely to have a negative effect on compliance because it will reduce the willingness to achieve compliance in a cooperative manner. This provides the starting point for much of the Chayes' work in *New Sovereignty*.

One concern expressed with the managerial approach to compliance is that it may actually in some cases hinder compliance. The concern is that making resources available to Parties unable to comply will actually provide an incentive not to comply in the first instance so as to be eligible for assistance.[35] The main criticism of managerialism is that it has not been shown to be effective in the context of deep cooperation.[36] Similarly, managerialists are concerned that sanctions may interfere with a general willingness to cooperate.

(d) Liberal Theories

Liberal theories of compliance challenge the notion that States behave as unitary actors in either making a rational decision whether to comply with international obligations or in being persuaded to comply through norms, ideas or social relations among States. Liberal theories focus on domestic actors and structures and their respective influence on the international behaviour of States.

In this context, liberal theory suggests for example that States with democratic domestic institutions are more likely to comply with decisions of international tribunals than illiberal States.[37] More generally, if a State has domestic institutions that are based on the rule of law, and if decision makers are accountable for breaching domestic rules, they are more likely to take international obligations seriously and take reasonable measures to comply. Furthermore, who stands to benefit and bear the cost of complying with an international obligation domestically relative to who makes decisions about implementation and compliance, according to liberal theories of compliance, will also impact on a State's commitment to meet its international obligations. One further example of how domestic structures can influence compliance with international obligations is the division of power in federal States. If there is a division between the power to enter into international agreements and the power to put in place measures necessary for effective implementation and compliance, this can be a significant compliance challenge in States where the level of government with the power to

[35] It is not clear that this concern is supported or refuted by any clear empirical evidence to date. See George W. Downs, David M. Rocke and Peter N. Barsoom, "Is the Good News about Compliance Good News about Cooperation?" (1996) 50 International Organization 379, at 387-99.

[36] See Section 3 for a more complete analysis of this debate.

[37] Laurence R. Helfer & Anne-Marie Slaughter, "Toward a Theory of Effective Supranational Adjudication" (1997) 107 Yale L. J. 273, at 331.

implement does not feel any sense of ownership or obligation toward the commitment taken on.

(e) Concluding Thoughts on Compliance Theories

These various theories about State behaviour generally and compliance with international obligations more specifically offer a wide range of explanations for why States comply or do not comply with international law. Two fundamental debates emerge from these theories. One is the debate between compliance based on a rational choice of the best interest of the State and compliance based on a desire to cooperate and a commitment to a joint effort to address a common problem. The other debate is between States as uniform entities that speak with a common voice and States as complex entities influenced by a great variety of interests and forces.[38]

It is not clear that these theories are necessarily mutually exclusive, and to what extent they each explain behaviour that States exhibit at some time. This raises the question whether a compliance system design has to choose among theories, or whether it can be designed on the assumption that all factors contribute to the understanding of why States behave the way they do under certain circumstances.

Alexander Thompson, for example, makes the point that you can explain any decision based on rational choice, but that does not necessarily mean that the behaviour in question is a result of rational choice, as opposed to a combination of factors or an irrational behaviour.[39] A key question in this regard is what these theories tell us about the likely effectiveness of different approaches to compliance. More specifically, what do they tell us about how to achieve compliance with obligations that require self-sacrifice by individual States in the interest of the global community, or "deep cooperation"? In the climate change context, what do these theories of compliance tell us about obligations such as the emission reduction obligations in the Kyoto Protocol, which are generally accepted to be onerous to implement, to require significant changes in behaviour, economic investment, or cultural adjustments?[40]

---

[38] Note that this debate is considered here only to a limited extent. A detailed analysis of the internal process of formulating positions on international agreements and compliance with international obligations is beyond the scope of this book. The need to consider, to the extent possible, the influence of non-state actors on compliance is, however, recognized. For a discussion of liberal theories of state behaviour, see Raustiala, *supra* note 8, at 409.

[39] Consider Alexander Thompson "Applying Rational Choice Theory to International Law: The Promise and Pitfalls" (2002) 31 L.S. 285. See also Robert Keohane, "Rational Choice Theory and International Law: Insights and Limitations" (2002) 31 J. Legal Stud. 307.

[40] See discussion on deep cooperation in Section 3 of this Chapter.

One way to possibly reconcile norm-driven theories of compliance with rational theories would be to consider the increasing importance of international cooperation and the increasing interdependence of States as the rational motivation for the observed cooperation referred to by norm-driven theorists. The question then becomes at what stage does the self-interest of States override the perceived benefit to cooperate. In many respects, we may already be seeing this line based on current value States place in international obligations. Following this line of reasoning, the value of cooperation would be reflected in what Parties are prepared to agree to in terms of substantive obligations in free rider situations. Presumably the same pressures that push countries to comply with an obligation in an effort to be seen as cooperating internationally can also motivate a State to participate in an international regime such as the Kyoto Protocol even though in isolation the State may not consider it to be in its best interest.

As we can see in the context of Kyoto, there are States who have decided to participate in spite of concerns about the impact of Kyoto on their economies. Other States with similar concerns have chosen not to participate. Presumably, following the reasoning of rationalism, some of these countries simply have crossed the line between the benefit of international cooperation and the cost of participation. For the US, for example, this may be because the cost of not cooperating internationally is seen by the current administration to be relatively low. For a State like Australia, the benefit of supporting the US may simply be seen as greater than the cost of not cooperating with other nations.

In the following Section, the compliance debate between norm-driven and rational choice theorists is considered in more detail in the form of the current debate between managerial and deterrence-based compliance in international law. This debate is considered here with a particular view to assessing the compatibility and conflict between these two views of compliance. The central question posed is whether compliance system design has to choose between these schools of thought, and if not, to what extent the relative influence of rational choice versus norms has to be resolved to be able to design effective compliance systems and to evaluate the effectiveness of existing compliance systems such as the one negotiated for the Kyoto Protocol.[41]

3. MANAGERIAL OR DETERRENCE BASED COMPLIANCE?

The debate between managerialism and deterrence as the basis for compliance in international law has particular relevance for Multilateral Environmental

---

[41] See, for example, Jutta Brunnée, "The Kyoto Protocol: A Testing Ground for Compliance Theories?" (2003) 63:2 Heidelberg Journal of International Law 255, for a discussion of the compatibility and potential for conflict between these approaches.

Agreements (MEAs) such as the UNFCCC and its Kyoto Protocol. Historically, MEAs have been short on enforcement measures, and have instead mainly relied on voluntary compliance with some facilitation. More recently, MEAs such as the Montreal Protocol on Ozone Layer Depleting Substances have started to use a managerial approach to compliance.

With the importance of effective, quick and "deep" international cooperation being highlighted in the environmental field through the emergence of numerous serious global environmental challenges such as climate change, threats to biodiversity, access to fresh water and supplies and the ability of voluntary and managerial approaches to compliance has been the subject of a particularly crucial debate in the context of MEAs. The following provides a closer look at this debate and what it may mean for compliance with the Kyoto Protocol.

(a) The Case for a Managerial Approach to Compliance

The starting point for much of the debate on compliance in international law invariably is the effectiveness of sanctions as the ultimate threat, the ultimate stick. On one side of the debate are the rationalists, who take the view that State behaviour in response to international obligations is dominated by rational choices made based on an assessment of the best interest of the State. The logical conclusion is a focus on a combination of deterrents and incentives as the heart of any compliance strategy.

On the other side of the debate are normative theorists. The predominant normative school of thought in the context of international compliance is managerialism, even though it also borrows from rational choice theory. As discussed above, managerialists suggest that deterrence-based compliance strategies are difficult to negotiate, even harder to apply in case of non-compliance, and are ineffective in most cases when applied because most incidents of non-compliance arise out of inadvertence, lack of capacity, and unforeseen circumstances, rather than rational choice.[42] Given that the enforcement versus management debate is currently the focus of the theoretical analysis of compliance with international obligations, a closer look at this debate is warranted here.

The debate was brought to the forefront in the mid 1990s with a book written by Antonia and Abraham Chayes entitled *The New Sovereignty*, in which the authors advocate for a managerial approach. It still provides the most detailed rationale for a managerial approach to compliance to date.[43] At the outset the Chayes' challenge the basic assumption of the rational choice theory of compli-

---

[42] New Sovereignty, *supra* note 3, at 29 to 33.
[43] *Ibid.* at 1 to 28.

ance that States only comply with international obligations when it is in their interest to do so. The authors instead argue that States have a general propensity to comply that arises out of a number of factors.[44]

The concept of the general propensity to comply is based on a number of observations. First, once a State ratifies a treaty or otherwise accepts an obligation, and directs its bureaucracy to implement, it takes effort to reach a decision not to comply. Compliance is therefore seen as the norm, and effort is needed to deviate from that norm. Secondly, the authors point to the fact that the obligations are assumed voluntarily, in that the agreements are negotiated on a consensual basis. This results in a presumption within a State that has accepted an obligation that the fulfillment of that obligation is in the State's best interest. Furthermore, the process of negotiating the agreement results in those involved developing a sense of ownership in the agreement and its objectives. In addition, the authors argue that there is a norm among States similar to a norm among individuals in a domestic context that laws are to be obeyed. This norm results in international obligations being fulfilled unless there is a strong "countervailing circumstance".

This leads the Chayes to a discussion of sources of non-compliance.[45] The starting assumption is that there is a general propensity to comply with international obligations as a result of involvement in the development of the obligation, the cost and effort involved in reversing a decision to make efforts to comply, and the general acceptance of the duty to obey international law. Given this, what accounts for non-compliant behaviour? At this point, the authors acknowledge that there are circumstances where the decision not to comply with an international obligation is a deliberate choice made by a given State. Iraq's invasion of Kuwait, and North Korea's refusal to permit International Atomic Energy Agency inspections[46] are offered as examples.[47] These are argued to be the exception rather than the rule. The more common sources of non-compliance offered by the authors are the following:

- Ambiguity and indeterminacy of treaty language;[48]
- Limitations in the capacity of Parties to carry out their undertakings;[49] and
- The temporal dimension of the social, economic, and political changes contemplated by regulatory treaties;[50]

---

[44] *Ibid.* at 3.

[45] *Ibid.* at 9.

[46] In accordance with the Nuclear Non-Proliferation Treaty (NPT) North Korea was Party to at the time. See Treaty on the Non-Proliferation of Nuclear Weapons, 21 UST 483, 1968, EIF Mar. 5, 1970.

[47] *Supra* note 3, at 9.

[48] *Ibid.* at 10.

[49] *Ibid.* at 13.

[50] *Ibid.* at 15.

Based on this, the Chayes propose an approach to compliance that focuses on overcoming these compliance problems. The proposal is to focus compliance systems on ensuring transparency, settling disputes, building capacity, and using persuasion to achieve compliance.

The foundation for the managerial approach proposed by the Chayes is their critique of deterrence-based compliance in Chapter 1 of *New Sovereignty*.[51] The critique is largely based on the authors' collection of evidence that the level of compliance with international obligations is high without resort to sanctions, that sanctions are rarely used in any event, and that they are often ineffective in improving compliance when they are used. Specific sanctions are considered in some detail, and the limits of their effectiveness exposed. Chapter two of Part I of *New Sovereignty*, for example, deals specifically with treaty based economic and military sanctions.[52] The focus is on the UN Charter and the OAS Charter; the only two treaties that formally authorize and have formal procedures for the international imposition of economic or military sanctions. The authors consider the following categories and events:

- Authorization of military force in response to large scale military action that threatens international peace and security (Korea, Cuba, Iraq);
- Economic sanctions to respond to racial oppression and colonialism in Africa; and
- Use and threat of sanctions to respond to a number of international crises since the Gulf War (Iraq, Yugoslavia, Somalia, Haiti).[53]

The Chayes discuss the crucial role played by the US in these cases,[54] the legitimacy challenge, and the influence this challenge has had on how the various sanctions were decided upon and implemented. The costs involved in the implementation of the sanctions, and the impact of the changing relationship between Russia and the US are also considered. Limitations identified include the need to rely on domestic decisions and resources for implementation, the consensus needed for legitimacy if not implementation of the sanctions, the difficulties encountered in making economic sanctions effective, and the internationally dominant role of the US in the decision making process and the implementation of sanctions.

New Sovereignty closely analyzes the effectiveness of treaty-based sanctions other than military and economic, categorized by the authors as membership

---

[51] *Ibid.* at 1 to 28.

[52] *Ibid.* at 34.

[53] *Ibid.* at 35.

[54] *Ibid.* at 40.

sanctions.[55] This refers specifically to sanctions used for the purpose of ensuring members of an international organization comply with the obligations imposed on them by the instrument which created the organization. The most obvious sanctions considered here are suspension, expulsion, or withdrawal of membership. In any of the three cases, it is a matter of weighing the cost to the member of leaving the organization against the loss of legitimacy of the organization itself as a result of the loss of the member. This means expulsions and withdrawals are rare. It also means that suspensions and expulsions are less likely, the more influence the member has internationally, and withdrawals are less likely, the less influence the member has.

The authors also assess and consider unilateral sanctions as an alternative to internationally coordinated action sanctioned by an international treaty.[56] Sanctions can range from the use of force to economic sanctions or changes in diplomatic relationships. Even though the sanctions considered here are unilateral, they are still considered in the context of enforcing treaty obligations, or otherwise responding to a Party's failure to comply with a treaty obligation. The authors consider the effectiveness of such sanctions in the context of GATT violations against the US, whaling, human rights, non-proliferation, and expropriation. The authors also consider the risk of unilateral action for the credibility of the treaty regime, the risk of "tit for tat", concerns about legitimacy of unilateral action as opposed to international action sanctioned under the treaty. The authors conclude that unilateral sanctions can be effective in specific situations, but are not an effective overall compliance strategy. It is important to note that the authors' focus in this part particularly, is very much on the role of the US as enforcer of international law.

The authors, in their final analysis, conclude that the threat of sanctions is not sufficient to ensure first order compliance with international obligations.[57] While rational choice theorists might offer second order compliance, and specifically enforcement through domestic institutions as a way to enhance compliance with international law,[58] the Chayes propose a more cooperative and participatory approach as an alternative. This is the approach referred to as the managerial approach to compliance and is discussed in Part II of *New Sovereignty*.[59] In this Part, the authors consider in more detail the role of cooperative means of managing

---

[55] *Ibid.* at 68.

[56] *Ibid.* at 88.

[57] *Ibid.* at 108.

[58] See, for example Fisher, *supra* note 8. For further consideration of second order compliance see also Laurence R. Helfer, Anne-Marie Slaughter, *supra* note 37. For discussion of the role of domestic intuitions, see Brett Frischmann, "Using the Multi-Layered Nature of International Emissions Trading and of International-Domestic Legal Systems to Escape a Multi-State Compliance Dilemma" (2001) 13 Geo. Int'l Envtl. L. Rev. 463.

[59] New Sovereignty, *supra* note 3, at 109.

compliance, including consideration of the role of norms in achieving compliance, the importance of transparency, the role of dispute resolution procedures, capacity-building, and the use of persuasion.

With respect to norms, the authors suggest that the process of establishing and refining norms is crucial in ensuring compliance.[60] The reference to norms in this context is about more than the initial agreement on the substance of an international treaty. It refers to an ongoing process of interpretation, application, and refining of the understanding of the nature and extent of the obligation taken on over time. Norms can be expressed in writing, or verbally, they can be formally agreed to, or develop as a common understanding either as background to a treaty, or by way of interpretation of a treaty obligation.

The authors discuss the role of justification of actions that may be seen as deviating from the norm (and the resulting dialogue on the norm itself). Other factors include the dependence of every country, even the most powerful, on the good will and support of others to achieve its objectives internationally, and the important role international organizations play in defining and refining the norms. Finally, the authors point to the importance of legitimacy of norms, including factors such as substantive and procedural fairness, and equal (or equitable) application of the norm.[61]

Issues of transparency, reporting, data collection, monitoring, and verification are also considered. Transparency is seen as important in improving coordination among Parties, to allow Parties to learn from each other how to implement the norm or obligation most effectively. Transparency also provides reassurance to Parties that others are meeting their obligations. This becomes more important as costs and other compliance constraints increase. Finally, transparency can act as a deterrent to discourage non-compliance through international embarrassment, by ensuring that the world knows if a country will not meet its obligations.[62]

Effective reporting and data collection serve as the foundation for transparency, and therefore in the end serve the same purpose. Obligations to report and collect data, at the same time, give rise to compliance challenges of their own. The authors discuss challenges involved in different approaches, including self reporting, advance notification and inspection, targeted inquiries, and the use of outside sources of information, including non-governmental organizations (NGOs), international organizations, and other Parties. Specific challenges of data collection considered include how to access the data, consistency and comparability of the data, and reliability of the data. Overcoming these challenges,

---

[60] *Ibid.* at 112.
[61] *Ibid.* at 127.
[62] *Ibid.* at 135.

itself, requires capacity-building, resources, and effective compliance tools to overcome unwillingness to share information where it is most needed, from Parties that are not complying with their obligations.[63]

One thing that is not clear is why this approach is offered as an alternative to sanctions. Domestically, prosecution is not offered as an alternative to monitoring, verification, or legitimacy of the rule for that matter, but they seem to go hand in hand. While domestic experiences do not necessarily apply to the international context, in this case it is not clear that the same would not apply internationally.

Up to this point, the authors focus more or less on what is required for an effective compliance alternative to sanctions. This is where credible norms, transparency, reporting, and verification are identified. The authors then turn to the question of how barriers to the creation of these conditions for compliance can be overcome. In this context, capacity-building, technical assistance, technology transfer, dispute settlement, and processes for treaty (or norm) adaptation are considered.[64]

Finally, the authors offer some practical suggestions on how to move forward. The potential roles of various players in moving toward an effective managerial, cooperative compliance strategy are identified.[65] The authors consider the International Monetary Fund (IMF)[66] as a case study to determine the relative impact of sanctions or the threat of sanctions, versus the use of managerial tools such as reporting, monitoring, and verification. Considered in this analysis are IMF compliance efforts such as surveillance of exchange rate policies, and conditions imposed on Parties before they can draw on the IMF's resources. While pointing out the potentially valuable role played by the use of managerial compliance tools, the analysis fails to make a convincing case that compliance with IMF rules is driven by the application of these tools as opposed to the ultimate threat of sanctions. The authors are not able to offer an effective empirical research method to separate these two influences on a Party's behaviour. Discussion of the application of similar tools in the context of the OECD, the GATT, the UNFCCC, and Long Range Transboundary Air Pollution (LRTAP) run up against the same limitation.[67]

---

[63] *Ibid.* at 154.

[64] *Ibid.* at 197.

[65] *Ibid.* at 229 to 285.

[66] *Ibid.* at 234.

[67] In considering the effectiveness of various tools, this limitation will have to be explored. No such research appears to have been done to date. To do this effectively, one would have to find examples that allow for an assessment of the effectiveness of managerial compliance tools separate from any threat of sanctions.

The role of NGOs and international organizations is considered as part of the analysis of the role of various players. With respect to NGOs, the authors conclude that they have played an important role, both in terms of pushing the envelope in the development of the norm, and in ensuring compliance. Much of the impact has been to enhance the credibility/legitimacy of the norm, to improve transparency, and to ensure reporting of non-compliance. Finally, in terms of international organizations set up to administer the implementation of the norm established, the authors point out the difficulties and limitations such organizations face over the long-term of retaining objectivity and autonomy on one hand, and retaining funding on the other.

This is particularly difficult with respect to the US and other powerful nations, who contribute significant resources without which these organizations cannot be effective. The challenge for international organizations in this context is how to fairly and effectively implement the norms in the face of this power imbalance, and in the face of the real danger of losing the support of these powerful nations. The authors conclude that only a managerial approach to compliance can hope to achieve this balance. The implication here is that some nations, including the US, are simply too powerful to be coerced, and that they therefore need to be convinced to comply, not only once, at the time of acceptance of the norm, but on an ongoing basis throughout the implementation of the norm.

In conclusion, the argument put forward in *New Sovereignty* is that compliance problems in international law are rarely a result of a deliberate choice not to comply with an obligation. Instead, non-compliance can be traced in most cases to ambiguity in the nature and extent of the obligation, lack of capacity, and unexpected intervening events that influence compliance with an obligation. The focus for compliance systems proposed by the authors is therefore to ensure transparency, settle disputes, build capacity, and the use of persuasion to achieve compliance.

(b) The Case for a Deterrence Based Approach to Compliance

Perhaps the sharpest critique of the managerial approach to compliance has come from George W. Downs and other supporters of the rational choice theory of compliance. Downs, in particular, has responded critically to the approach advocated by the Chayes in *New Sovereignty*. Downs rejects the managerial approach based on his conclusion that it only works when the obligation in question is a minor or modest deviation from what a State would have done without the agreement. In his view, as the level of effort or deviation from business-as-usual increases, the managerial approach fails, and strong, deterrence-based enforcement is needed to ensure States meet their obligations.[68]

---

[68] For a detailed review of deterrence-based enforcement, see George W. Downs (1997/98), *supra*

At the heart of the critique is that *New Sovereignty* ignores the level of cooperation needed to meet the obligation in assessing the level of compliance achieved with respect to a given international agreement. Downs argues that managerialists really only have been able to show that compliance is high when the effort needed to meet obligations is modest or low. Down argues that there are an ever increasing number of challenges facing the international community that will require more than modest efforts by States, if the international community is to have any hope of solving these problems. He refers to this higher level of cooperation as "deep cooperation".[69] Examples of areas where deep cooperation will be needed include human rights, arms control, and global environmental challenges such as climate change, biological diversity, and natural resource depletion.

Downs raises the question of whether deep cooperation can be achieved without enforcement. Can the managerial approach achieve cooperation and compliance where the effort required of individual States is high? As a starting point, Downs makes the case that the examples used in *New Sovereignty* to demonstrate States' propensity to comply are examples of shallow cooperation. One can debate whether the cases of deliberate non-compliance, such as the Iraqi invasion of Kuwait and North Korea's rejection of its obligations under the Non-proliferation Treaty are cases of deep or shallow cooperation. Regardless, there appears to be a general recognition by all involved in this debate over the managerial versus enforcement approach to compliance that there is simply not enough evidence to empirically determine whether the managerial approach is effective in case of obligations requiring deep cooperation.

Downs argues generally that the level of effort and the tools required to secure compliance depend on the level of effort a State has to make to come into compliance. Specifically, Downs accepts that the managerial approach of building capacity, ensuring transparency, and ensuring effective communication among the Parties can be an effective compliance strategy in case of shallow cooperation or modest obligations on Party States. As the level of effort and cooperation needed to address an issue increases, enforcement has to become more and more the focus of the compliance system.

In the end, Downs is convincing in arguing that the evidence presented by the Chayes is not enough to prove that a managerial approach alone can be effective in ensuring deep cooperation. This does not necessarily mean, however, that a managerial approach would not work for deep cooperation. To reach that conclu-

---

note 26, George W. Downs (1996), *supra* note 35, and George W. Downs, Kyle W. Danish and Peter N. Barsoom, "The Transformational Model of International Regime Design: Triumph of Hope or Experience?" (2000) 38 Colum. J. Transnat'l L. 465.

[69] For a detailed discussion of the definition of deep cooperation and how it might be measured, see Downs (1996), *supra* note 35, at 382, 383.

sion, one would expect to find evidence that the managerial approach failed in case of deep cooperation. Downs' overall point appears to be that deep cooperation has been rare, and therefore there is little evidence to test whether non-compliance in a deep cooperation context results from rational choice or from inadvertence. This leaves theoretical reasoning as the only basis for a conclusion on the question of whether a managerial approach can be effective in case of deep cooperation, and whether it can be as effective as or more effective than an enforcement approach.

Similarly, Downs is unable to show, based on empirical evidence, that enforcement is a more effective way of achieving compliance with obligations that require deep cooperation than a managerial approach. Downs' answer is to demonstrate that strong enforcement is needed for deep cooperation based on game theory, based on the assumption that States make rational choices about their interests in deciding whether to comply with an obligation. This, of course begs the question whether, in case of deep cooperation, States are more likely to make rational choices because the stakes are higher, or whether States are still more likely to be influenced by a desire to cooperate, or an acceptance of the rule of law. In other words, as the effort increases, does the likelihood increase that a State will fail to comply deliberately? If so, can a managerial approach nevertheless deal with such instances of non-compliance, or is an enforcement approach needed to prevent or punish a deliberate choice of non-compliance?

Intuitively, Downs' view may be compelling. It seems reasonable to expect that as the level of effort and the level of cooperation required goes up, so does the likelihood that the propensity to comply will be overshadowed by a decision that the cost of compliance is too high and not in the interest of a given State. Without empirical evidence to support this view, it is difficult to predict under what circumstances and at what point this transition might take place. It is reasonable, however, to assume that under some circumstances, for some States, that point does arise. In fact, some of the examples provided in *New Sovereignty* illustrate that States do get to a point where a deliberate choice is made not to comply. The question of what motivates such a deliberate choice, and whether it is possible to predict the circumstances under which such a deliberate choice may be made is not resolved by either *New Sovereignty* or by Downs in his critique. It is important to remember that Downs assumes that States are not influenced in any significant way by a norm that international law is to be followed, but rather will focus on what is in the self-interest of the State. In other words, both sides assume a clear, one sided answer to the most basic question: What is the relative influence of self-interest and a desire to conform to international norms on State behaviour?

Finally, one other issue not resolved is the connection between compliance and the obligations of a State. In most cases in the past, compliance systems have been negotiated either simultaneously or following agreement on the substance of the treaty in question. In those cases, States had the opportunity to consider the level of cooperation or effort required to meet the obligation in deciding what type of compliance system to agree to. Furthermore, States have generally been able to consider the combination of substantive obligation and consequences for non-compliance before deciding whether to ratify or otherwise agree to be bound by a treaty. This raises the question of the impact an enforcement based compliance system with a strong deterrence focus would have on the willingness of States to either agree to onerous obligations or to accept an obligation if it combined deep cooperation with strong enforcement. One possible outcome would be that Parties would be less willing to agree to accept onerous obligations because of a fear of sanctions. The other possibility is that States may be more willing to accept onerous obligations because they had a higher level of comfort that other States would cooperate rather than defect. This issue is also left open by both managerialists and proponents of deterrence-based compliance.[70]

(c) Implications for Compliance System Design

In the end, both schools of thought offer valuable insights into compliance, but both so far lack the empirical evidence needed to resolve key questions about their relative importance in understanding specific cases on non-compliance, and both are incomplete in their analysis in dismissing the influence of the other.[71] The Chayes, for example, start out with the proposition that while compliance measures for the implementation of many international norms at least in theory include economic and military sanctions, these measures are generally considered to be ineffective. The basis for this conclusion is that they are infrequently applied, and ineffective in restoring compliance when applied.

What is not addressed is whether they are designed to be effective at restoring compliance or preserving compliance. Could sanctions be more effectively designed to overcome a Party's motivation to violate a given rule of international law? While it is relatively straight forward to determine when sanctions are ineffective at restoring compliance, it is more difficult to empirically separate out the deterrent effect of these sanctions. Can sanctions be designed to serve both a deterrent function and a restorative function? While the Chayes appear to assume their ineffectiveness, or draw this conclusion from practical experience, others have actually considered these questions in a bit more detail, for example, by separating the effectiveness of sanctions on States and individuals acting on behalf of States.[72]

---

[70] Consider Downs (2000), *supra* note 68.

[71] See, for example, Jutta Brunnée, et al, "Persuasion and Enforcement: Explaining Compliance with International Law" (2002) 13 Finnish Yearbook of International Law 1.

[72] Fisher, *supra* note 8.

Downs starts out with the proposition that the effectiveness of the managerial approach is limited to shallow cooperation, and that the limitations of the managerial approach are hidden by the relative absence of treaties that require deep cooperation. Downs then proceeds to claim that enforcement is essential to ensure compliance in case of deep cooperation. In the end, neither Downs nor the Chayes is able to support their respective positions sufficiently with empirical evidence to allow for any firm conclusions about the relative effectiveness of either approach where it counts the most, in circumstances where a high level of cooperation is needed to address a pressing international challenge, such as the climate change challenge facing the international community today.

In the end, these two schools of thought on State compliance provide two enlightening and plausible explanations for State behaviour in response to international obligations. Neither school has demonstrated that it provides a complete explanation for when and why States comply and don't comply. In addition, there is nothing to suggest that these are alternative theories, rather it appears to be a question of which theory provides the better explanation of a particular State's behaviour in a specific context. From a compliance system design perspective, therefore, both schools of thought offer valuable insights into possible motivations to comply and possible challenges and barriers to compliance as well as a range of tools that can be used to secure compliance in various circumstances.

It would seem that there are a range of influences on State behaviour, some of which will be more normative, others will be more rational choice based. A compliance system designed exclusively based on enforcement or management of compliance is therefore not likely to be effective in ensuring compliance on its own. Rather, a system that recognizes the value of deterrence-based influences without compromising the ability of cooperative, managerial approaches is most likely to ensure the highest level of compliance by the greatest number of Parties in the long-term. An effective compliance system therefore should anticipate both a rational choice not to comply and non-compliance due to inadvertence, change in circumstances, and lack of capacity, and develop compliance tools to encourage compliance in any and all of those circumstances.

The challenge for compliance system design is to develop an overall system that takes into account as many possible influences for compliance and non-compliance as possible. This means developing a range of mechanisms and tools that allow the system to anticipate and respond to a range of motivations and circumstances. The various categories of circumstances that could lead to non-compliance and the tools available to overcome those compliance challenges are considered below.[73]

---

[73] See Section 5 of this chapter.

4. THE BINDING/NON-BINDING DIVIDE

Another current debate that is tied to the rational choice versus norm-driven theory discussion is the question of whether more progress can be made internationally with binding or non-binding commitments. The link is the potential link between a given approach to compliance and the willingness of States to take on onerous obligations. Do strong compliance systems make States more reluctant to make commitments to take substantively meaningful action out of fear of being forced to comply at a high cost? Alternatively, will States be more likely to accept the obligation in the comfort that others will also meet their obligations?

It is also particularly relevant to the following discussion about compliance with commitment and obligations in the Kyoto Protocol, as the binding nature of the commitments was the subject of considerable debate in the negotiations leading up to the Marrakech Accords. In the end, as discussed in the next Chapter, Kyoto contains both binding obligations and non-binding commitments, though the exact legal status of the binding obligations is yet to be fully resolved.[74] To gain a better understanding of the connection between the nature of obligations and commitments on the one hand, and compliance on the other hand, this debate is reviewed here.

The issue of binding versus non-binding obligations, in terms of effectiveness in achieving environmental objectives as well as in terms of achieving compliance with the specified change in behaviour, has received particular attention in the international environmental law field since the Earth Summit in Rio in 1992.[75] This is in large part due to the fact that Rio in many ways was a watershed moment for the development of international environmental law. It followed on the heels of the Brundtland Report, and took place at a time when public awareness of the global nature of the environmental problems we are facing had reached new heights. Issues such as acid rain, ozone layer depletion, climate change and loss of biodiversity were seen as enormous challenges to the international community and to international environmental law.

At Rio, the world community faced a fundamental choice: address these issues through binding obligations, through motivation, incentives, education, development of norms and principles, and other non-binding mechanisms, or both. At Rio, the international community rejected the idea of a world governing body to

---

[74] See discussion about the status of the compliance text in Section 3 of Chapter 2 and Sections 2-4 of Chapter 4.

[75] Rio Declaration on Environment and Development, June 13, 1992, 31 I.L.M. 874 [hereinafter Rio Declaration]. The first global environmental agreement to take a soft law approach was actually the Declaration of the United Nations Conference on the Human Environment, June 16, 1972, 11 I.L.M. 1416 [hereinafter Stockholm Declaration]. See also: Report on the United Nations Conference on the Human Environment at Stockholm, 11 I.L.M. 1416 (1972).

protect the environment and ensure these global environmental issues were tackled. Instead, while leaving open the door to future binding obligations, world leaders predominantly chose to take the route of agreeing to non-binding principles that were to guide future national and international efforts to address these issues. This choice in 1992 has raised the obvious question of the effectiveness of non-binding commitments. It has also raised the question of how to measure and enhance compliance with these non-binding commitments.

In a recent book on compliance, edited by Dinah Shelton,[76] 27 scholars, mainly from the United States and Europe, look at various aspects of compliance with international norms, with a particular focus on non-binding norms. The starting point for the work is a stock-taking of the status and role of non-binding obligations in international law, and some discussion of what distinguishes soft law from hard law. The book then presents numerous studies of compliance with specific non-binding norms to test the hypothesis that countries comply with non-binding norms as much as with binding norms. Beyond this, the studies attempt to identify what motivates countries to comply with non-binding norms. In the analysis, the focus is on the influence of the institutional setting, on regional diversity, and on the type of obligation, including whether the obligation is general and vague or specific and clear.[77]

The following summary of conclusions reached on compliance extracts the results of this work that have broader application to compliance with international norms generally. As pointed out by Peter Haas in Chapter 2 of the book, the distinction between international law and domestic law is much greater from a compliance theory perspective than the distinction between hard and soft law at the international level.[78] Much of this work on non-binding norms is therefore considered by Haas to be generally relevant to any understanding of what motivates countries and non-state actors to comply with international norms.

Peter Haas suggests in his contribution that a State's capability, its willingness or motivation, and the cost of compliance are the main factors that influence a country's level of compliance with an international obligation.[79] A State that has the capacity and the motivation to comply, and is able to comply without significant cost, is most likely to comply with an international norm. On the other hand, a State that does not have capacity, is not willing or motivated, and faces significant costs, is unlikely to comply. Haas then considers how a country's compliance

---

[76] Dinah Shelton (ed.), Commitment and Compliance: The Role of Non-Binding Norms in the International Legal System (Oxford and New York: Oxford University Press, 2000).

[77] Shelton, *Ibid.* at 3.

[78] See Chapter 2, Choosing to Comply: Theorizing from International Relations and Comparative Politics, *Ibid.* at 43.

[79] For a more detailed list of factors, see Shelton, *Ibid.* at 117- 118.

record with respect to a particular norm, and the effect of various possible factors identified can actually be evaluated. He proposes the following at page 51:

> First, process tracing for a given country will focus over time on the decision to comply with a given treaty or set of treaties, seeking to determine if the institutional factor correlates with the decision to comply and if there is a plausible causal mechanism between the institution and the decision. Secondly, aggregate analysis of treaties with these characteristics correlated with compliance levels offers a statistical appraisal of the propositions. Thirdly, and more ambitiously, multivariate analyses taking account of changes in institutional factors over time and of state choices over time offer a more convincing appraisal of the propositions, capable of taking account of changes in state behavior and the presence or absence of each factor, as well as identifying where more factors correspond with higher levels of compliance. Finally, counterfactuals applied to individual state choices could also contribute to an appraisal of these propositions.

Haas then proceeds to discuss various tools that can be used to enhance motivation and capacity to comply. They include compliance monitoring and verification measures, linkages among institutions involved in related issues, an understanding and use of connections between issues in motivating compliance, capacity-building, concern about the issue within Member States, and the institutional profile of the institution overseeing the substance, the compliance process or both. Haas then makes the point that each nation will have a unique sensitivity to certain motivations and incentives. He points out, for example, that democracies are more likely to be susceptible to international influence to implement environmental obligations.[80] Haas concludes by arguing that compliance is a matter of State choice, subject to institutional and constructivist forces.

Bilder, in his contribution,[81] focusses on the meaning of compliance, and in particular, the different meanings it may have to academics, Member States, and non-state actors. He makes the point that States' understanding of what should be done to comply with the norm established often evolves over time. He suggests that Parties would be less willing to accept the risk of being bound by a norm, if the norm was seen as rigid and unable to evolve with changing circumstances. To Bilder, the important thing is to develop techniques and signals that allow States to determine how seriously a given norm is to be taken, and what level of reliance on the norm is justified. Finally, Bilder suggests that compliance cannot be considered in isolation, but must be considered in the broader context of the norm and surrounding circumstances.

---

80 Shelton, *Ibid.* at 61, one point not addressed of course is the relative environmental impact democracies have on the global environment, and the potential connection between the form of government and the environmental impact.

81 See Chapter 2, Beyond Compliance: Helping Nations Cooperate, *Ibid.* at 65–73.

Reinicke and Witte[82] explore the impact of interdependence, globalization, and sovereignty on compliance with non-binding norms. As a starting point, the authors draw a distinction between economic interdependence, and globalization. They see the first as a process between States, whereas globalization is seen as a process structured by private actors. The authors consider the threat globalization poses to state sovereignty, and offer the development of international norms among States as a way to limit or counter this loss of sovereignty. This approach suggests that globalization has left States with the choice of giving up sovereignty to private actors, or to retain it in the form of global sovereignty, in terms of international agreements as the basis for internal sovereignty. In other words, international agreement to protect biodiversity, for example, may allow States to take internal action to protect biodiversity where globalization otherwise makes this impossible.

It is in this sense that international norms are seen as a means of preserving internal sovereignty against the threat of globalization. This, in turn, may become an increasing source of motivation on behalf of States to accept and implement international norms, to cooperate in their development and enforcement. In the context of compliance with norms related to environmental and natural resources issues, Kiss distinguishes between compliance at the international level, compliance within domestic legal orders where international norms have been incorporated into domestic law, and compliance with norms created by non-state actors.[83]

Edith Brown Weiss provides valuable insights into compliance with soft law.[84] She asks what difference it makes to compliance whether the norm is binding or not. Her conclusion is that it depends upon the circumstances. Based on the case studies in the book, in some cases binding norms evoke much greater compliance whereas in others a non-binding norm may invoke better compliance. Weiss then proceeds to reflect on the hypothesis of the overall study, which is that the context in which soft law develops does affect compliance. Weiss emphasizes a number of factors identified in the various studies, including connection to other international obligations, more generally the international context in which the norm is placed, the degree of consensus that develops on the issue during the course of negotiations and implementation, and the relationship between the context and

---

[82] See Chapter 3, Interdependence, Globalization, and Sovereignty: The Role of Non-binding International Legal Accords, *Ibid.* at 75.

[83] See Chapter 5, Commentary and Conclusion, *Ibid.* at 228-237. Examples of norms created by non-state actors would be the Code of Conduct for Responsible Fisheries developed by the FAO (though negotiated by States), and various standards issued by the International Standards Organization (ISO). One interesting issue here, though beyond the scope of this book, is the impact this has on the application of precaution, given that it assumes a loss of sovereignty over environmental protection in the absence of international agreement. It will be interesting to see if precaution will continue to get lost in the process.

[84] Chapter 9, Conclusion, *Ibid.* at 535.

compliance. In the context of the influence of social norms, Weiss proposes the following six factors:[85]

- If there is a continuing long-term relationship among participants, this will enhance compliance, and make it less crucial to have binding norms;
- If there is sufficient concern about reputation (i.e. concern over reputation outweighs barriers to compliance, such as lack of capacity, willingness, or high cost of compliance), binding norms may be unnecessary to achieve compliance;
- Compliance can be enhanced if the norm is consistent with established social norms. This suggests that it will be more difficult to ensure compliance when the norm is a deviation from existing social norms. This would suggest, perhaps, that compliance with environmental norms may be more difficult than with human rights;
- A shared desire to maximize welfare and minimize transaction costs may encourage compliance without the need for legal enforcement. This refers to the process of agreeing to efficient informal ways to ensure compliance as a means of jointly avoiding the more formal compliance process and possibly even the imposition of binding norms. An interesting issue left unresolved is the effect of participant's definition of welfare on their willingness to act. In other words, how do countries decide whether compliance increases the welfare of its citizens?
- The threat of sanction may increase likelihood of compliance (especially if the threat is credible); and
- Institutional structures that enhance transparency and accountability further compliance.

Weiss concludes with the suggestion that the binding or non-binding nature of an international norm is one of many factors that can influence compliance, and that all the circumstances surrounding an international norm should be considered in determining the most effective way of insuring implementation. The overall message from this study of compliance in the context of non-binding norms is that there are no short cuts. What is needed is a thorough understanding of the motivations and barriers of the participants, generally, and with respect to the specific norm to be implemented. Only with this base of information can appropriate compliance tools be selected and implemented.

The debate over the role of soft law tools in addressing global problems is closely related to the question of compliance. It is not clear that this debate has advanced sufficiently to offer any firm conclusions on what is essentially a debate between norm based and rational choice based theories of State behaviour. On the issue of concern over reputation, for example, Weiss suggests that there may

[85] *Ibid.* at 539.

not be any need for binding norms. From a practical perspective, however, the more important question might be whether States are likely to be more concerned about the impact on their reputation of non-compliance with a binding or a non-binding norm. The work that has been done on soft law suggests that non-binding norms do influence State behaviour. The relative importance of such normative influences remains an open question.

5. FROM THEORY TO SYSTEM DESIGN[86]

The preceding Sections on various aspects of compliance theory lead to the following conclusions. One is that there are a range of possible influences and explanations for State behaviour in response to international obligations. Another is that there is insufficient empirical evidence to determine the relative impact of these factors, especially as the level of effort required to meet an obligation increases. This Section considers the implications of these influences for compliance system design.

Ronald B. Mitchell[87] provides a useful overview of compliance that assists with this transition from theory to compliance system design in Chapter 1 of *Improving Compliance with International Environmental Law*.[88] In it, he considers various sources of compliance and non-compliance with international environmental obligations. On the compliance side, Mitchell distinguishes between compliance that results from independent and interdependent self-interest. Independent self-interest refers to the situation where the required behaviour itself is in the self-interest of the Party. There can be numerous reasons for this, ranging from no change in behaviour required in the first place, to changes that the Party had already decided to make for reasons independent of the international obligation taken on. The bottom line in the context of independent self-interest is that a Party will make its decision to comply based on the benefits it will derive directly from complying with the obligation.

In case of interdependent self-interest, Parties will consider the impact their own behaviour will have on others, either in the context of the specific obligation or treaty in question, or in a broader context. These factors can range from a net benefit of mutual compliance with the specific obligation to a desire to uphold

[86] The term "Compliance System" is used here in the same sense as it is used by Mitchell, *supra* note 10, as consisting of the primary rule system, the compliance information system, and the non-compliance response system, all in a treaty context.

[87] For Mitchell's recent work on compliance systems in the context of climate change, see also Ronald B. Mitchell, "Flexibility, Compliance and Norm Development in the Climate Regime" in Jon Hovi, Olav Schram Stokke, and Geir Ulfstein, eds., Implementing the Climate Regime: International Compliance, (London: Earthscan Press, 2005).

[88] Mitchell, *supra* note 10.

the integrity of international law generally because of the benefits a Party derives from other States meeting other existing or expected future obligations. Mitchell suggests that the concept of the net benefit from mutual compliance with a specific obligation has not become a significant motivating factor to date because of the realistic fear of free riders. This suggests a catch 22 that may play an important role in the renewed interest in compliance, especially in the context of international environmental law.[89]

Assuming that mutual compliance with meaningful and effective international obligations to address environmental issues is in the common interest of all, and assuming there is no independent self-interest in compliance in the absence of global compliance, a precondition for individual compliance may be a high level of comfort that there will be mutual or global compliance. It is important to note that the lack of comfort on global compliance may not only impact on a Party's compliance, but may also affect the negotiation of the obligation in the first place. This raises an interesting question in the context of the negotiation process. When are compliance systems negotiated relative to the negotiation of the obligations? How does one affect the other? If Parties were able to first design a compliance system that ensured mutual or global compliance, how would that affect the negotiation of the obligation as well as the individual Party's motivation to comply with the obligation? If, as Mitchell suggests, this issue of mutual compliance versus individual compliance in terms of the benefits of compliance is a crucial factor in compliance behaviour, then finding an effective way to ensure compliance may be a crucial step in ensuring more effective international action on environmental issues.[90]

Mitchell also considers a broad range of categories of non-compliance. In this context he considers non-compliance as a preference, non-compliance due to incapacity, and non-compliance due to inadvertence.[91] Mitchell's approach differs from the Chayes in that he accepts the positive influence enforcement measures can have, and he accepts the possibility that a rational choice about the self-interest of a State can lead to a decision not to comply in spite of the mutual benefit of international cooperation. Mitchell's structure for analysis therefore provides a useful basis for compliance system design that seeks to combine the managerial and deterrence-based compliance approaches.

---

[89] For a slightly different discussion of some of these issues, see Brown Weiss, *supra* note 12, at 6 - 8. Brown Weiss considers as categories for her discussion of compliance the following: characteristics of the activity involved, characteristics of the treaty, the international environment, and factors involving the country.

[90] The question posed here is essentially whether effective compliance offers a way out of the problem sometimes referred to as the multiple persons prisoners dilemma or free riding. For further discussion of the issue of collective action see: M. Olson, The Logic of Collective Action: Public Goods and the Theory of Groups (Cambridge, Mass.: Harvard University Press, 1965).

[91] Mitchell *supra* note 10, at 11 to 13.

As a starting point, it is perhaps stating the obvious that in the absence of compliance measures there will be circumstances where a Party concludes after taking on an obligation that the costs of non-compliance outweigh the benefits and therefore makes a decision not to comply.[92] This, of course, leaves unanswered for now the question of what factors into this decision, what benefits are considered, and how far outside the specific benefits in the context of the given agreement the Party will look to determine what is in its best interest. Examples of how a Party may get to this point could include that their needs were not adequately considered in the negotiation of the obligation, that the Party was unprepared at the negotiations and did not understand the implication of taking on the obligation until later, and that circumstances changed after the negotiation of the obligation to reduce the benefit of compliance or increase the benefit of non-compliance. Collectively, non-compliance in this category is likely to be most difficult to address, at least conceptually.

The second category Mitchell discusses is non-compliance due to incapacity. Violations can be due to financial, administrative or technological incapacity. There are numerous examples of this, most notably with respect to developing countries and economies in transition. One interesting consideration is the overlap between this category and non-compliance by preference. When to consider a Party's decision to put resources elsewhere a question of capacity versus preference is a reflection of the relative importance placed on the objective of the obligation on the one hand, and the domestic issue competing for the resource on the other. If a country chooses to protect its economy rather than prevent environmental harm for future generations, when is that a preference? When is it an issue of capacity? The concept of sustainable development might suggest that it depends on whether we are protecting needs at both ends, and that a current want compared to a future need would make the non-compliance a preference, whereas a current need as a competing interest might make it a capacity issue, regardless of whether the future right to be protected is a want or a need.

Thirdly, Mitchell refers to non-compliance due to inadvertence.[93] This category has much to do with the nature of the obligation and the implementation options. There are many obligations that come with inherent uncertainties on the compliance side. Furthermore, even if the obligation does not necessarily carry with it this uncertainty, there may be reasonable implementation options that carry with them this type of uncertainty. The Kyoto targets are perfect examples of this kind of uncertainty. First, due to variation in the climate and the level of economic activity from year to year, energy consumption and resulting GHG emissions will vary to some degree from year to year. This provides some level of uncertainty for countries who are working toward compliance. Furthermore, if a country

[92] *Ibid.* at 11.
[93] *Ibid.* at 13.

chooses to make use of sinks, CDM projects and joint implementation, these compliance policies carry with them additional uncertainty.[94] A forest counted on for a significant removal of GHGs could be destroyed by natural causes shortly before the end of the commitment period under Kyoto, resulting in inadvertent non-compliance.

These three broad and general categories of non-compliance are considered in the context of the complete absence of any measures, systems, or processes to achieve compliance, but rather based on what might happen after a Party has committed to some change in the context of an international agreement. In other words, the focus has been on a Party's predisposition to comply rather than the ultimate behaviour in response to the obligation or commitment. This leads Mitchell to consider how compliance might be elicited in these various circumstances of predisposition not to comply.[95] He considers positive and negative inducements, and then turns to compliance system design.

Under the heading of positive rewards are a number of compliance tools to address capacity and inadvertence. They include educational programs to build capacity and raise awareness, other capacity-building programs, building of partnerships, financing programs, and other similar initiatives. Positive inducements can also address a predisposition to non-compliance based on preference; however, if the inducement comes from other Parties, any significant monetary inducement may amount to a renegotiation of the obligations taken on. Inducements in other forms may include linking compliance with one to compliance with other obligations that may benefit different Parties, thus creating a network of interlinked commitments that would fall apart if one Party failed to comply with one obligation. This raises reciprocity as a possible basis for Parties to comply in the absence of self-interest.[96] Inducements from an international body such as a compliance body would likely be fairly limited in their ability to motivate a Party that is predisposed to non-compliance as a result of a self-interested cost benefit analysis.

Negative sanctions offer an alternative to positive sanctions, and are most likely to be applied to prevent a predisposition to non-compliance, out of pref-

---

[94] See Ronald B. Mitchell, "Institutional Aspects of Implementation, Compliance, and Effectiveness" in Urs Luterbacher and Detlef Sprinz, eds., International Relations and Global Climate Change, (Cambridge, Mass.: MIT Press, 2001), at 221-244.

[95] Mitchell *supra* note 10, at 13.

[96] Self-interest here includes the self-interest to bow to pressure from a more powerful State discussed below. This raises the question whether reciprocity, if applied as an incremental strategy to build a track record of compliance with gradually "higher stakes obligations" that are less and less in the interest of individual States but more and more in the interest of the common good is a way to overcome the prisoners dilemma/ tragedy of the common problem.

erence, to turn into actual non-compliance.[97] To be effective, a negative sanction must be credible and change the balance to ensure that compliance is now in the self-interest of the Party. An interesting question in the context of sanctions (but also relevant to the positive inducement option) is who hands out the sanction? The sanction could be issued by other Parties, by a compliance body, or by some other international institution. In certain circumstances, it could also originate from a non-state actor such as ENGO, the public, or a corporation. A related question is who receives the sanction, the Party who is failing to meet its obligations or the entity within that Party responsible for the non-compliance, such as a non-state actor.

One particularly interesting issue in the context of who hands out and who receives the sanction is the role of dominant States or blocs of States such as the US and the European Union. The relationship between relative power and level of compliance in terms of who can influence the compliance of others and who can be influenced by others to comply is particularly relevant here. This can be an issue directly if the inducements are in the hands of State Parties. The dominant State may choose to tolerate the violations, force other States to comply directly, or force other States or institutions to put pressure on a State to comply. The role of dominant States, therefore, is still relevant even if the inducements are in the hands of international institutions, as long as effective implementation, in the absence of world government and global enforcement, still depends on or benefits from pressure from dominant States.

Having considered various influences on compliance, Mitchell then turns to the role of the design of the compliance system in enhancing compliance.[98] A compliance system in this context would be any part of the treaty or regime rules that has the potential to influence compliance with the rule, other than the substantive rule itself. The general assumption made here is that there are choices to be made in the design of a compliance system. These choices can range from reporting obligations to access for NGOs and general transparency provisions. The question is whether and how these choices influence Parties' compliance.

Mitchell proposes three basic types of compliance systems.[99] More accurately, he proposes three distinct aspects to an overall compliance system. The primary rule system includes the Parties to the substantive obligation, the rules established for meeting the substance and the processes set up for the implementation of these rules. In other words, this part of the system is about setting the obligation and the process for meeting it. The compliance information system refers to the development, collection, analysis and dissemination of information regarding

---

[97] Mitchell *supra* note 10, at 14.

[98] *Ibid.* at 16.

[99] *Ibid.* at 18 to 20.

level of compliance with the obligation. Finally, the non-compliance response system deals with the question of how to respond to non-compliance. Much of the discussion on this last part of the compliance system complements the discussion above on inducements and sanctions.

The primary rules system is about the relationship between the obligation and compliance. It considers the compliance implication of what obligations are imposed on which Parties, what action is expected from States in implementing the obligation, and what the relationship between the expected actions of States and compliance is. In other words, is compliance an automatic result of States taking the expected action, or does compliance depend on the effectiveness of those actions, cooperation from other Parties, non-Parties, or non-state actors? Any specific roles and responsibilities in meeting a State's obligations identified for non-state actors are considered in this context.

The nature of the obligation is considered here, specifically whether the expected change is technically difficult, is behaviourally difficult and/or is difficult to enforce. Does it affect different States differently? What non-state actors does it affect and how? A further factor to consider here is how specific the obligation is. The more vague the obligation is, the more likely there will be different levels of compliance, whether due to self-interested interpretations of the obligation, or inadvertence. A related but distinct issue is to what extent the obligation itself is in the public eye, and the nature of the obligation is publicly understood. This is distinct from the issue of transparency of the compliance record of a Party; the focus here is on whether the nature of the obligation is publicly understood and what level of interest there is in the obligation.[100]

The compliance information system is all about transparency of compliance with the obligation. While we often think of transparency in terms of the general public or NGOs, transparency is a crucial component of compliance regardless of whether the pressure to comply ultimately comes from a dominant State, the public and ENGO, or an international institution. It is furthermore important regardless of whether the predisposition to non-compliance results from lack of capacity, inadvertence, or preference. Those in a position to influence the predisposition must have access to information about the level of compliance and the causes of non-compliance in order to be able to exercise their role in the compliance system effectively. The most common form of information gathering in this context is self reporting, with some roles for ENGOs, international institutions, the public, and industries in generating information and verifying compliance.

This leads to the third part of the compliance system, the non-compliance response system. As a starting point, in spite of its name, the compliance response

[100] These issues are also considered by Brown Weiss, *supra* note 12, at 6–12.

system as defined by Mitchell[101] can address non-compliance in a preventative manner. It can pro-actively seek to inform Parties of their obligations to prevent inadvertent violations. It can seek to identify capacity gaps and work pro-actively to fill them. It can identify failure to make demonstrable progress and put pressure on Parties to take measures to comply. Preventative measures can either be taken in response to early warning signs, or generically to create additional motivation to comply.[102] Secondly, the response can originate from a compliance body, from some other international institution, from a Party, or from a non-state actor. A preventative approach requires careful consideration at the design stage of opportunities to coerce compliance, eliminate opportunities to avoid compliance, to improve transparency of the obligation and compliance with it, and to focus on obligations that are suitable to effective enforcement action.

In terms of the actual response system to non-compliance in the narrower sense of the term, this takes us back to facilitation and sanctions, or the carrot and the stick. As pointed out by Mitchell, inducements tend to be expensive for the entity trying to influence a Party's compliance, whereas sanctions are usually often costly only if they do not work.[103] As a result, one would expect a tendency to rely on sanctions over inducements in motivating compliance.[104] While this may be the case generally, the picture changes somewhat in the context of non-compliance due to lack of capacity. In a number of recent environmental treaties, there has been an up front recognition that developing countries and economies in transition lack the capacity and sometimes the moral duty to reach the same level of response to an environmental issue as wealthier nations. This has been addressed in part by placing different obligations on countries with less capacity and moral responsibility. In recognition that nature does not distinguish environmental effects of human activity on this basis, however, there has also been a recognition that environmental improvements have to be made in countries with less capacity or moral responsibility. In such situations, there has been a formal recognition that the way to achieve compliance in developing countries is through inducements, not through sanctions.[105]

The role of Parties versus non-state actors and international institutions is perhaps the most challenging issue in the context of the compliance response system. On one hand, the challenge for Party control over sanctions is one of

---

[101] Mitchell *supra* note 10, at 16 to 23.

[102] An example would be the announcement by the US that it would not allow vessels into US ports that had not installed segregated ballast tanks to come into compliance with MARPOL; see Ronald B. Mitchell, "Regime Design Matters: International Oil Pollution and Treaty Compliance" (1994) 48 International Organization 425.

[103] Mitchell, *supra* note 10, at 21.

[104] Especially in cases of non-compliance by preference.

[105] Two recent examples where these concepts have been implemented to varying degrees are the Montreal Protocol on Ozone Depleting Substances, and the Kyoto Protocol on Climate Change.

motivation to impose a sanction on another Party to force compliance. Just as a Party is considered to be motivated to comply in part due to its dependence on others to meet their international obligations, a Party may be unwilling to impose sanctions for fear that this will result in sanctions imposed on it should it fail to comply with another international obligation at some point in the future. It would be interesting to relate this reluctance to the relative dominance of a particular State to see if this fear of becoming the recipient of a sanction in the future is as much of a factor for more dominant States.

On the other hand, leaving the imposition of sanctions in the hands of international institutions or non-state actors raises the question of the level of influence those actors are likely to have over a non-complying Party. The obvious conclusion, at least on the surface, would appear to be that compliance may depend predominantly on the actions taken and motivation of the most dominant States. The test for the effectiveness of any compliance system may be whether it can demonstrate compliance beyond or independent of the efforts of such States, or whether it is merely facilitating those efforts.

What is proposed here is that the effectiveness of a compliance system should be assessed based on its ability to overcome the various motivations for non-compliance identified, and at the same time, its ability to enhance member States' propensity to comply. This issue will be considered in more detail in the next Chapter in the context of the Kyoto compliance system.

4

The Kyoto Compliance System

1. INTRODUCTION

This Chapter assesses the Kyoto Compliance System based on the theoretical context provided in the previous Chapter. As discussed in Section 5 of Chapter 3, for present purposes, the compliance system consists of the primary rule system, the compliance information system and the non-compliance response system. Before considering in detail the likely effectiveness of the Kyoto compliance system, however, it is worth reflecting on the potential importance of compliance with the Kyoto Protocol.[1] The role of compliance can be considered from a variety of perspectives, ranging from its importance to the climate change issue itself, to its role in shaping environmental law, compliance, and international law more generally.

What then is the role of compliance in the success of the Kyoto Protocol?[2] Compliance is generally considered crucial for the proper functioning of the trading mechanisms in the Kyoto Protocol.[3] Beyond this, given that Kyoto is primarily about the reduction of GHG emissions in developed States as a modest first step toward more meaningful global action, a low level of compliance undermines the ability of the Kyoto Protocol to achieve its main objective of setting the stage for future action. In short, a high level of compliance is crucial for the success of the Kyoto Protocol itself.

---

[1] *United Nations Framework Convention on Climate Change*, Intergovernmental Negotiating Committee for a Framework Convention on Climate Change OR, 5th Sess., Annex, UN Doc. A/AC.237/18 (PartII)/Add.1 (1992), 31 I.L.M. 849, online: UNFCCC <http://unfccc.int/resource/docs/a/18p2a01.pdf> [UNFCCC or The Framework Convention], and *Conference of the Parties to the Framework Convention on Climate Change: Kyoto Protocol*, 10 December 1997, U.N. Doc. FCCC/CP/1997/L.7/add. 1, 37 I.L.M. 22 (1998), [hereinafter the Kyoto Protocol]. For the substantive discussion about the implications of compliance for the climate change regime and the future of environmental law, see Chapters 9 and the Epilogue.

[2] Success is seen here as an effective and constructive step toward an effective climate change regime, or an effective internationally coordinated response to climate change. See, for example, David G. Victor, "Enforcing International Law: Implications for an Effective Global Warming Regime" (1999) 10:3 Duke Envtl L. & Pol'y F. 147, and Jacob Werksman, "Responding to Non-compliance under the Climate Change Regime" 1999 OECD Information Paper, (Paris, 1999) online: OECD Documents <http://www.olis.oecd.org/olis/1999doc.nsf/LinkTo/env-epoc(99)21-final>.

[3] See Section 3 of Chapter 2.

In light of this, the level of compliance with Kyoto will have implications beyond the Protocol itself. A related issue is the role of Kyoto compliance for the future of the climate change regime. In other words, what does Kyoto compliance mean for the future of the climate change regime beyond 2012? This is a crucial question in the context of the climate change regime, because the Kyoto Protocol is recognized to be only the first modest step in mitigating climate change. Is the level of compliance likely to affect what happens after 2012? One of Kyoto's main functions is to serve as a signal to developing nations that developed States are taking the issue of climate change mitigation seriously, that they are taking the lead on this issue, and that GHG emission reductions can be made in developed States without sacrificing quality of life. Without a high level of compliance with some of the key obligations in the Kyoto Protocol, it cannot achieve this objective, and future negotiations to convince developing countries to take on emission reduction obligations will inevitably be affected.

While a detailed discussion is not warranted or possible here, the absence of a clear alternative and the challenges involved in getting the Kyoto process off the ground also point to potentially serious consequences for the climate change regime as a whole, should the Kyoto process fail as a result of low compliance.[4] High levels of compliance on the other hand can help develop a level of trust among Member States that would allow for more meaningful targets in the future, and thus has the potential to lead to an effective international response to climate change beyond 2012.

A third consideration of the importance of compliance with Kyoto would be its importance for international environmental law. By all accounts, as inadequate as the Kyoto obligations are in addressing the climate change challenge, it seems undisputed that the level of cooperation and effort required of Annex I Parties in the Kyoto Protocol is unprecedented for Multilateral Environmental Agreements.[5] High levels of compliance with the Kyoto obligations therefore have the potential of setting a new standard of cooperation for addressing global environmental challenges. In other words, high levels of compliance could raise international environmental law, arguably for the first time, to the level of deep cooperation.[6] On the other hand, low levels of compliance may serve to re-enforce the perception that deep cooperation without global governance may not be possible in inter-

---

[4] See Chapter 9 for a discussion of alternatives to the Kyoto process and on the future of the climate change regime more generally.

[5] Ronald B. Mitchell, "Institutional Aspects of Implementation, Compliance, and Effectiveness" in Urs Luterbacher and Detlef Sprinz, eds., *International Relations and Global Climate Change*, (Cambridge, Mass.: MIT Press, 2001), at 221- 226.

[6] See discussion in Chapter 3. Deep cooperation generally refers to a high level of effort by a State to comply with an international obligation, one that goes beyond what a State might do if it only considered its short-term self-interest.

national environmental law. These issues are considered in more detail in Chapters 9 and 10.

Finally, in the context of the previous Chapter's discussion of competing theories of compliance in international law, it is clear that developing effective compliance tools is crucial for the future of international law. As one of the first testing grounds for a combination of enforcement and management of compliance, the importance of compliance with Kyoto obligations to our understanding of and the evolution of compliance theory cannot be overstated. The implementation of the Kyoto compliance system has the potential to be invaluable in shedding light on the debate over the roles of management and enforcement, and over the influences of rational choice versus normative influences on compliance.[7]

2. THE KYOTO COMPLIANCE SYSTEM AND OBLIGATIONS DIRECTLY RELATED TO ARTICLE 3.1

This Section considers the compliance system as it relates to Article 3.1 of Kyoto, specifically the obligation on developed States to meet State specific emission reduction targets by 2012. The analysis starts with an identification of possible motivations for non-compliance with this obligation, under the broad categories of preference, capacity, inadvertence and ambiguity. Ambiguity is treated as a separate category here because its influence on compliance may, at times, be linked to either preference or inadvertence, but it does not fit neatly into either category. The compliance system as it relates to this obligation is then considered in three parts, the primary rule system, the compliance information system, and the non-compliance response system. The overall ability of the Kyoto compliance system to respond to various compliance challenges for the Article 3.1 obligations are considered through this process. This process is then repeated for obligations indirectly linked to Article 3.1, and those independent of Article 3.1.

(a) Nature of the Article 3.1 Obligation

The first Kyoto obligation considered is the obligation to reduce GHG emissions in accordance with national or regional targets assigned in Annex I. Specifically, Article 3.1 provides:

> The Parties included in Annex I shall, individually or jointly, ensure that their aggregate anthropogenic carbon dioxide equivalent emissions of the greenhouse gases listed in Annex A do not exceed their assigned amounts, calculated pursuant to their qualified

---

[7] For a good discussion of the potential for the Kyoto compliance experience to enlighten the compliance theory debate, see Jutta Brunnée, 'The Kyoto Protocol: Testing Ground for Compliance Theories?' (2003) 63:3 Heidelberg Journal for International Law 255.

> emission limitation and reduction commitments inscribed in Annex B and in accordance with the provision of this Article, with a view to reducing their overall emissions of such gases by at least 5 percent below 1990 levels in the commitment period from 2008 to 2012.[8]

This obligation applies to Annex I countries, which are developed countries and economies in transition. Parties can meet this obligation in a number of ways,[9] including through domestic emission reductions, removal of greenhouse gases through sinks, through support of emission reductions in other countries, and through the purchase of emission reduction credits. This obligation is generally considered to be at the heart of the Kyoto Protocol.

There has been considerable debate about the adequacy of this obligation, and the role it is likely to play in achieving the objectives of the climate change regime as expressed in Article 2 of the UNFCCC. In this regard, the reduction levels, the limited number of countries with reduction targets, and the fact that some of the highest emitters are currently not participating in the Kyoto process are all cause for concern. The fact that a number of countries, particularly the economies in transition, have business-as-usual projections well below their emissions targets has also been the subject of considerable controversy.

Nevertheless, for all other Annex I countries with binding reduction targets, it is generally accepted that the Article 3.1 obligation can only be met with considerable effort, and at a significant cost. In fact, the level of effort required to comply with Article 3.1 is generally accepted to be unprecedented for Multilateral Environmental Agreements. In this sense at least, Article 3.1 is likely to test the commitment to international cooperation on climate change mitigation, and must be considered an obligation that requires "deep cooperation".[10]

(b) Possible Reasons/Motivations for Non-compliance[11]

As was discussed in some detail in the previous Chapter, there is considerable debate among international relations theorists, international lawyers and others about the relative importance of various influences on State behaviour in response to international obligations such as those in Article 3.1 of the Kyoto Protocol. Much of the disagreement is about their relative influence rather than over the

---

[8] Kyoto Protocol, *supra* note 1, Article 3.1.

[9] See Chapter 2.

[10] See, for example David G. Victor, "Enforcing International Law: Implications for an Effective Global Warming Regime" (1999) 10:3 Duke Environmental Law & Policy Forum 147.

[11] Categories are taken from Ronald B. Mitchell, "Compliance Theory: an Overview" in James Cameron, Jacob Werksman, and Peter Roderick, eds., *Improving Compliance with International Environmental Law* (London: Earthscan Publications Ltd., 1996) 3, at 11 to 13.

existence of a wide range of influences on State behaviour. In this Section, the various possible influences are therefore identified separately from any discussion about their relative influence on the actual behaviour of specific States in response to their obligations under Article 3.1.

The intention here is to identify possible influences, not to debate their relative weight in the context of a specific obligation and a specific State. This is done on the assumption that whatever the relative weight of these various influences, and whatever the actual influence on a given State during the course of implementation, at the design or assessment stage of a compliance system, one would be best served to consider as many potential influences on compliance as can be anticipated.

The issue of the relative weight of various influences, it is argued here, becomes relevant when choices have to be made in the design or implementation of a compliance system, such as a choice due to a conflict between two compliance responses, where the application of one response reduces one risk while it increases another. In other words, the question of which influence will more likely lead to non-compliance becomes most relevant when the compliance tools available to reduce the influence of one factor, such as reputation, have a side effect of increasing the likelihood that another influence on non-compliance increases, such as a sense of distrust among the Parties. The question of conflict or compatibility of compliance responses or tools can arise at two stages, the compliance system design stage and the implementation stage. For these reasons, the analysis will return to the question of the relative importance of the various influences only where there is a concern that various compliance tools or response options available may be incompatible with each other.

The various reasons for non-compliance with international law have been described and categorized for as long as academics and practitioners have debated the role of international law in influencing State behaviour, and are discussed in the previous Chapter. For purposes of this analysis, following the general approach used by Mitchell, these various factors are considered below in four categories: preference, capacity, inadvertence, and ambiguity.

(i) *Preference*

There are two categories of preference leading to an inclination not to comply with an obligation. One is a situation where a State had an intention to comply at the time it ratified the treaty or otherwise took on the obligation. Assuming an intention to comply at the time the commitment is made, the issue in the context of preference would appear to be a change in circumstances from what the Party expected at the time of ratification. The alternative scenario is one where a State had no intention to comply at the time it accepted an obligation or made a commitment. Under this scenario, the additional issue that arises is why a State may be motivated to take on an obligation that it does not intend to meet, or that it does not consider in its interest to meet. For purposes of this analysis, it is generally assumed that States have an intention to comply with obligations they take on, but the possibility of the alternate scenario will be considered as it becomes relevant to the analysis of the Kyoto compliance system.

In terms of factors that may lead to a preference to fail to meet the obligations under Article 3.1 of the Kyoto Protocol, the most plausible factors would be various unexpected changes in circumstances. Such changes could include changes of a scientific nature, changes of a social and political nature, and changes of an economic nature.

In terms of changes of a scientific nature, there are three possible scenarios. One is that new scientific discoveries suggest that climate change will either not take place at the rate currently predicted, or that the changes are not linked to the emission of greenhouse gases to the extent currently thought. This could lead States to conclude that the commitment they made has lost its relevance, and they may no longer feel obliged to comply. If the scientific conclusion is undisputed, this outcome will likely be uncontroversial, and not considered a compliance problem, given that the failure to comply will not hinder the effectiveness of the regime in achieving the objective of preventing human interference with the climate system.

Alternatively, new science could conclude that climate change will take place much more quickly and severely than predicted. In this case, two responses are possible. States may conclude that the current targets are inadequate and that more significant reductions are necessary to address climate change. On the other hand, it is also conceivable that some States will react by concluding that the climate is changing so rapidly that it is too late for mitigation, and that efforts will have to shift to adaptation. Therefore, even under the scenario of scientific evidence that the problem is worse than currently thought, there is a possibility that the new science will create an incentive for some States not to comply with their obligations under Article 3.1.

Thirdly, a more localized factor could be that the climate change threat to the territory of a particular State is less or more than predicted. For example, if predictions about the collapse of the Gulf Stream or the relative higher increases of temperature in polar regions turn out to be false, northern countries and European countries might be less willing to accept the cost of complying with their Article 3.1 obligations.

There is a range of circumstances that could lead to changes in the economic cost of complying with Article 3.1. The decision of the Bush Administration to pull out of the Kyoto Process, for example, reduced the predicted price of international credits for emission reductions because the US was generally recognized to be a significant net buyer of international credits.[12] Technological advances could disproportionately reduce the cost of compliance for specific countries. If policies are not as effective as predicted, countries may find the cost of compliance significantly higher than anticipated. Unforeseen events, such as forest fires can also impact on the credits needed to comply, and can impact on the cost of compliance.[13] Higher than predicted economic activity can also result in a higher cost of meeting the Kyoto target, though in most cases one would expect that the economic benefit from this unexpected boom would more than offset any additional cost of meeting the Kyoto target. Finally, whether other States comply with their obligations, and whether enforcement action is taken against those countries can also have an impact on the cost of compliance for other States.[14]

At the end of the day, whatever the cause, there are two basic scenarios that can lead to unexpected economic costs in complying with Article 3.1. Either the price of credits is higher than expected, or a State is in a position of having to purchase more credits than contemplated. In the latter scenario, the State will either have higher than predicted emissions, or efforts to offset emissions would not be as successful as predicted.

In terms of social and political changes, there is again a range of circumstances one could anticipate that might result in a State's preference to change from compliance to non-compliance. Events such as the emergence of terrorism as a major international threat to security, or prolonged armed conflict, for example, could affect States' perspectives on the relative importance of addressing climate change, making them less willing to invest in meeting their emission reduction obligations under Article 3.1. Other plausible scenarios in this category would

---

[12] The US pull out, in turn, decreased the cost of compliance to most other Annex I states. It also reduced the benefit of joining the process for States who expected to be net sellers of credits, such as the Russian Federation, contributing to the delay in Russia's decision to ratify the Kyoto Protocol.

[13] See Section 3 of Chapter 2. In short, a forest fire will result in the release of GHG gases, and may result in the release of carbon a State may have sought credit for under Articles 3.3 or 3.4 of Kyoto. Depending on the size of the fire, this could significantly alter the state of compliance of a State.

[14] See Jon Hovi & Steffen Kallbekken, "The Price of Non-compliance with the Kyoto Protocol: The Remarkable Case of Norway" (2004) Centre for International Climate and Environmental Research (CICERO Working Paper 2004:07) online: CICERO <http://www.cicero.uio.no/media/2773.pdf>, and Catherine Hagem *et al*, "Tough Justice for Small Nations: How Strategic Behaviour can Influence the Enforcement of the Kyoto Protocol" (2003) Centre for International Climate and Environmental Research (CICERO Working paper 2003:01), online: CICERO <http://www.cicero.uio.no/media/2186.pdf>.

include a change in political leadership in specific States leading to a change in position with respect to the obligation. In situations where the change in leadership comes with a full opposition to Kyoto, such as that of the official opposition in Canada, a change in leadership might result in a withdrawal from the Kyoto Protocol. An alternative, however, depending on the circumstances, would be a change in domestic policy without a formal withdrawal from the process.

Other factors could influence a State to prefer not to comply with its obligations. For example, the fact that States such as the United States and Australia refused to participate in the Kyoto process could influence a State who ratified Kyoto before the position of these States was known. An evolving perception that the Article 3.1 obligations are not being taken seriously by some Annex I Party States could lead to a snow ball effect resulting in general refusal to comply by Annex I States. It is also conceivable that a perception could develop in some countries that the target taken on turns out to be unfairly onerous compared to targets taken on by other Annex I Parties, or unfair relative to activity in developing countries without Article 3.1 obligations.[15] In other words, the internal or external free rider problem has the potential to lead to State preferences to not comply with their Article 3.1 obligations.[16]

(ii) *Capacity*

The line between lack of capacity and preference based on economic factors will frequently be somewhat difficult to draw. Whether a State failed to comply with an obligation because it lacked the capacity to implement renewable technology, energy conservation technology, change behaviour, or purchase credits or whether this was a deliberate choice made by the State to spend the resources to achieve the target on other priorities can be a difficult distinction to make. At one extreme, the choice may be between spending the resources on compliance with an international obligation and providing for the basic short term needs of a State's population. At the other extreme, it may be a choice between compliance with an international obligation or funding an aggressive armed conflict, or a choice between further economic growth in a developed country and meeting an international commitment.

In the context of most if not all Annex I States it is suggested that lack of capacity is not likely to be a factor with respect to Article 3.1. Barring unforeseen

[15] This potential compliance problem is at least in part a result of the pledge based rather than principle based negotiation of the Article 3.1 targets in the Kyoto Protocol.

[16] Internal free riders are States who are Parties to Kyoto, but decide not to take the necessary action to meet their obligations. External free-riders are States who decide not to participate in the Kyoto process. Both internal and external free-riders share in the environmental benefit without sharing the cost.

circumstances, the only States that may lack the resources or technology to reduce GHG emissions or purchase credits to meet their article 3.1 obligations are economies in transition (EITs), all of which have an assigned amount well above projected emissions. Economies in transition may, however, have capacity challenges with regard to reporting and monitoring obligations under Articles 5 and 7 of the Kyoto Protocol. Other than these reporting requirements, given the absence of any quantitative limit on a State's reliance on the Kyoto mechanisms for compliance, lack of capacity in the context of Article 3.1 really equates to lack of revenues to purchase the credits necessary to come into compliance.

Lack of capacity would therefore not be expected to be a significant compliance issue in this group of obligations, other than for Article 5 and 7 obligations in EITs and as a result of unforeseen events for Article 3.1. Such unpredicted events could include economic collapse within a State or region, or a price for international credits that puts their purchase out of the reach of even the wealthiest States.

(iii) *Inadvertence*

There are clearly a number of unforeseen events that could result in initial non-compliance with the Article 3.1 obligation. It is important to keep in mind, however, that it is very likely that credits will be available for purchase during and even after the first commitment period. Given that States have access to these credits to compensate for the unexpected right up to the end of the 100 day true-up period[17] sometime round 2015, it must be considered very unlikely that the unforeseen events will make compliance impossible. On this basis, the only true non-compliance factor that fits into the inadvertence category would be a shortage of international credits available. Most likely, in response to all other unexpected events, the State will still have a choice to make, whether to respond to these events with the purchase of credits or to accept non-compliance with its Article 3.1 obligation.

Circumstances in the category of inadvertence that can create compliance challenges include a number of unforeseen events, such as forest fires, extreme heat or cold, etc., resulting in higher than predicted emissions. Similarly, States may encounter compliance challenges as a result of the underperformance of domestic policies and actions to reduce emissions. A third category might be the

---

[17] See *Report of the Conference of the Parties on its Seventh Session*, Conference of the Parties, UnitedNations Framework Convention on Climate Change (UN FCCC), 29 October – 10 November 2001, FCCC/CP/2001/13/Add.1(Decisions 1/CP.7 - 14/CP.7), FCCC/CP/2001/13/Add.2(Decisions 15/CP.7 - 19/CP.7), FCCC/CP/2001/13/Add.3(Decisions 20/CP.7 - 24/CP.7), FCCC/CP/2001/13/Add.4(Decisions 25/CP.7 - 39/CP.7 & Resolution 1/CP.7 – 2/CP.7), online: UN FCCC <http://unfccc.int/2860.php> [hereinafter COP 7 or Marrakesh Accords], Procedures and Mechanisms Relating to Compliance under the Kyoto Protocol, 24/CP.7., at 74.

purchase of credits on a buyer liability basis that turn out to be bad credits. A similar type of event would be the underperformance of a CDM project in terms of the certified emissions reduction credits it actually generates relative to expectations.

It is impossible to provide a comprehensive list of categories of how inadvertence might influence compliance. Another example of a possible change in circumstances that could be considered inadvertent and lead to compliance challenges would be public pressure that affects the political freedom of a State to meet its obligations. The emergence of other pressing domestic or international issues could also make it more difficult for a State to meet the Article 3.1 obligations.

Finally, there may be links between other obligations and the Article 3.1 obligation. For example, a State that is not able to meet certain accounting obligations with respect to sinks will not be able to use credits it may otherwise be eligible to use. If the failure to meet the accounting obligations are as a result of lack of capacity, as some fear with respect to economies in transition, the resulting compliance problem under Article 3.1 could be considered to be as a result of inadvertence.

(iii) *Ambiguity*

This category consists mainly of disputes over the nature of the legal obligation, and whether the obligation has in fact been met. In the case of Article 3.1, the obligation itself is actually quite clear, the ambiguity comes in through the estimation of emissions in a given State on one hand, and any dispute over the validity of the assigned amount units, certified emission reductions, emission reduction units, and removal units held by a State at the end of the first commitment period on the other hand[18]. The extent to which this will be a concern depends on the clarity of the rules, and the ability of the compliance system to anticipate areas of ambiguity and resolve them sufficiently early to prevent them from becoming compliance issues.

---

[18] For a good discussion of various forms of uncertainty introduced into the Kyoto process through the use of flexibility mechanisms, see, for example, Mitchell (2001), *supra* note 5, at 221. See also Peggy Rodgers Kalas *et al*, "Dispute Resolution under the Kyoto Protocol" (2000) 27 Ecology L.Q. 53.

(c) Compliance Mechanisms and Tools under the Kyoto Compliance System[19]

This Section considers the range of tools and other influences available to try to prevent non-compliance and to respond to it appropriately so as to enhance future compliance. As with the various influences on State compliance considered in the previous Section, there is considerable debate about the respective roles of the primary rule system, the compliance information system and the compliance response system in enhancing compliance. The purpose here is not to debate their relative effectiveness in specific circumstances, but to outline more generally what tools are available within the overall compliance system under the Kyoto process. It is important to note that outside influences to enhance compliance or punish non-compliance are not considered in this Section.[20] The focus in this Chapter is rather on tools and influences that either are part of the Kyoto compliance system or could be incorporated into the negotiations and design of primary rule and compliance systems in the context of an international instrument such as the Kyoto Protocol.

In very general terms, it is safe to conclude from the previous Section that preference, capacity, inadvertence, and ambiguity can all play a role in whether Annex I countries will meet their Article 3.1 obligations. In considering the components of the compliance system, therefore, its ability to prevent and respond to motivations in each of these categories must be considered. While individual components of the compliance system may be particularly suitable to addressing individual factors, in the end, the effectiveness of the compliance system as a whole will have to be measured based on its ability to prevent and respond to non-compliance whatever the motivation.

(i) *The Primary Rule System and Compliance*[21]

The connection between the primary rule system and compliance in the international law context is well established. To be effective at achieving the substantive objectives of an international instrument, the primary rule system must consider whether and how compliance with the primary rules under consideration

---

[19] Categories are adopted from Mitchell (1996), *supra* note 11.

[20] They are considered in Chapter 9 and the Epilogue after the discussion about linkages between climate change and other international regimes, norms and institutions.

[21] Consider Ronald B. Mitchell, "Regime Design Matters: International Oil Pollution and Treaty Compliance" (1994) 48 International Organization 425, for a good illustration of the influence of the primary rule system on compliance. See also Scott Barrett, "International Environmental Agreements: Compliance and Enforcement: International Cooperation and the International Commons" (1999) 10 Duke Env. L. & Pol'y F. 131 for a discussion about the connection between primary rules, compliance and participation by States in a given regime.

can be ensured. At the same time, the effectiveness of the primary rule system in the context of the substantive objectives of the international instruments and the extent to which the rules are acceptable to negotiating States are likely to be crucial. The first relates directly to the effectiveness of the regime, the other to the level of participation. The narrower question asked here is to what extent the primary rules system will assist or hinder compliance efforts.

The focus in this Section of the analysis is on Article 3.1. However, it is important to consider Article 3.1 in the context of the overall primary rules system as it affects the Parties Article 3.1 obligations. The obvious starting point would be the emission reduction targets taken on by Annex I countries for the first commitment period. There are a number of characteristics and circumstances surrounding these targets that are worth noting for their potential impact on compliance.

First, a limited number of countries have obligations under Article 3.1. There is certainly a potential that the limited number of States bound by these targets will negatively affect compliance. Furthermore, the failure to extend targets to developing countries has also been a factor in the United States and Australia's decision not to participate in the Kyoto process. On the other hand, the participation of many if not all developing countries in the Kyoto process was dependent on this focus on action by developed countries, at least for the first commitment period. In terms of the impact on compliance with Article 3.1, the fact that targets are limited to developed countries and States with economies in transition is in accordance with the UNFCCC principle of common but differentiated responsibility and is generally recognized as an equitable way forward.[22] This should reduce any negative compliance implications, as should the fact that States with targets agreed to them and ratified Kyoto in the full knowledge that developing countries would not be required to accept binding emission reduction targets for the first commitment period.

A second factor worth noting is that the targets are essentially pledge based rather than principle based, meaning that they are based on what individual countries agreed to do relative to other States, rather than based strictly on some formula or set of principles, such as historical contribution, current emissions, or capacity to make reductions. On the one hand, the fact that States came forward and voluntarily accepted individual targets may enhance compliance, because this may make it difficult for States to suggest that the targets are unfair. On the

---

[22] UNFCCC, *supra* note 1, Article 4.1. See also Harris, Paul G., "Common but Differentiated Responsibility: the Kyoto Protocol and United States Policy" (1999) 7 N.Y.U. Envtl. L.J. 27, and Ward, Halina, "Common but Differentiated Debates: Environment, Labour and the World Trade Organization" (1996) 45 I.C.L.Q. 592.

other hand, the absence of a set of clear principles to explain and justify the targets also has the potential to become a compliance challenge.

It has already become clear in the context of the debate over ratification in Canada, for example, that the absence of a clear set of principles as the basis for the targets negotiated can undermine the domestic credibility of the targets, and thereby make it more difficult to attain the collective support needed to comply with the obligations taken on. In Canada, this issue was raised with respect to the relative targets between Canada and the US before the US pulled out of the process, and more recently was raised in the context of the ratification debate in Canada with respect to the relative burden on Canada and Australia. Should Canada find itself in a position of having difficulties meeting its Article 3.1 obligations, it is likely that the fairness of the target, and Canada's burden relative to other Kyoto Parties will be raised again. In the meantime, any question about the fairness of the obligation can negatively affect the willingness of governments, citizens and industries in Canada to take the steps necessary to meet Canada's obligations.

A third compliance factor related to the primary rule system is that the highest total contributor, and one of the highest *per capita* emitters, the United States, is currently not participating in the Kyoto process. Less critically, Australia, the highest *per capita* emitter, but a relatively low emitter in terms of total contribution to global emissions is also no longer part of the Kyoto process. The fact that there are two external free-riders is mitigated by the fact that all other Annex I countries ratified with the full knowledge of the formula for Kyoto's coming into force and the full knowledge of the decision of the US administration not to participate in the Kyoto process.

Fourthly, it is generally recognized that the targets set for the first commitment period are inadequate to address climate change, and that more significant reductions in the 50 to 80% range for Annex I countries will be necessary in the future. There is a risk that the inadequacy of these targets will result in the obligation to reach these targets not being taken seriously. Conversely, as negotiations start in the next couple of years on emission reduction targets, it must be considered likely that the need for much deeper cuts after 2012 will make any failure to meet the first commitment period target less acceptable, and thereby increase the pressure on Annex I countries to meet their Article 3.1 obligations. Furthermore, the Montreal Protocol, while clearly in a very different context, does provide a precedent for the acceleration of emission reduction targets even during the current commitment period. While accelerated reduction schedules for GHG emissions must be considered unlikely here given the lead time needed to achieve reductions, on balance the inadequacy of the Article 3.1 obligations in meeting the objectives of the UNFCCC should have a net positive, not a negative impact on compliance.

Another issue in the context of the effect of the primary rule system on compliance is the ability of Member States to accurately and consistently estimate emissions. This issue can fall into the category of capacity, inadvertence and ambiguity, depending on the State involved and the circumstances that lead to problems estimating emissions accurately. Given the States who have accepted Article 3.1 obligations, capacity is likely only going to be an issue for some of the economies in transition. The extent to which inadvertence and ambiguity will be factors depends upon the clarity of the rules for estimating and reporting emissions. Other issues related to the ability of the compliance information and response systems to identify and prevent problems related to reporting on emissions during the first commitment period will be discussed below. One factor clearly in favour of compliance is that the UNFCCC process already requires States to report on emissions,[23] making it easier to identify and remedy problems before the start of the first commitment period.

The Kyoto mechanisms are clearly a central part of Kyoto's primary rule system with tremendous implications for compliance. As a starting point, and perhaps most significantly, the Kyoto mechanisms collectively operate as a release valve to ensure States who have difficulty meeting their Article 3.1 obligations through domestic action have a number of alternatives that are designed to level the playing field in terms of cost of compliance on a per ton carbon basis. The mechanisms therefore have the potential to address the preference factors related to unexpected challenges in making reductions domestically. Most importantly, the Kyoto mechanisms will significantly reduce the cost of compliance for countries that will have the most difficult time making domestic reductions. In terms of addressing issues of preference, inadvertence and capacity, these mechanisms must therefore be considered one of the most important compliance tools included in the Kyoto Protocol.

At the same time, in the context of ambiguity as a compliance factor, the Kyoto mechanisms have the potential to introduce a level of uncertainty to the question whether a given State is in compliance with its obligations under Article 3.1 at the end of the first commitment period. There is some potential for uncertainty about whether credits claimed by a State for compliance purposes were obtained in accordance with the rules established for the various mechanisms. For example, there may turn out to be disagreement over whether a clause in the Marrakech Accords under which Parties agree to "refrain from using" CDM credits obtained through the promotion of nuclear technology is in fact a binding agreement, or whether such credits could potentially be used for compliance purposes. More likely, there could be compliance implications arising out of certain buyer liability provisions. On one side buyer liability provides an incentive for the buyer to ensure that the credits sold are valid credits. On the other, the uncertainty sur-

[23] UNFCCC, *supra* note 1, Article 4.1.

rounding the validity of the credits in the hands of the buyer can result in inadvertent non-compliance. On balance, however, given their effect on the cost of compliance, mechanisms must be considered to enhance opportunities for compliance considerably.[24]

The inclusion of sinks[25] is another part of the Kyoto primary rule system with potential implications for compliance. As with the Kyoto Mechanisms, the fact that sinks offer an alternative to domestic emission reductions does provide an additional release valve for countries with difficulties meeting their Article 3.1 obligations through domestic emission reductions, thus providing additional opportunities to reduce the cost of compliance.[26] Sinks also create compliance challenges in a number of ways. For example, the ability to generate sinks credits domestically is not evenly distributed among States with Article 3.1 obligations. If some countries are perceived to get a windfall through sinks credits, this may serve to undermine the credibility of the targets allocated to Annex I Parties. As discussed in Section 3 of Chapter 2, this concern is mitigated in the Marrakech Accords through limits placed on the credits States can generate, limits that were negotiated before Annex I States began the ratification process. Another compliance concern is that countries who were counting on sinks credits, such as the Russian Federation, may find it too hard to meet accounting and reporting requirements to get sinks credits. In short there are a number of compliance challenges created by the primary rules dealing with the generation of sinks credits.

Another aspect of the primary rule system with potential implications for compliance is the provision for revocation of ratification with one year notice. Theoretically, this would allow a State to sell its assigned amount close to the start of the first commitment period and then revoke ratification and avoid the consequence of non-compliance. While this must be considered very unlikely, the theoretical possibility may influence the level of comfort States will have going into the first commitment period that other States will meet their obligations. This in turn could influence the willingness of States to take costly measures to meet their own Article 3.1 obligations.

---

[24] See, for example, Glenn Wiser, "Hybrid Liability Revisited: Bridging the Divide Between Seller and Buyer Liability" (2000) Center for International Environmental Law (CIEL, 2000) online: CIEL <http://www.ciel.org/Publications/HybridLiabilityCOP6.pdf>; Ronald B. Mitchell, "Flexibility, Compliance and Norm Development in the Climate Regime" in Jon Hovi, Olav Schram Stokke, and Geir Ulfstein, eds., *Implementing the Climate Regime: International Compliance*, (London: Earthscan Press, 2005), and Brett Frischmann, "Using the Multi-Layered Nature of International Emissions Trading and of International-Domestic Legal Systems to Escape a Multi-State Compliance Dilemma" (2001) 13 Geo. Int'l Envtl. L. Rev. 463.

[25] Kyoto Protocol, *supra* note 1, Articles 3.3, 3.4.

[26] Assuming that the State in question has the potential for sinks credits and plans for the use of those credits well enough in advance of the first commitment period. In other words, sinks do not provide the same last-minute release valve as the Kyoto Mechanisms, but rather require longer term planning.

Finally, provisions in Kyoto for cooperation in developing policies, technology,[27] and with respect to research and development on causes and solutions to climate change can impact on compliance. Such efforts, if taken seriously, can enhance capacity in all participating States, reduce the cost of compliance, develop trust and cooperation, and further help level the playing field.

(ii) *Kyoto's Compliance Information System*

The Kyoto system is closely linked to and builds upon the UNFCCC information system.[28] It broadly consists of requirements for Member States to estimate emissions of greenhouse gases from all relevant sources in their own countries, to report annually on those emissions in accordance with modalities set out mostly in the Marrakech Accords, and to comply with requirements and rules for the accounting of the use of the various mechanisms in the Kyoto Protocol, such as sinks, emissions trading, joint implementation and the clean development mechanism.[29]

Both the enforcement and managerial schools of compliance see an important role for the compliance information system as part of an overall compliance strategy. Where the two schools differ is in the specific roles that they see for the information gathering process. While the enforcement school considers this stage to be more of a preparatory stage for the response system, the managerial school focuses much more on the ability to use the information to prevent non-compliance through facilitation and on the direct impact the information gathering process may have on compliance by addressing ambiguities, preventing non-compliance by inadvertence, and identifying capacity-building needs. In the Kyoto compliance system, both objectives are pursued.[30] As suggested by the Center for International Environmental Law (CIEL) in 1998, the focus initially is on gathering information to facilitate compliance. As the first commitment period comes to a close, the enforcement part of the compliance system takes over, using much of the same information previously used for early identification and prevention of non-compliance through facilitation. The seeds for this hybrid

---

27 Victor talks about the connection between the primary rules and compliance in his 1999 article, and actually offers as an alternative a primary rule system that is designed to help key technologies penetrate. See David G. Victor, "Enforcing International Law: Implications for an Effective Global Warming Regime" (1999) 10:3 Duke Environmental Law & Policy Forum 147.

28 See Farhana Yamin, *The International Climate Change Regime: A Guide to Rules, Institutions and Procedures* (Sussex: Cambridge University Press, 2005), Chapter 11, Reporting and Review, comparing UNFCCC and Kyoto rules for reporting and review.

29 For background, see Molly Anderson, "Verification under the Kyoto Protocol" in Trevor Findlay and Oliver Meier, eds., *Verification Yearbook 2002* (London: The Verification Research, Training and Information Centre (VERTIC), December 2002) 147.

30 See Jutta Brunnée, "A Fine Balance: Facilitation and Enforcement in the Design of a Compliance Regime for the Kyoto Protocol" (2000) 13 Tul. Envtl L. J. 223, at 229 to 240.

compliance system can be found in the text of the Kyoto Protocol itself, which includes reference to the multilateral consultative process (MCP) provided for in the UNFCCC,[31] and at the same time suggests the need for a non-compliance response system with consequences for non-compliance in addition to the facilitative approach inherent in the reference to the MCP.[32]

Considering this dual role of the Kyoto compliance information system, it can serve a wide range of objectives, from providing the information needed to manage compliance to providing the information base for enforcement. In the process, the compliance information system may improve transparency, reassure Parties that there are no free riders, and build upon Parties' concern for their international reputation. It can provide information for compliance assessment and non-compliance response, and identify opportunities to resolve disputes. It can further serve to clarify obligations, build capacity, and otherwise enhance compliance.[33] The bottom line is that the information system is an integral part of any effective compliance system, regardless of whether the focus is on enforcement, facilitation or both.[34]

There are a number of options available in designing the compliance information system.[35] As a starting point, information can be gathered by Member States and submitted. An alternative or possible supplement could be some form of in-country inspection carried out by individuals or institutions so mandated under the compliance information system. A third approach involves the use of NGOs[36] as a source of information about State behaviour in response to its obligations.[37] Other options include questioning of Parties' representatives, and otherwise obtaining additional information either directly from Parties or from

---

[31] Kyoto Protocol, *supra* note 1, Article 16.

[32] *Ibid.* Article 18.

[33] See Yamin, *supra* note 28, Chapter 11, Introduction, where the author makes the point that the reporting and review system "make up the backbone of the climate change regime".

[34] See Donald Goldberg *et al*, "Building a Compliance Regime under the Kyoto Protocol", (1998) Center for International Environmental Law & EuroNatura (CIEL, 1998), online: CIEL <http://www.ciel.org/Publications/buildingacomplianceregimeunderKP.pdf>, at 6 for a discussion about the dual role of the compliance information system for both facilitation and enforcement. The idea is to facilitate compliance up to the end of the commitment period, then move to enforcement. The overlap is discussed at 7. See also Brunnée, *supra* note 30, at 229 to 240.

[35] For a good overview of the design options for reporting, see Glenn Wiser, "Compliance Systems under Multilateral Agreements: A Survey for the benefit of Kyoto Policy Makers" (1999) Center for International Environmental Law (CIEL, 1999) online: CIEL <http://www.ciel.org/Publications/SurveyPaper1.pdf>, at 3.

[36] See Steinar Andresen *et al*, "The Role of Green NGO's in Promoting Climate Compliance" FNI report 4/2003 (Lysaker: The Fridtjof Nansen Insitute, 2003).

[37] For a good overview of options for information system design generally, see Glenn Wiser 1999, *supra* note 35, at 2-15.

other sources.[38] Other design issues include questions about whether tools can be used at the discretion of the body in charge of the review, of the Parties being investigated, or some other entity, such as the COP. The Center for International Environmental Law points to the importance of consistency in methodologies and reporting format as another crucial design consideration. Consistency is essential for a meaningful review of the state of compliance, and in case of the imposition of consequences for non-compliance, clarity and precision of methodologies will be crucial from a credibility perspective.[39]

The Kyoto compliance information system consists of two main stages, the reporting stage and the review and verification stage.[40] As the focus here is on Annex I Party obligations under Article 3.1, this Section is concerned primarily with the collection and review of information on Annex I countries, particularly their emissions and emission reduction efforts. In this regard, Parties are required to put in place by 2007 a national system for the estimation of GHG emissions. The point is to ensure capacity to follow the methodologies agreed to and accurately report on GHG emissions before the commencement of the first commitment period in 2008.[41]

Specifically, Article 5 of Kyoto requires Annex 1 Parties to put in place a system for national emissions estimations on an annual basis in accordance with agreed upon methodologies. Article 7 requires Parties to use those national systems to report annually on emissions by source and removal by sink, again, in accordance with agreed upon methodologies. Finally, Article 8 provides for review, verification and adjustment of the information provided, by expert review teams, to ensure that Parties' annual reporting on emissions and sinks is accurate, consistent, and complies with the agreed upon methodologies.

The Marrakech Accords authorize expert review teams to make conservative adjustments to the emissions estimations filed by parties under some circumstances. Specifically, such adjustments can be made if the methodologies are not followed or a party has submitted insufficient information. This provides an incentive to Parties to follow the methodologies and provide sufficient information to allow for proper verification of the estimations made.

Reporting obligations can be usefully broken up into obligations that precede the start of the first commitment period, and those that follow the start of the

---

[38] *Ibid.* at 12, 13, discussing the independent in country inspection process under the Nuclear Non-Proliferation Treaty.

[39] *Ibid.* at 5 regarding the LRTAP example.

[40] See Xueman Wang and Glenn Wiser, "The Implementation and Compliance Regimes Under the Climate Change Convention and Its Kyoto Protocol" (2002) 11:2 RECIEL 181, see also Yamin 2005, *supra* note 28, Chapter 12.

[41] See Kyoto Protocol, *supra* note 1, Article 5.1. See also Yamin, *supra* note 28, Chapter 11, at 17.

commitment period in 2008.[42] Reporting obligations before 2008 focus on ensuring Parties have the required reporting capacity, that they collect the appropriate information, and that compliance problems can be identified early, before the commitment period actually commences. The first reporting obligation designed specifically to encourage Parties to implement measures to come into compliance early is the requirement to report on demonstrable progress by 2005.[43] Specifically, Parties have to file a report by January 1, 2006 to the COP/MOP.[44]

Other reporting and related obligations designed to manage and facilitate compliance before the start of the commitment period include the establishment of a national registry to record transactions involving the generation or trading of units obtained through the use of emissions trading, joint implementation, the clean development mechanism, or land use or land use change projects. A report formally establishing a Party's assigned amount and its capacity to estimate its emissions is required within a year of the entry into force of the Kyoto Protocol or before January 1, 2007. The more general National Communications will be due periodically as decided by the Parties to the Kyoto Protocol, with the first communication due on January 1, 2006. Annual reporting is otherwise voluntary until 2008.

Once the first commitment period commences, the focus of reporting is on verifying compliance. As a starting point, the GHG emission inventory for 2008 is due in April 2010, and yearly thereafter for the other 4 years of the first commitment period.[45] In addition, at the conclusion of the true up period,[46] which runs for 100 days after the final report filed by the Expert Review Teams, each Party has to file a final report which will form the basis for the final assessment of compliance and the imposition of consequences in case of non-compliance.[47]

The Kyoto reporting guidelines[48] require annual reporting during the first commitment period. Included in these reporting requirements is the following information:

---

[42] See Molly Anderson, *supra* note 29, at page 10, and Yamin, *supra* note 28, Chapter 11, at 22.

[43] Kyoto Protocol, *supra* note 1, Article 3.2.

[44] Yamin, *supra* note 28, Chapter 12, at 23.

[45] Methodologies and best practice guidance has been developed by the IPCC and adopted by the Parties for purposes of their reporting requirements under Articles 5 and 7 of the Kyoto Protocol, See Yamin, *supra* note 28, Chapter 11, at 18.

[46] See Section 3 of Chapter 2.

[47] Yamin, *supra* note 28, Chapter 11, at 18.

[48] See Marrakech Accords, Decision 22/CP.7, Guidelines for the Preparation of the information required under Article 7 of the Kyoto Protocol, and draft decision -/CMP of the same title; and decision 22/CP.8, Additional sections to be incorporated in the guidelines for the preparation of the information required under Article 7, and in the guidelines for the review of information under Article 8 of the Kyoto Protocol.

- The Party's GHG emissions data for the year in question;
- Detailed accounting of land use, land use change, and forestry (LULUCF) activity and resulting credits and debits claimed;
- Detailed accounting of various emissions units ranging from assigned amount units to credits generated through the use of the Kyoto mechanisms held in the Party's national registry at the start and the end of the year;
- Detailed accounting of all transactions the Party was involved in during the course of the year;
- Reporting on the Party's commitment period reserve at the end of the year;
- Reporting on the Party's efforts to comply with Article 3.14 by taking steps to minimize adverse impacts of mitigation measures on developing countries;[49] and
- Reporting on any changes to national registries or national systems.

Another central part of the compliance information system is the review of the information submitted. The information received prior to the commencement of the commitment period is primarily intended for use by the facilitative branch with a view of facilitating compliance, but it is also received by the COP/MOP. In addition, the review of some of the pre-commitment period reporting will serve to train experts for the expert review process and work out any unforeseen difficulties with the expert review process. Information on activities during the first commitment period is formally reviewed by Expert Review Teams for each Annex I country to determine compliance with reporting and emission reduction obligations for the first commitment period. The focus of these Expert Review Teams is the review of information submitted by the Parties for the five commitment period years from 2008 to 2012. It is important to note that the Expert Review Teams do not make compliance decisions, but the reports filed by these Expert Review Teams provide the basis for compliance assessment and non-compliance response.[50]

Expert Review Teams are authorized to put questions to Parties, are expected to conduct in country visits with the consent of the Party, offer advice, and generally conduct a detailed technical assessment of compliance with reporting and emission reduction obligations. If reporting problems are identified, Parties are to be given an opportunity to correct the problem. If the compliance problem is not limited to reporting, but extends to the Article 3.1 obligation to meet the

---

[49] For an Annex II Party, additional information must be filed on how the Party is phasing out market imperfections, cooperating in developing non-energy uses for fossil fuels, and assisting developing countries to diversify their economies.

[50] For the detailed rules under which the Expert Review Teams operate, see the Marrakech Accords, and a subsequent COP 8 decision: Decision 23/CP.7, Guidelines for review under Article 8 of the Kyoto Protocol, and draft decision -/CMP.1 of the same title; Decision 22/CP.8, Additional sections to be incorporated in the guidelines for the preparation of the information required under Article 7, and in the guidelines for the review of information under Article 8 of the Kyoto Protocol.

Party's emission reduction target, the Party will have the 100 day true-up period following the filing of the final report of the Expert Review Team to make up any deficiency through the purchase of credits.[51]

The ability of the Kyoto compliance information system to function effectively for facilitation and enforcement of compliance depends on a number of factors, some of which cannot be assessed completely until after the compliance system is in full operation. Transparency, for example, is an open question, given that there is considerable discretion to keep information confidential.[52] The effectiveness of various guidelines, modalities and guidance documents prepared to ensure accurate and consistent reporting of information also remains to be seen. What can be said at this point is that an unprecedented effort was made in the negotiations leading to the Marrakech Accords to establish rules for consistent reporting of information.

Another factor in the effectiveness of the information system as a basis for facilitation of compliance is timing. Valuable years of collecting information about compliance problems due to lack of capacity, inadvertence and ambiguity have been lost between the signing of the Marrakech Accords and the entry into force of the Protocol. This is time lost for the facilitative approach.

Subject to these considerations, the compliance information system should serve as a valuable tool in the overall compliance system. It should provide a solid basis for both the facilitation of compliance leading up to the first commitment period, and for the enforcement of compliance once the work of the Expert Review Teams is done in 2015.

The Center for International Environmental Law (CIEL), in 1999,[53] made three specific recommendations with respect to the compliance information system. First, it recommended that review teams be given the opportunity to request additional information and conduct in-country fact finding. With some limitation, this recommendation has been adopted in the compliance information system negotiated. Second, CIEL recommended that review teams be able to raise compliance concerns with Parties and identify cases of non-compliance. Review teams are able to raise questions of implementation, but are not able to make findings of non-compliance. This is left to the enforcement branch of the compliance committee. Nevertheless, the substance of this recommendation has been adopted by allowing the review teams to identify compliance concerns. Third, CIEL recommended that the process take advantage of the contribution civil society can make to the compliance information gathering and review process. To do so

[51] See *supra* note 17, at 74.

[52] See discussion in Section 3 of Chapter 2.

[53] See Wiser (1999), *supra* note 35, recommendations/observations at 14, 15.

effectively requires access to information, resources, and opportunities for involvement in the information gathering and review process. On this point, it is too early to tell whether the compliance information system will be implemented so as to take advantage of these opportunities.

(iii) *Kyoto's Non-Compliance Response System*

The third part of the compliance system has been termed by Mitchell the non-compliance response system.[54] Included in this component of the compliance system are both facilitative and enforcement responses to non-compliance, both preventative and reactive measures. This Section is therefore essentially about how the information gathered and generated under compliance information systems is used to respond to non-compliance, and how effective its use is for the management and enforcement of compliance.

Given the dual role of the non-compliance response system under Kyoto to facilitate and enforce compliance, there are a number of objectives to consider in evaluating the effectiveness of this component of the compliance system. On the facilitation side, even though the focus is now more on the actual commitment period and beyond, the objectives still include the building of capacity to comply, facilitation of compliance, resolving ambiguities, and avoiding inadvertent non-compliance. On the enforcement side, the focus is on encouraging compliance, ensuring that there are no free riders, ensuring no one benefits from failing to comply, and ensuring non-compliance is made known to other Parties and the public to use reputation as a motivator.[55] Where necessary violators may be punished.

These objectives raise a number of design questions.[56] Questions about the design of a non-compliance response system generally fall into the following categories:

- Initiation of the non-compliance response process;
- Determination of non-compliance;
- Response to non-compliance.

With respect to initiation, the main issue is who can initiate. Options range from initiation by Parties themselves only, by other Parties, by institutions under

[54] See Mitchell, *supra* note 11, at 20.

[55] For a more detailed discussion on the role of reputation in compliance, see George W. Downs & Michael A. Jones, "Reputation, Compliance, and International Law" (2002) 31 J. Legal Stud. 95.

[56] For a good overview of design considerations considering the dual role of a compliance response system under Kyoto, see Wiser (1999), *supra* note 35, at 15 to 38.

the regime, or by non-state actors.[57] A purely facilitative approach would perhaps require initiation by Parties themselves only, based on the assumption that the non-compliance response system is there to assist non-complying Parties, not to punish them. Even then, because of concerns about reputation, a non-complying Party may be reluctant to initiate non-compliance procedures, unless the incentive in terms of access to aid and resources outweighs the concern about reputation.

Any response system that seeks to enforce compliance in addition to facilitating it has to consider avenues of initiating the process other than by Parties themselves. Other Parties, however, may also not provide a sufficient and consistent trigger for the non-compliance response system, given that there may be reluctance either generally or with respect to specific States to initiate the process. International institutions under the climate change regime are dependent upon the Parties for funding and other support. In other words, any one group of entities may at times be an inappropriate decision maker on whether the process should be initiated.

Furthermore, each of these groups may lack the specific knowledge required to identify circumstances where an initiation of the process is warranted. The more complex the obligations, the broader the range of entities that may be needed to ensure that compliance issues are identified consistently and as early as possible. The more open the process is to initiation by a broad range of entities, including non-state actors, the more likely it is that instances of non-compliance will be identified for both facilitation and enforcement purposes. The benefit of allowing a broad range of entities to initiate the process may have to be weighed against any reluctance of the Parties to subject themselves to such scrutiny.

With respect to determinations of non-compliance, the central issue is who makes the determination that a Party is not complying with its obligations. A related question is how the determination is made, and specifically what level of discretion is left to the decision maker to decide whether to make a finding of non-compliance.[58] One central issue here is whether a finding of non-compliance should be reserved for instances where facilitation has failed or is considered to be inappropriate as a result of a finding that a Party's failure to comply was due to self-interest or preference.

In other words, should there be discretion to determine the cause of the problem before a determination is made whether a Party failed to comply, or is the determination on compliance a technical question as to whether a Party has met the strict obligations under Article 3.1, with the question of motivation and appropriate response as a follow-up step? In the absence of evidence that a finding of non-compliance makes facilitation more difficult, it is argued here that the finding

---

[57] *Ibid.* at 21.

[58] *Ibid.* at 22-26.

of non-compliance should be a technical issue, to be determined by an independent body following due process and based on clear criteria without political influence. Such a process would enhance the credibility of the finding made, and provide a strong base for any follow-up action, be it facilitative or enforcement oriented.[59]

Once a determination of non-compliance is made, the next design issue is who decides on the appropriate response to the non-compliance, including facilitation of future compliance and the imposition of consequences of non-compliance,[60] and what range of responses are available. This is where the choice between facilitation and enforcement comes to the surface. Up to now, the impact of this choice has been mainly a question of whether the system can be designed to accommodate both functions. At the non-compliance response stage, however, a choice has to be made whether to respond to the compliance problem identified with facilitation, enforcement or both. Depending on the answer to this general question, there are then a number of more specific choices to be made in terms of how to most effectively facilitate compliance and/or how to most effectively sanction non-compliance.

In terms of the range of responses available, these have been discussed in the previous Chapter.[61] On the facilitation side, they range from funding and capacity-building to dispute resolution. On the enforcement side, response options range from publication of non-compliance, loss of treaty privileges,[62] financial penalties,[63] and requirements to make up emissions in future commitment periods, to trade sanctions.[64]

---

[59] Particularly in cases, such as the Kyoto compliance system, where the consequences of non-compliance are not punitive, but rather are designed to prevent non-complying Parties from benefiting from their failure to comply as well as to minimize any environmental harm suffered as a result.

[60] See Glenn Wiser (1999), *supra* note 35, at 27.

[61] See discussion on compliance tools in Chapter 3.

[62] See, regarding enforcement in the context of emissions trading, Donald M. Goldberg *et al*, "Responsibility for Non-Compliance Under the Kyoto Protocol's Mechanisms for Cooperative Implementation" (Washington, D.C.: The Centre for International Environmental Law and Euronatura-Centre for Environmental Law and Sustainable Development, 1998), online: CIEL <http://www.ciel.org/Publications/pubccp.html>, Wiser (2000), *supra* note 23, and Brett Frischmann, *supra* note 23, at 491.

[63] For a discussion of the compliance reserve options, see Goldberg (1998), *supra* note 34 at 18, 19. For a discussion of the compliance fund option, and other financial penalty options, see Donald Goldberg, Glenn Wiser, "The Compliance Fund: A New Tool for Achieving Compliance under the Kyoto Protocol" (1999) Center for International Environmental Law & EuroNatura (CIEL, 1999) online: CIEL <http://www.ciel.org/Publications/ComplianceFund.pdf>.

[64] See Goldberg (1998), *supra* note 34. See also Jennifer Morgan, *et al*, "Compliance Institutions for the Kyoto Protocol: A Joint CIEL/WWF Proposal" (1999) (Discussion draft prepared for the Fifth Conference of the Parties) online: CIEL <http://www.ciel.org/Publications/complianceinstitutions.pdf>, and Jacob Werksman, "Responding to Non-compliance under the Climate Change Regime" 1999 OECD Information Paper, (Paris, 1999) online: OECD Documents <http://

Against this backdrop, we can now consider the actual non-compliance response system established for the *Kyoto Protocol* in Marrakech.[65] As described above,[66] the *Marrakech Accords* provide for a compliance committee that will function in the form of a plenary, a bureau, and two branches, one for the purpose of facilitating countries efforts to comply with obligations under the *Kyoto Protocol*, and the other to enforce compliance with specific obligations. Each branch is composed of one member from each of the five regional groups of the United Nations, one member representing small island States, and two members each from Annex I countries and Non-Annex I countries. Decisions are to be made by consensus whenever possible. If consensus is not possible, a majority of three-quarters is required for any decision of the committee or one of its branches.

The bureau serves the function of receiving and reviewing questions of implementation brought to the compliance committee and determines which branch of the compliance committee is responsible for responding to the issue raised. Questions of implementation can be brought before the committee by any Party, either with respect to itself or another Party, or as a result of the review carried out by the expert review teams with respect to article 3.1 obligations. This means the compliance committee and its two branches are not in a position to facilitate or enforce compliance at their own initiative, nor are non-state actors able to initiate such a process.

If a matter is referred for facilitation by the bureau, the facilitative branch is generally responsible for assisting Parties in their efforts to meet their obligations under the *Kyoto Protocol*. This includes providing advice, and otherwise facilitating compliance with respect to commitments under Articles 3.1, 5.1, and 5.2, and 7.1, and 7.4. With respect to these provisions, the mandate of the facilitative branch overlaps with that of the enforcement branch, which has a mandate to determine compliance and impose consequences of non-compliance with these provisions.

The jurisdiction of the enforcement branch is limited to matters brought before it by a Party or as a result of the expert review process by way of referral from the bureau. Substantively, the enforcement branch is limited to considering compliance with obligations that have a clearly accepted link to the emissions reduction target under Article 3.1.[67] Decisions of the enforcement branch regarding

---

www.olis.oecd.org/olis/1999doc.nsf/LinkTo/env-epoc(99)21-final>, on the range of available non-compliance consequences.

[65] See Marrakech accords, *supra* note 17, Procedures and Mechanisms Relating to Compliance under the Kyoto Protocol, Decision 24/CP.7 (Annex) at 64.

[66] See Section 3 of Chapter 2.

[67] More precisely, this includes meeting its emissions reduction target, complying with monitoring and reporting obligations, and questions of eligibility to use the Kyoto mechanisms. See Wang, *supra* note 40, at 189.

compliance with Article 3.1 will generally follow the review of the final reports submitted by a Party under Article 8 some 15 months after the end of the commitment period, unless otherwise initiated by a Party. Before a determination of non-compliance is made at this point, Parties are provided with an opportunity to come into compliance by purchasing the necessary credits from another Party. Under Part XIII of the Annex to the compliance decision text,[68] a Party may buy credits for compliance purposes up to 100 days after the expert review process for the commitment period under Article 8 is declared by the conference of the Parties to be concluded.[69]

The importance of the distinction between the two branches becomes clear when one compares the consequences applied by each of the branches. The facilitative branch, under Part XIV of the Annex on compliance, can apply the following consequences:

- Provision of advice and facilitation of assistance;
- Facilitation of financial and technical assistance, including technology transfer and capacity-building; and
- Formulation of recommendations to a Party on what could be done to address concerns about a Party's ability to comply with its obligations.[70]

The work of the facilitative branch with respect to Article 3.1 and related obligations is clearly most useful early in the process, before and during the early stages of the first commitment period. As a result of the information gathered through the various reporting requirements discussed previously in the context of the compliance information system,[71] the facilitative branch should have adequate information to identify compliance issues and identify them early. Of note is that the ability of the facilitative branch to actually provide the resources needed to assist a State in addressing a compliance issue is limited. For example, the facilitative branch has the ability to facilitate financial and technical assistance, not to provide that assistance.

The enforcement branch is authorized to apply the following consequences:

- Make a declaration of non-compliance;
- Require a Party to submit a compliance action plan, which would include an

[68] As is the general practice, the Annex contains the draft text on compliance for adoption at the first meeting of the Parties to the Kyoto Protocol.

[69] See Marrakech accords, *supra* note 17, Part XIII, Additional Period for Fulfilling Commitments at 74, also referred to as the "true-up period".

[70] Marrakech Accords, *supra* note 17, Part XIV, Consequences Applied by the Facilitative Branch at 75.

[71] See compliance information system, Section 2 above.

analysis of the causes of non-compliance, measures to be taken to return to compliance, and a timetable for implementing the measures;

- Suspend eligibility to use the mechanisms, if a Party is found not to meet one of the eligibility requirements; and
- In case of failure to meet its emissions reduction target under Article 3.1, deduct from the Party's assigned amount for the second commitment period 1.3 times the amount of excess emissions from the first commitment period.[72]

The application of these consequences of non-compliance is not discretionary. The substantive penalty for not meeting the first commitment period target is therefore an automatic reduction of the assigned amount in the second commitment period with a multiplier or penalty rate.[73] The question of what to do at this stage of the process was the subject of considerable disagreement during the negotiations. The challenge from a compliance perspective is that by 2012, it will be too late to actually reduce emissions during the commitment period, either through domestic or international action. A Party that has not been able to gather enough credits to offset its emissions has few options left at this point, and no control over whether they are available. The general options left at that stage are to require the Party to restore the reductions in the second commitment period, to impose a penalty, or both.

The compliance challenge at the end of a commitment period, on one hand, highlights the important role of the facilitative branch to identify these concerns early enough to assist a country to avoid non-compliance at the end of the commitment period. On the other hand, recognizing that the facilitative process may not prevent non-compliance with article 3.1 obligations in all cases,[74] it raises the question of how to effectively respond at this stage, when emissions have already taken place and it is too late to facilitate compliance through assistance with policy implementation directed at emission reduction. If at least some Parties have exceeded their targets, a non-complying Party may be able to use emissions trading as a means to come into compliance. Without the U.S. in the Kyoto process, and given the amount of "Hot Air" and sinks credits expected to be available from countries such as Russia and Ukraine, it is likely that this will be

[72] Marrakech Accords, *supra* note 17, Part XV, Consequences Applied by the Enforcement Branch at 75.

[73] Also referred to as borrowing or restoration.

[74] Especially considering the uncertainty introduced through the expanded use of sinks. Parties are not required to do qualitative reporting until they try to issue sinks credits. This could mean considerable differences between a Party's expectation and credits issued. In addition, any area included under articles 3.3 or 3.4 could lose its sinks credits in the case of a forest fire. This means considerable uncertainty over the amount of sinks credits a Party will have at the end of a commitment period. As a result, a Party could be out of compliance in spite of reasonable efforts to meet its target. Similar uncertainties would be introduced if a Party decided to purchase credits from a Party that was not complying with its commitment period reserve requirement, as it might again not know early enough whether those credits will be available for compliance purposes.

sufficient to allow countries to come into compliance at the end of the first commitment period, even if best efforts have not resulted in expected reductions in domestic emissions.[75] Given that the negotiations understandably were conducted on an assumption that the US would not remain outside the Kyoto process indefinitely, and that other unforeseen events could also cause international credits to be unavailable,[76] the compliance system cannot rely exclusively on this mechanism for compliance. The restoration rate was therefore included as an additional means of restoring the balance and motivating compliance.[77]

The general process in the compliance system is that a question is brought to the bureau for a determination of which branch has jurisdiction. There are three ways issues can come before the compliance committee: as a result of a review of a country's submissions by an expert review team under Articles 5 and 7, at the initiative of a Party that realizes it requires assistance in meeting one of its obligations, or at the request of another Party that questions compliance of a Party with one of its obligations.[78] In cases of issues over which both branches have jurisdiction, such as reporting obligations under Articles 5 and 7, the two branches are expected to work in parallel, one assisting the Party in its efforts to correct the problem, the other to apply consequences associated with the breach.

In terms of transparency, both branches are able to consider submissions from outside sources, and decisions are generally to be made public and are to include conclusions and reasons for reaching those conclusions.[79] Both branches are authorized to seek expert advice. Beyond this, there is little in the Marrakech agreement on compliance to guide the facilitative branch on transparency. It is important to consider here that transparency in the context of facilitation may be a double-edged sword. While transparency clearly aids the enforcement agenda, it has the potential to hinder the facilitation effort in that it may make States defensive rather than cooperative. On the other hand, being seen to be cooperating may also be useful to States who are unable to meet their target, as it will show a genuine effort to come into compliance and thus diminish reputational damage.

The process set up for the enforcement branch is generally open. However, there are provisions in the compliance agreement[80] that can reduce or eliminate

---

[75] See Section 3 of Chapter 2 for discussion on likely availability of "Hot Air".

[76] Such as little or no effort domestically by a large number of Annex I countries.

[77] See Chapter 2, Section 3 on compliance with respect to the options considered, including the compliance fund. See also Jutta Brunnée, "A Fine Balance: Facilitation and Enforcement in the Design of a Compliance Regime for the Kyoto Protocol" (2000) 13 Tul. Envtl L. J. 223, and Goldberg, *supra* note 63.

[78] Marrakech accords, *supra* note 17, Part VIII, General Procedures, para. 3 at 70.

[79] *Ibid.* para 6 at 70.

[80] *Ibid.*, paras. 4-6 at 70; *Ibid.*, Part IX, Procedures for the Enforcement Branch, para. 2 at 71; *Ibid.*, Part X, Expedited Procedures for the Enforcement Branch, para. 1 at 72.

the transparency of the process to a point where it risks losing its credibility. There are broad powers, for example, to prevent information from being made public until after the conclusion of the process,[81] but such powers were intended to be exercised only in exceptional circumstances. Similarly, there is provision for the hearings of the enforcement branch to take place in private. These powers, if applied more broadly than intended, could undermine the progressive provisions in the compliance agreement, which allow for submissions from intergovernmental and non-governmental organizations.

Depending on the exercise of discretion, particularly by the enforcement branch, the compliance process could range from open and transparent to being cloaked in secrecy. Most Parties to the negotiations seemed to value transparency and participation by non-Parties, and it was mainly due to the strong bargaining position of Russia, as a key Party for the coming into force of the *Protocol* without ratification by the U.S., that these powers to restrict access were included. It would be reasonable to expect, particularly given the composition of the two branches, that these powers will not be exercised.

The agreement provides for an appeal process, but it is limited to due process issues. The Conference of the Parties (COP) will serve as the appeal body, and the decision being appealed will stand pending the appeal.[82] This will ensure that the appeal process, which can take some time given that the COP generally only meets once a year, is not used as a way to delay application of consequences of non-compliance.

Overall, it would appear that the Kyoto compliance system is well suited to ensuring compliance with Annex I Parties' obligations under Article 3.1 and related obligations with respect to monitoring, reporting, and the use of the mechanisms. The mechanisms should ensure that the cost of compliance is reasonable, and that the relative burden placed on Parties is fairly balanced. Furthermore, both the compliance information and the non-compliance response systems offer a good balance between facilitation and enforcement, ensuring that both objectives can be pursued without significant interference, as long as efforts are well coordinated.

The compliance information system should work to provide the necessary information to identify compliance problems, to allow for early and effective facilitation of compliance and for application of non-compliance responses as appropriate. The process for determining non-compliance would appear to be designed to be free of political interference. Similarly, the application of non-compliance responses on the enforcement side appears clear, and without unnec-

---

[81] *Ibid.* para 6 at 70.
[82] *Ibid.* Part XI Appeals, para 4, at 74.

essary discretion. Appropriately, the application of response on the facilitation side appears discretionary and able to adapt to circumstances.

There are nevertheless some concerns about the ability of the compliance system to respond to non-compliance with Article 3.1 and related obligations. The late coming into force of the Kyoto Protocol will leave less time to identify compliance problems, especially with respect to Article 5 and 7. This could become a problem for EITs. In addition, the 1.3 restoration rate for each ton of carbon by which a State misses its target at the end of the true-up period may not be sufficient to prevent non-compliance by preference, as it may still turn out to be cheaper to make up the 1.3 tons in the second commitment period than to purchase the credits, depending on the cost of compliance in future commitment periods. This concern is somewhat dampened by the fact that other factors are likely to combine with the straight financial consideration to reduce or eliminate any benefit of non-compliance. Such other factors include the relatively low cost of international credits expected, and the influence concerns over reputation and the integrity of the regime is likely to have on Annex I countries commitment to their obligations.

On the facilitation side, the ability to actually address the problem may be limited and outside the control of the facilitation branch, as it does not have independent access to resources and technology to address the problem, but rather it is limited to facilitating the access. The complexity of the monitoring, reporting and verification rules, while necessary, will create compliance challenges due to inadvertence, as it will take time for States to become familiar with the rules of the game. As long as the overall credibility of the Kyoto process is retained and enhanced in the time between its coming into force and the end of the first commitment period, the combination of these factors should be enough to overcome any self-interested decision to refuse to purchase the credits needed to comply.

In the end, the effectiveness of the compliance system will depend significantly on its implementation. If the process is able to stay clear of political interference, if the compliance committee ensures the principle of transparency is preserved, if the cost of international credits stays reasonable through to the true-up period, and if sufficient resources are allocated to assist with the facilitation of compliance, particularly in the context of Articles 5 and 7 in eastern European countries, the Kyoto compliance system stands a good chance of dealing effectively with compliance issues related to Article 3.1 obligations.

3. THE KYOTO COMPLIANCE SYSTEM AND OBLIGATIONS INDIRECTLY RELATED TO ARTICLE 3.1

As discussed in Chapter 2, there are a number of obligations in the Kyoto Protocol that are not directly linked to the obligation in Article 3.1 to reduce GHG emissions, but they are indirectly related to that core obligation.[83] These obligations include the requirement to demonstrate progress under Article 3.2, an obligation to minimize the adverse impact of mitigation measures on developing countries under Article 3.14, a commitment that use of emissions trading and the clean development mechanism will be supplemental to domestic emission reduction efforts under Articles 6 and 17, and the obligation to cooperate in the development and implementation of policies and measures in a number of specific areas under Article 2. These obligations, while they do have some connection to the Article 3.1 obligation to reduce emissions, are not subject to review, determination or response by the enforcement branch of the compliance committee. The task of ensuring compliance with these obligations is left exclusively to the facilitative branch.

The requirement to demonstrate progress by the end of 2005, while included mainly as a result of concerns expressed by developing countries over the long time lag between the signing of the Kyoto Protocol and the start of the first commitment period in 2008, may prove to be a crucial step in the overall effort to ensure compliance with the Kyoto obligations. It is likely to be the focal point of the early efforts of the facilitative branch, with respect to Article 3.1 obligations, but also with respect to obligations under Article 5, 7, 3.14, and the issue of supplementarity under Article 6 and 17. It is important to note that for many of these obligations, there is no detailed agreement on the nature and extent of the obligation, making facilitation of compliance a challenge. At least, by including these obligations in the requirement to demonstrate progress, the facilitative branch will have some information on the basis of which to comment on progress in these areas and perhaps formulate recommendations for future progress.[84]

Given the timing of reporting on demonstrable progress, well before the start of the first commitment period, it is not surprising that the implementation of this obligation was in the end left to the facilitative branch in spite of considerable effort by developing countries to make it subject to enforcement. The requirement to demonstrate progress through reporting will enable the facilitative branch and Parties themselves to develop a general sense of whether Parties are on track to meeting their Kyoto obligations.

---

[83] See Section 3 of Chapter 2.

[84] Molly Anderson, "Demonstrable Progress on Climate Change, Prospects and Possibilities" in Trevor Findlay and Oliver Meier, eds., *Verification Yearbook 2003* (London: The Verification Research, Training and Information Centre (VERTIC), December 2003) 191.

In the context of reviewing demonstrable progress reports, the facilitative branch is likely to be challenged to consider all issues directly or indirectly related to the emission reduction obligations in Article 3.1. The review of demonstrable progress should include whether States are relying too heavily on the use of Kyoto mechanisms,[85] and whether sufficient efforts are made to minimize impacts on developing countries[86] and to develop and implement policies and measures to reduce GHG emissions.[87]

If taken seriously, the requirement to report on demonstrable progress should provide States with valuable feedback on where they are relative to other States in implementing policies and measures to reduce emissions, in establishing national systems under Article 5.1, and in developing national registries under Article 7. This objective can be achieved through the review and possible feedback on Parties' reports on progress by the facilitative branch. In addition, this goal would be enhanced significantly simply through the process of making public the relative effort put forward by States to meet their Kyoto obligations, though there appears to be uncertainty over the transparency of these reports.[88]

As pointed out by others,[89] the rules for reporting on demonstrable progress are not sufficiently detailed and precise to ensure consistency in the methodologies applied in estimating progress toward meeting obligations. This means the facilitative branch will be somewhat limited in its ability to point out clear compliance problems with the obligation to report on demonstrable progress. It is important to keep in mind, however, that it is not the role of the facilitative branch to make determinations of non-compliance in any event. Rather, the facilitative branch is expected to provide advice and assistance, to facilitate financial and technical help, and to make recommendations. While it may have to be careful not to overstate its case, the facilitative branch should be able to provide some clear signals if States are not on track to meeting their targets or related obligations. So while the facilitative branch will not be able to do much if States refuse to take the obligation to report on demonstrable progress seriously, it will be able to use the information provided to do its work to facilitate compliance with the Article 3.1 and related obligations leading up to the start of the commitment period in 2008.

There are at least two reasons to think that Parties will take their obligations to report on demonstrable progress seriously. One is that this is not an onerous

---

[85] That is, whether they are complying with the supplementarity requirements in Articles 6 and 17.

[86] Article 3.14.

[87] Articles 2, 3.

[88] See Molly Anderson, *supra* note 84, at 185, indicating that the UNFCCC Secretariat will be asked to provide an overall synthesis of the submissions made, which might indicate an unwillingness to make public the reports filed by individual Parties.

[89] *Ibid.* at 178.

obligation, and compliance theorists seem to agree that a managerial approach is effective at ensuring high levels of compliance with obligations that do not require deep cooperation.[90] The second is that Parties know there are more serious consequences associated with failing to meet these obligations a few years later, so reporting on them early and subjecting their efforts to the review of the facilitative branch should provide States with a better sense of whether or not they are on track to getting ready for 2008.

If this turns out to be the case, the ability of the facilitative branch to identify potential compliance problems early will be tested here, both with respect to emissions reduction efforts and with requirements related to monitoring and reporting under Article 5 and 7. The effectiveness of the facilitative approach in addressing various compliance problems, identifying the underlying cause, and overcoming the compliance challenges in time for the first commitment period will undoubtedly be tested here. The review and response to the demonstrable progress reporting process should therefore provide much opportunity to assess the facilitative approach in the Kyoto compliance process.

With respect to the other obligations with indirect links to emission reduction obligations of Annex I Parties under Article 3.1, the ambiguity over the exact nature and extent of the obligations are likely to create significant challenges for the facilitative branch. At best, it may be able to use the relative effort made by some Parties to try to motivate others who are doing less to increase their efforts.

4. THE KYOTO COMPLIANCE SYSTEM AND COMMITMENTS NOT RELATED TO ARTICLE 3.1

Finally, there are a number of obligations under the Kyoto Protocol that have no connection to Annex I Parties' obligations under Article 3.1. These are mainly commitments made by developed countries to assist developing countries in some form. There are also a few commitments made by developing countries, mainly in the context of monitoring and reporting, but since the overall focus of the Kyoto Protocol is so heavily on developed country commitments, they are not considered here for purposes of assessing the compliance system.

In terms of specific requirements for Annex I countries to assist Non-Annex I countries, such commitments generally relate to facilitation of access to and transfer of technology that may be of assistance to developing countries in addressing climate change, cooperation in scientific and technical research, education, training, and reference to similar requirements under Article 4 of the UNFCCC.[91] Measures to achieve these objectives focus on the commitment of

---

[90] See Section 3 of Chapter 3.

[91] Kyoto Protocol, *supra* note 1, Article 10.

additional funding resources. Such measures can be made available globally, regionally, or bilaterally.[92] To date, these general commitments have resulted in the establishment of three new funds under the Kyoto process, in addition to extending the use of the Global Environmental Facility (GEF) from the UNFCCC to Kyoto. The three new funds are the Least Developed Countries Fund, the Special Climate Change Fund, and the Kyoto Protocol Adaptation fund.[93] Collectively, these funds are designed to provide financial assistance to developing countries in their efforts to adapt to and mitigate climate change. The Adaptation Fund will be funded through a percentage of revenues generated from the use of the clean development mechanism. Otherwise, the general commitment made by developed country Parties is to provide additional funding through these funding mechanisms and to facilitate the transfer of technology.

These commitments are also not subject to review, determination or response by the enforcement branch of the compliance committee. The task of ensuring compliance with these commitments is again left exclusively to the facilitative branch. Given the general nature of the commitments, and the mandate of the facilitative branch, the process of facilitating compliance with these commitments will likely take the form of reporting on funding levels and use of the funds made available. Compliance with these commitments comes down to the willingness of individual States to make funding available either through one of the four funds or on a bilateral basis to assist developing countries.

Reporting on the funding provided on a Party by Party basis is likely to be the most effective tool available to the facilitative branch with respect to these commitments. Given the nature of the commitments and the fact that the commitments are made only by Annex I Parties, it must be considered unlikely that lack of capacity, inadvertence or ambiguity will be significant compliance issues.[94] In other words, preference must be considered to be the dominant influence on whether and how much assistance a Party will offer to developing countries. With publication of donations as the main tool to influence compliance, it is highly questionable if compliance tools within the Kyoto Compliance System will be effective in improving compliance with Article 10 and 11 commitments.

---

[92] *Ibid.* Article 11.

[93] See Farhana Yamin, *supra* note 28, Chapter 10 at 25-30, 60.

[94] This is assuming an absence of a major change in the economic strength of Annex I Parties, such as took place after the collapse of the former Soviet Union.

5. CONCLUSION: IS THE KYOTO COMPLIANCE SYSTEM PREPARED FOR THE CHALLENGES AHEAD?[95]

In this Chapter, Kyoto compliance issues have been considered for Article 3.1 obligations, obligations with an indirect link to Article 3.1, and commitments independent of Article 3.1. Possible State behaviour in response to these various Kyoto obligations leading to non-compliance has been considered in the context of preference, capacity, inadvertence, and ambiguities. The ability of the Kyoto process to deal with each of these potential causes of non-compliance was then assessed, with a focus on obligations related to Article 3.1, but with some consideration of other obligations as well. What is left to be done is to take stock, to consider the overall effectiveness of the Kyoto regime to anticipate and prevent non-compliance with its central obligations.

On the enforcement side, the agreed upon consequences for non-compliance were clearly a compromise to ensure States would not be deterred from participating in the Kyoto process, and that those in the process would meet their emission reduction obligations. The restoration rate of 1.3, at best, will prevent a financial benefit to Parties who refuse to purchase the credits needed at the end of the compliance process for the first commitment period. Parties stopped short of ensuring that there would be a penalty associated with failure to hold sufficient credits to offset emissions. In addition, options for financial penalties and trade sanctions were also rejected as possible consequences of non-compliance.

The resilience of the enforcement side of the compliance system will likely only be tested in unusual circumstances, given that the cost of international credits are likely to be much lower than they were for much of the negotiation process, during which the US was still a full participant in the process.[96] While the ability of the enforcement process to deter non-compliance based on economic considerations alone is therefore somewhat limited, there is no current indication that the ability of the enforcement branch to ensure compliance will be seriously challenged.

The use of reputation to encourage compliance is limited. It is somewhat troubling that there is considerable uncertainty over the transparency of compliance procedures given the discretion left to the compliance committee to decide

[95] See Jørgen Wettestad, "Enhancing Climate Compliance: What Are the Lessons to Learn from Environmental Regimes and the EU?" FNI report 2/2003 (Lysaker: Fridtjof Nansen Institute Publications, 2003).

[96] The withdrawal of the US from Kyoto has eliminated the biggest potential buyer of international credits, given that the US is the single largest contributor of GHG emissions, and given that it was generally recognized even as early as 1997 that the US was not likely to meet its target through domestic emissions reductions alone. See also discussion on the availability of Hot Air in Section 3 of Chapter 2.

in the future how transparent the process will be. This is unfortunate considering that the pressure to keep compliance issues confidential will undoubtedly increase as actual compliance issues arise. It will be important for the compliance committee to resist this pressure and establish a precedent of transparency early in its development. Only then can the use of reputation to motivate compliance be fully explored.

The ability of the facilitative branch to overcome problems of capacity is also limited. The main concern is that it does not have direct access to the resources needed to address capacity issues, other than to provide advice and facilitate access to funding, expertise and technology. The effectiveness of the facilitative branch will therefore depend on its ability to bring compliance problems to the attention of the international community early and convince it to take the necessary steps to address them. The fact that most Annex I countries should have the capacity to comply means that the compliance problems to be expected are already fairly well known to deal mainly with monitoring and reporting obligations in EITs and should greatly assist the work of the facilitative branch in this regard.

Some observers have raised concerns about potential collateral impacts of enforcement action under the Kyoto compliance system on complying States. The concern is essentially that the imposition of the restoration rate on certain States will have a negative economic impact on States that have close economic ties to those countries, particularly in the energy sector.[97] Such unexpected effects, if they turn out to be significant and perceived as unfairly affecting some States, have the potential to undermine the credibility of the Kyoto process. Such concerns highlight the importance of considering the first commitment period as a test phase, so that by the time States are ready to commit to more meaningful targets, important lessons about the relative fairness of the obligations and the compliance system will have been learned.

The overall conclusion of this Chapter is that the Kyoto compliance system is the first serious attempt at combining the deterrence of non-compliance and the facilitation of compliance in the multilateral environmental context. Within limits, the compliance system stands a good chance at addressing effectively the compliance challenges that lie ahead. This leaves the climate change regime in a position of having a fair chance of high levels of compliance with inadequate targets, and with many of the largest emitters either outside the Kyoto process or without binding obligations to make emission reductions.

The combination of facilitation and enforcement is a central part of the compliance regime, and neither alone is likely to ensure high levels of compliance.

---

97 Consider, for example, the collateral damage concerns raised in two separate reports by Hovi and Hagem, both at *supra* note 14.

At the same time, there is some potential for conflict between facilitation and enforcement. It remains to be seen, for example, how willing Parties will be to share information for facilitation purposes, if that information can be used for enforcement. On the other hand, the potential for the two approaches to complement each other appear to outweigh the risk of conflict. It will be interesting to see whether Parties accept punitive penalties to encourage compliance with obligations requiring deep cooperation in the future. Much may depend on whether the non-punitive measures, such as the 1.3 restoration rate, the compliance action plan, and the loss of ability to trade plus normative influences work together to ensure compliance in a deep cooperation context.

In the end, the answer to the question of compliance with the Kyoto Protocol cannot be found within the Kyoto Compliance System alone. In the next four Chapters, outside influences on compliance and the future of the climate change regime are therefore considered.[98] Specifically, these Chapters will consider the extent to which there are international influences outside the climate change regime. Influences will be considered on compliance with Kyoto obligations and on future participation in the climate change regime, particularly by the US. The extent to which external influences have the potential to motivate States to take on more meaningful targets in the future will also be assessed. The international influences considered in this context include the World Trade Organization, International Human Rights, The UN Law of the Sea Convention, and MEAs using the UN Convention of Biological Diversity as an illustration.

[98] See, for example, Olav Schram Stokke, "Trade Measures, WTO, and Climate Compliance: The Interplay of International Regimes" FNI report 5/2003 (Lysaker: Fridtjof Nansen Institute Publications, 2003), on opportunities to use trade sanctions as an outside influence on Kyoto compliance.

PART II

EXTERNAL INFLUENCES ON THE CLIMATE CHANGE REGIME

In this Part, four key external influences on the climate change regime are considered. The influences selected are all international law influences. They include the World Trade Organization, the United Nations Law of the Sea Convention, International Human Rights Norms, and Multilateral Environmental Agreements. For each of these potential influences, a number of key questions are posed. They are:

- Does the regime in question offer opportunities to motivate States to take climate change mitigation measures generally?
- Does the regime offer opportunities to motivate States to ratify and comply with Kyoto?
- Does the regime offer co-benefits, in other words, are there areas where compliance with the requirements of the regime also has or can have climate change mitigation benefits?
- Are there possible or inevitable clashes, where compliance with one means violating the other?

In the context of these questions, each of the following four Chapters considers the potential for normative and enforcement influences on the climate change regime.

5

The WTO and Its Potential Impact on the Climate Change Regime

1. INTRODUCTION

This Chapter explores the role of trade in motivating action on climate change, using the specific example of developments within the World Trade Organization (WTO). Both enforcement and normative influences are possible. However, due to the binding dispute settlement process incorporated into the WTO in 1994, the focus here is on opportunities to motivate State action on climate change through the use of the WTO dispute settlement process. The fundamental issue considered is to what extent the rules governing the WTO regime can assist a nation's efforts to address climate change. Specifically, the following questions arise:

- Can a State take measures to protect industries that are adversely affected by the State's efforts to implement the Kyoto Protocol? (Against Annex I Parties to the Kyoto Protocol? Against Non-Annex I Parties?)
- Can a State take measures to protect industries that are adversely affected by efforts to address climate change more generally? (i.e. what if a State takes steps which are more stringent than those required under Kyoto?)
- Can a State take measures to influence the climate change impact of products imported into that State?
- Can a State take measures to prevent a competitive disadvantage resulting from efforts to address climate change in the case of exports?[1]

The WTO was created as a result of the Uruguay Round of trade negotiations, which concluded in 1994 in Marrakech, Morocco. The negotiations were built upon the original global trade agreement, the GATT 1947,[2] which had remained

---

[1] A likely means of achieving these objectives would be the imposition of Border Tax Adjustments on products that are considered to contribute to climate change relative to domestic or other imported products. See, for example, Christian Pitschas, "GATT/WTO Rules for Border Tax Adjustment and the Proposed European Directive Introducing a Tax on Carbon Dioxide Emissions and Energy" (1995) 24 Ga. J. Int'l & Comp. L. 479, and A. J. Hoerner "The Role of Border Tax Adjustments in Environmental Taxation: Theory and US Experience" (working paper, presented at the International Workshop on Market Based Instruments and International Trade of the Institute for Environmental Studies Amsterdam, 19 March 1998).

[2] *General Agreement on Tariffs and Trade* (GATT 1947), reprinted in *The Results of the Uruguay Round of Multilateral Trade Negotiations: The Legal Texts* (Geneva: GATT Secretariat, 1994), at 485 to 558.

essentially unchanged for almost 50 years. Perhaps the most significant change resulting from the Uruguay Round was the establishment of a binding dispute settlement process, including a panel process, an appeal process, and a process for the implementation of rulings of these WTO bodies. In addition, there were adjustments made to certain provisions of GATT 1947 (in the form of GATT 1994),[3] and a number of specific agreements dealing with certain aspects of the GATT were concluded as part of the Uruguay Round. Most notable, from an environmental perspective, are the Agreement on Technical Barriers to Trade (TBT Agreement),[4] and the Agreement on the Application of Sanitary and Phytosanitary Measures (SPS Agreement).[5]

2. TREATMENT OF TRADE ISSUES UNDER THE UNFCCC AND THE KYOTO PROTOCOL

In considering the impact of the WTO on compliance with the climate change regime under the UNFCCC, it is important to first consider the specific connections made in the various agreements that make up the two regimes. As a starting point, the UNFCCC makes general references to the need to consider the economic cost of measures to address climate change, the principle of State sovereignty, and the need for a cooperative approach to climate change among other indirect references to the relationship between trade and climate change. More directly, the principles of the UNFCCC include a reference to the need to ensure that measures to address climate change do not 'constitute a means of arbitrary or unjustifiable discrimination or a disguised restriction on international trade'.[6]

---

[3] Multilateral Trade Negotiations Final Act Embodying the Results of the Uruguay Rounds of Trade Negotiations (Marrakech, 15 April 1994), reprinted in The Results of the Uruguay Round of Multilateral Trade Negotiations: The Legal Texts (Geneva: GATT Secretariat, 1994) 1.

[4] *Agreement on Technical Barriers to Trade* (TBT), reprinted in The Results of the Uruguay Round of Multilateral Trade Negotiations: The Legal Texts (Geneva: GATT Secretariat, 1994), at 138 to 162.

[5] *Agreement on Sanitary and Phytosanitary Measures* (SPS), reprinted in The Results of the Uruguay Round of Multilateral Trade Negotiations: The Legal Texts (Geneva: GATT Secretariat, 1994), at 69–84.

[6] *United Nations Framework Convention on Climate Change*, Intergovernmental Negotiating Committee for a Framework Convention on Climate Change OR, 5th Sess., Annex, UN Doc. A/AC.237/18 (PartII)/Add.1 (1992), 31 I.L.M. 849, online: UNFCCC <http://unfccc.int/resource/docs/a/18p2a01.pdf> [UNFCCC or The Framework Convention], Articles 1-3. See specifically Article 3(5). Note that the UNFCCC secretariat in its 1999 submission to the WTO took the position that the measures in the Convention and the Kyoto Protocol do not constitute trade measures, but that trade implications may arise from national implementation. See Communication from the UNFCCC Secretariat (1999), WT/CTE/W/123.

The Kyoto Protocol includes an obligation on Parties to strive to implement climate change measures so as to minimize adverse effects on international trade.[7] On the one hand, it is clear from these provisions that the drafters intended there would be some limit on the ability of Parties to choose any measures they considered appropriate to address climate change and meet their obligations under the UNFCCC and the Kyoto Protocol. On the other hand, it seems equally clear that they anticipated that some restrictions to international trade would result from measures to address climate change and comply with the UNFCCC and Kyoto Protocol. The provisions within the climate change regime leave open the question of when measures may be justified, and when they may not be. As we will see, the language under the WTO regime is more specific. In this context, the general language in the UNFCCC and Protocol can be reasonably interpreted in such a manner as to be consistent with the more specific language within the WTO regime.

3. TREATMENT OF ENVIRONMENTAL ISSUES UNDER THE WTO

The interaction between trade and environmental issues under GATT and the WTO is essentially a result of a concern that States' measures to protect the environment in some form or another are either intended to, or have the effect of, restricting trade rather than just protecting the environment. In considering possible conflicts between a Party's environmental protection measures and the objective of unrestricted trade, a number of possible perspectives arise.

One consideration is what a Party's purpose is in putting forward the environmental protection measure. Is the purpose ultimately to protect the environment or to protect domestic industries? What if it is both? What if you cannot tell what the purpose is? Should the motivation matter, or should the focus be on the effect?

An alternative consideration is therefore to focus on the effects of the measure. Does the environmental protection measure have the effect of restricting trade? Does it have the effect of restricting trade in a manner that benefits domestic industries? If so, is the environmental objective legitimate? Is the environmental objective being met through the measure? Are there fewer trade restrictive ways of meeting the environmental objective? Who determines what a legitimate objective is, what an effective measure to meet the objective is, and whether other measures that are less trade restrictive are as effective? How are conflicts between legitimate environmental protection measures and trade restrictions resulting from their implementation resolved?

---

[7] *Conference of the Parties to the Framework Convention on Climate Change: Kyoto Protocol*, 10 December 1997, U.N. Doc. FCCC/CP/1997/L.7/add. 1, 37 I.L.M. 22 (1998), [hereinafter the Kyoto Protocol], Article 2(3).

A crucial question is to determine where the balance can be struck (in terms of environmental protection and trade) between State sovereignty to act in the interest of its citizens and the 'treaty obligation' to consider extra territorial effects of domestic action. The following overview of the relevant provisions of the various WTO agreements as they relate to the environment should give some insight.

A good starting point for the consideration of how the WTO regime treats environmental issues is the preamble to the Marrakech Agreement establishing the WTO, which includes the following provision:

> Recognizing that their relations in the field of trade and economic endeavour should be conducted with a view to raising standards of living, ensuring full employment and a large and steadily growing volume of real income and effective demand, and expanding the production of and trade in goods and services, while allowing for the optimal use of the world's resources in accordance with the objective of sustainable development, seeking both to protect and preserve the environment and to enhance the means for doing so in a manner consistent with their respective needs and concerns at different levels of economic development . . .[8]

The relevance of this provision was considered by the WTO Appellate Body in the Shrimp Turtle Ruling (No 1). The Appellate Body concluded that this provision is relevant to the interpretation of the provisions of the various WTO agreements that were in place at that time. The Appellate Body concluded that the preamble:

> . . . demonstrates a recognition by the WTO negotiators that optimal use of the world's resources should be made in accordance with the objective of sustainable development. As the preambular language reflects the intentions of negotiators of the WTO Agreement, we believe that it must add colour, texture and shading to our interpretation of the agreements annexed to the WTO Agreement, in this case the GATT 1994. . .[9]

The general GATT rules most relevant in the environmental context are Articles I (most favoured nation treatment), III (national treatment), and XX (exceptions). Articles I and III set out the general rules for treatment of foreign products compared to like domestic products. Article XX is perhaps of most interest from an environmental perspective, as it provides for exceptions to the general GATT rules in the pursuit of certain environmental objectives. Article XX exempts measures necessary to protect the life or health of humans, animals or plants, as long as those measures do not arbitrarily or unjustifiably discriminate between

[8] As applied in the Appellate Body Shrimp Turtle Ruling No. 1, WTO AB 12 October 1998, *United States – Import Prohibition of Certain Shrimp and Shrimp Products*, WT/DS58/AB/R (AB-1998-4), at para. 152 (hereinafter 'Shrimp Turtle (No.1)').

[9] *Ibid.*, at paragraph 153.

countries and the measures do not constitute a disguised restriction on international trade (Article XX(b)). A similar exception is included in Article XX for the conservation of natural resources, if there are similar domestic restrictions on the production or consumption of the resource in question (Article XX(g)). As indicated, two agreements added to the GATT during the Uruguay Round of negotiations, the TBT Agreement and the SPS Agreement, are of particular interest from an environmental perspective[10]. They are briefly reviewed below.

(a) The Agreement on Technical Barriers to Trade (TBT)[11]

The TBT Agreement is particularly relevant to the consideration of linkages between trade and climate change as it provides considerable detail on the interpretation of provisions of the GATT dealing with technical barriers to trade. Given that the most likely challenge to State action with respect to climate change mitigation is that it amounts to a technical barrier to trade, this agreement is likely to play a central role in any dispute over the trade implications of climate change mitigation measures. Furthermore, the TBT Agreement provides a new basis for the interpretation of GATT rules that have otherwise been in place, essentially unchanged for 50 years. This provides the WTO panels and appellate bodies with the opportunity to take a fresh look at the relationship between trade and environment. A key aspect of this is the extent to which the provisions of this agreement may change the status quo, and whether the rulings of the WTO demonstrate any change in application since the adoption of the TBT Agreement.

The TBT Agreement came into force with the establishment of the WTO on 1 January 1995. All Member States of the WTO are bound by the terms of this Agreement. The TBT Agreement provides an interpretative tool for Articles I and III of the GATT. Specifically, it provides a basis on which to distinguish legitimate technical regulations and standards for such purposes as human health, environmental protection and national security from those that favour domestic products over those imported from other countries (Article 2). All products, including industrial and agricultural products, are subject to the Agreement, with two exceptions: products purchased under government procurement are not subject to

---

[10] See WTO Secretariat, *Guide to the Uruguay Round Agreements* (Boston, Kluwer Law International, 1999). For the legal texts from the Uruguay Round, see *supra* note 3. For an overview of the history of trade negotiations prior to the Uruguay round, see G. R. Winham, *International Trade and the Tokyo Round of Negotiation* (Princeton University Press, 1986). See also G. R. Winham, "The World Trade Organization: Institution-Building in the Multilateral Trade System" (1998) 12 *The World Economy* 349.

[11] See *supra* note 4. For a more detailed assessment of the TBT Agreement, see C. Thorn *et al*, "The Agreement on the Application of Sanitary and Phytosanitary Measures and the Agreement on Technical Barriers to Trade" (2000) 31 Law & Pol'y Int'l Bus. 841.

this Agreement, but are subject to the Agreement on Government Procurement;[12] and pursuant to Article 1, the Agreement does not apply to sanitary measures and phytosanitary measures as defined in Annex A of the Agreement on the Application of Sanitary and Phytosanitary Measures.

The preamble recognizes that no country should be prevented from taking measures to protect human, animal or plant life or heath or from protecting the environment, as long as those measures do not arbitrarily or unjustifiably discriminate between countries. The Agreement therefore seeks to strike a balance between preserving countries' rights to protect their citizens and the environment on the one hand, and preventing such measures from restricting international trade on the other.

Generally, the Agreement encourages the country which seeks to implement measures to protect its citizens or its environment to do so in a manner that is consistent with international standards. In the absence of suitable international standards, the State is obliged to ensure that the standard of technical regulation is designed for the purpose of achieving the human health or environmental objective and has the effect of achieving the objective in the least trade restrictive manner.

To this end, the Agreement establishes general obligations designed to ensure standards and technical regulations do not discriminate against international trade. It then develops processes to ensure this objective can be met and a dispute resolution mechanism in case of dispute over the purpose and effect of specific measures. The Agreement recognizes the special circumstances of developing countries in meeting the obligations of the Agreement. Article 12.7 specifically obligates Members to provide technical assistance to developing countries to ensure that technical regulations, standards and conformity assessments do not create unnecessary obstacles for exports from developing country Members. Article 12 further recognizes that developing countries face institutional and infrastructure problems, and obligates Members to assist developing countries to overcome those challenges.

The fundamental approach of the Agreement puts it at odds with the precautionary principle in the sense that it requires countries to justify action to protect the environment, rather than requiring justification of inaction, as contemplated under the precautionary principle.[13]

---

[12] See Marrakech Agreement Establishing the WTO, Annex 4: Plurilateral Trade Agreements: Agreement on Government Procurement (Marrkech, 15 April 1994), in *The Results of the Uruguay Round of Multilateral Trade Negotiations: The Legal Texts*, *supra* note 3 above, at 438.

[13] See, for example, Articles 2.5 and 2.9. For a detailed analysis of precaution, see N. de Sadeleer, *Environmental Principles: From Political Slogans to Legal Rules,* (Oxford, Oxford University Press, 2002), at 91.

(i) *Structure*

The Agreement consists of a preamble, 15 Articles and three annexes. The preamble places the Agreement in the context of the Uruguay Round of Multilateral Trade Negotiations and the 1994 GATT.

Articles 1 to 4 set out the scope of the Agreement and the general obligations on Parties. The objective of these obligations is to ensure that technical regulations and standards adopted are not more trade restrictive than necessary and do not accord favourable treatment to domestic products. These provisions apply to central government bodies, and Parties agree to take reasonable steps to ensure compliance by local government and non-governmental bodies as well. Articles 5 to 9 provide for mechanisms to ensure that domestic standards and technical regulations conform to the obligations under this Agreement. To this end, the Agreement establishes procedures for assessing conformity. Articles 10 to 12 provide for information sharing and assistance among Parties, with a particular emphasis on developing countries in Article 12. Articles 13 to 15 deal with dispute settlement and administrative matters.

The three annexes are considered integral parts of the Agreement. Annex I defines a number of terms for purposes of the Agreement.[14] Annex II sets out procedures for Technical Expert Groups set up under Article 14 to assist with dispute settlement. Annex III provides a code of good practice for the development of standards. It applies to national and regional standardizing bodies.

(ii) *Obligations*

The TBT agreement imposes a number of key obligations on Member States. They include the following:

- No Less Favourable Treatment
- Remove Technical Regulations
- Use International Standards
- Justify Technical Regulations
- Ensure Compliance
- Provide Enquiry Point
- Assist Other Countries

[14] Annex I defines "technical regulation", one of the central terms in the Agreement. The definition provides that a document that "lays down product characteristics or their related processes and production methods" is considered a technical regulation under the agreement if compliance is mandatory. Annex I also defines "standard", "conformity assessment procedures", "international body or system", "regional body or system", "central government body", "local government body", and "non-governmental body".

(A) No Less Favourable Treatment

At the core of the Agreement is the obligation to ensure that imported products are not treated less favourably than domestic products (Art. 2.1). Related to this is the obligation to ensure that technical regulations[15] are not introduced for the purpose of creating unnecessary obstacles to trade (Art. 2.2).

(B) Remove Technical Regulations

Pursuant to Article 2.3, Members agree to remove technical regulations once the circumstances giving rise to them have changed.

(C) Use International Standards

Members will use international standards for technical regulations except where they are ineffective or inappropriate to meet the legitimate objective pursued (Art. 2.4).[16]

(D) Justify Technical Regulations

Countries agree to give notice and justify technical regulations that are not in accordance with international standards and that may have a significant impact on trade.[17]

(E) Ensure Compliance

Articles 2 and 3 oblige Members to ensure compliance by central government bodies, local government bodies, and non-governmental bodies. Members are also expected to ensure that standardizing bodies comply with the code of good practice in Annex III (Art. 4). In addition, conformity assessment procedures have

---

[15] "Technical regulations" under the Agreement include any regulation to protect the environment unless they are covered under the WTO Agreement on Sanitary and Phytosanitary Measures.

[16] There are several international organisations that either have or are currently developing standards that may have relevance to this Agreement. They include the Codex Alimentarius, the World Organisation for Animal Health (OIE), the World Health Organisation, and the International Plant Protection Organization. A current list of Codex Alimentarius standards can be found at <http://www.codexalimentarius.net/STANDARD/ standard.htm> (accessed: 23 March 2001). Standards developed to date address food hygiene, food labelling, food import and export inspection and certifications systems, and organically produced food. For current OIE standards, see <http://www.oie.int/eng/en_index.htm> (accessed: 23 March 2001). The OIE has, for example, developed an International Aquatic Animal Health Code.

[17] Justification is required upon request by another Member, see Article 2.5.

to conform to requirements in Articles 5, 7 and 8. Countries are expected to accept such conformity procedures in other countries.

(F) Provide Enquiry Point

Each Member State must provide an enquiry point to ensure access to relevant information for other countries and interested Parties (Art. 10).

(G) Assist Other Countries

Members shall assist, where requested, other countries in the development of technical regulations (Art. 11). Members are especially expected to provide favourable treatment to developing country members (Art. 12).

(b) The Agreement on Sanitary and Phytosanitary Measures (SPS)[18]

The SPS Agreement is the other main agreement concluded as part of the Uruguay Round of negotiations with direct relevance to trade and environment issues. It deals specifically with sanitary and phytosanitary measures,[19] and as such is not as likely to be directly applicable to climate change mitigation measures. It is, however, relevant to this discussion, given that many of the initial rulings of the WTO on environmental matters have dealt with the SPS Agreement. The following brief overview should provide the context to consider these rulings and the extent to which the findings are going to be applied more broadly by the WTO.

Like the TBT Agreement, the SPS Agreement entered into force on 1 January 1995 and all WTO Member States are bound by it. (It is essentially an interpretative tool for Article XX(b) of the GATT).[20] Its primary objective is to clarify the balance struck between the rights of Member States to protect human, animal and plant life or health and the overall objective of the WTO to promote international trade (Article 2). The SPS Agreement seeks to identify mechanisms to distinguish between appropriate measures and measures that are intended or have

---

[18] See *supra* note 3, above. For a more detailed assessment of the SPS Agreement, see C. Thorn *et al*, *supra* note 11, at 841; J. M. Wagner, "The WTO's Interpretation of the SPS Agreement Has Undermined the Right of Governments to Establish Appropriate Levels of Protection Against Risk" (2000) 31 L. & Pol'y. Int'l. Bus. 854, and L. Hughes, "Limiting the Jurisdiction of Dispute Settlement Panels: The WTO Appellate Body Beef Hormone Decision" (1998) 10 Geo. Int'l Envtl. L. Rev. 915.

[19] 'Sanitary' measures address food safety issues and 'phytosanitary' measures deal with plant and animal protection.

[20] As such, the SPS Agreement applies only if a trade measure is in some way inconsistent with GATT Rules, most commonly Articles I and III of the GATT.

the effect of restricting international trade unnecessarily. It does so in part by separating the interests of the exporting and importing countries. The issue in this context is how far the exporting country has to go to justify the adequacy of its measures and how far the importing country can go in imposing conditions on the exporting country before it has to accept the product.

Most of the provisions in the Agreement provide rules on how to distinguish between necessary and unnecessary measures. The preferred solution in dealing with this struggle is through internationally negotiated standards. All Members would ideally accept these standards. The process becomes more complicated when countries either have different views on what level of risk is acceptable, when States disagree over the information that forms the basis for risk analysis, or there are national or regional issues that make an international standard difficult or impossible to apply.

The Agreement establishes an equivalency process to allow Members to alter international standards (Art. 4). In addition, where no appropriate international standards exist, members can apply a risk assessment process to identify an appropriate response that balances the human health and environmental risks against the trade restrictiveness of the measures proposed (Art. 5). The Agreement does not completely resolve the fundamental question of what level of risk is appropriate and how to value that risk against the restriction of trade associated with proposed measures.

Annex A offers the following definition of appropriate level of sanitary or phytosanitary protection: "The level of protection deemed appropriate by the Member establishing a sanitary or phytosanitary measure to protect human, animal or plant life or health within its territory". It is left to the dispute settlement process and negotiations on future standards and risk assessment processes to resolve what the limits of a Member's powers to impose measures are under this definition. The extent to which Members will be able to take a precautionary approach to the protection of the health of humans, plants and animals should be clarified in the process.

(i) *Structure*

The Agreement consists of a preamble, 14 Articles and three annexes. The preamble and the first two Articles provide the context and general provisions on the relationship between trade and the protection of human health, plants and animals from pests and disease. Articles 3 to 6 deal specifically with the mechanisms to establish acceptable limits on trade through international standards, equivalent national standards,[21] and accepted risk assessment processes.

---

[21] Pursuant to Article 4, the equivalency provisions allow for differences in SPS standards if the

Articles 9 and 10 deal with developing country issues. The remainder of the Agreement deals with administrative matters, transparency, and dispute settlement. Annex A provides definitions for key terms in the Agreement, Annex B deals in some detail with the issues of transparency raised in Article 7, and Annex C sets out control, inspection and approval procedures.

(ii) *Obligations*

The SPS agreement imposes a number of key obligations on Member States, including the following:

- General
- Base Measures on International Standards
- Accept Standards of Other Members
- Apply Internationally Accepted Risk Assessment
- Adapt Measures to Regional Characteristics
- Transparency
- Take Account of Special Needs of Developing Countries

(A) General

The general obligation in the Agreement is for Parties to ensure that sanitary and phytosanitary measures are taken either in accordance with international standards, or otherwise in such a way as to minimize their impact on international trade (Arts. 1 and 2).

(B) Base Measures on International Standards

Members shall base their sanitary and phytosanitary measures on international standards where possible (Art. 3). Exceptions include circumstances where there is no international standard and where there is scientific justification for a higher standard or the higher standard is otherwise based on a Party's decision to apply an appropriate level of protection that is different from the level on which the international standard is based.[22]

---

exporting country can demonstrate to the importing country that its measures meet the importing country's standards for SPS protection. Members are encouraged to enter into bilateral or multilateral agreements to formalise these equivalency arrangements.

[22] There are several international organizations that either have or are currently developing standards that may have relevance to this Agreement. They include the Codex Alimentarius, the World Organisation for Animal Health, and the International Plant Protection Convention (IPPC). These organizations are recognized in Article 3.4 as playing an important role in developing international standards. The role of the IPPC in setting standards under the SPS Agreement has been formally

(C) Accept Standards of Other Members

Standards that differ from international standards are to be accepted by importing countries if the exporting country can demonstrate that the standard meets the importing country's appropriate level of protection (Art. 4).

(D) Apply Internationally Accepted Risk Assessment

Members' measures are to be based on risk assessment, taking into account internationally recognised assessment techniques, available scientific evidence, inspection, testing and sampling methods, and balancing the risk to humans, animals and plants against the economic cost associated with the trade restrictive effect of the measures being considered (Art. 5).[23]

(E) Adapt Measures to Regional Characteristics

Members are required to ensure that measures applied to the export and import of a product are adapted to the area from which the product in question originates and to the area of destination (Art. 6). This means consideration of the level of prevalence of specific pests and diseases, the existence of eradication and control programmes, and appropriate criteria or guidelines.

(F) Transparency

Annex B establishes processes for notification and access to information on sanitary and phytosanitary measures of a Member State (Art. 7). They include the requirement for one entry point to be set up by each Member State to answer all questions from interested Members. The Annex provides for the availability of documentation through these entry points and includes a notification procedure to be followed by Members.

---

recognized through changes to the IPPC in 1997. A current list of Codex Alimentarius standards can be found at <http://www.codexalimentarius.net> (accessed: 23 March 2003). Standards developed to date address food hygiene, food labelling, food import and export inspection and certifications systems, and organically produced food. For current OIE standards, see <http://www.oie.int/eng/en_index.htm> (accessed: 23 March 2003). The OIE has, for example, developed an International Aquatic Animal Health Code.

[23] One thing not clear from Article 5 is whether and why risk assessment is required in circumstances where a Party determines that the appropriate level of risk is no risk. See also the Beef Hormone Rulings below.

(G) Take Account of Special Needs of Developing Countries

Members are required to take account of the special needs of developing countries, in particular the least developed countries (Art. 10). This requirement includes an allowance for phased introduction of measures and extended time frames for compliance. The Committee on Sanitary and Phytosanitary Measures[24] is empowered to grant developing countries specific exemptions from the obligations under the Agreement.

4. KEY RULINGS WITH ENVIRONMENTAL IMPLICATIONS

In the field of international trade rules, there has been more reliance on formal dispute resolution processes than is the case in many other areas of international law. As a result, there are a number of rulings before and after the Uruguay round of trade negotiations that have dealt with issues related to the questions posed. Before considering the application of the current WTO trade rules on climate change, it is therefore important to assess the interpretative value of these rulings as they relate to the questions posed. To this end, key trade rulings with environmental implications are therefore briefly assessed here.

These rulings do not have the same formal interpretive value as court rulings in many domestic legal systems. In particular, the WTO panels and appellate bodies established to resolve disputes among WTO Members States are not bound by precedent in the same manner as many domestic courts. Nevertheless, there is every indication from rulings to date that the WTO panels and appeal bodies will consider carefully previous rulings, and will make a considerable effort to develop a consistent body of law interpreting the various provisions of the WTO agreements. In the context of the questions raised here, therefore, previous rulings on similar issues are likely to be the best indicators available for predicting how the WTO might deal with disputes involving climate change mitigation measures. The following survey of key decisions is broken into two groups: those decisions made before the establishment of the WTO; and, those made under the new WTO rules (after 1994).

There were six panel reports prior to the establishment of the WTO that dealt directly with human health and the environment. They are the Canada-US dispute on Tuna,[25] the Canada-US dispute on unprocessed Herring and Salmon,[26] the

[24] Established under SPS Agreement, *supra* note 5, Article 12.

[25] *United States – Prohibition of Imports of Tuna and Tuna Products from Canada*, adopted on 22 February, 1982, BISD 29S/91. The US import prohibition on Canadian tuna was found to violate Article XI:1 of GATT 1947, and not justified under Articles XI:2 or Article XX(g).

[26] *Canada – Measures Affecting Exports of Unprocessed Herring and Salmon*, BISD 35S/98 (1988)

Thailand-US dispute over cigarette taxes,[27] two Dolphin Tuna disputes involving the US, Mexico, and the European Community,[28] and the Corporate Average Fuel Economy regulation (CAFE) dispute between the US and the European Community.[29] In each case, the environmental regulation challenged was found to violate the GATT rules and was found not to be justified under Article XX of the GATT.

Perhaps the most relevant pre-WTO rulings are the two Dolphin Tuna disputes, given their striking similarities to the two Shrimp Turtle Rulings discussed below. In the first Dolphin Tuna dispute, the Marine Mammal Protection Act of the United States was challenged under Articles III, XI, and XIII of GATT. Specifically, Mexico challenged a provision in the Act, which prohibited the importation into the US of yellowfish tuna, which was caught using technology that results in the incidental killing or serious injury of ocean mammals in excess of US standards. The Act essentially prohibited the import of tuna, unless a determination was made that the exporting country had measures in place to protect marine mammals that were comparable to those in the US. The panel concluded that the import restrictions were not internal regulations in accordance with Article III, were inconsistent with Article XI, and not saved by Article XX. A similar ruling was made in response to an EU complaint in the second Dolphin Tuna dispute. In the process, these decisions seem to close the door on non-product related process and production methods (PPMs) as a basis for determining that two products are not like products for purposes of Article III of the GATT.[30]

Since the conclusion of the Uruguay Round and the establishment of the WTO with its binding dispute settlement procedure, there have been six key disputes involving environmental issues. The final Appellate Body rulings in each case are briefly reviewed below.

(hereinafter 'Herring and Salmon Case'). Canadian export restrictions were found to violate Article XI:1 of GATT 1947, and not justified under Articles XI:2(b) or Article XX(g).

[27] *Thailand – Restrictions on Importation of and Internal Taxes on Cigarettes*, BISD 37S/200 (1990) Thailand import restrictions were found to violate Article XI:1 of GATT 1947, and not justified under Articles XI:2(b) or Article XX(g).

[28] *United States – Restrictions on Imports of Tuna*, BISD 39S/155 (1991), and WTO DS 16 June 1994, *United States – Restrictions on Imports of Tuna*, DS29/R (hereinafter 'Tuna-Dolphin Case'). US import prohibitions for tuna were found to violate Article XI:1, were found not to be internal regulations under Article III, and were found not to be justified under Article XX(b), (d), or (g).

[29] WTO DS 11 October 1994, *United States – Taxes on Automobiles*, DS31/R. The separate fleet accounting for foreign fleets violated Article III:4 and was not justified under Article XX(b), (d), or (g).

[30] See Section 5 below.

(a) Reformulated Gasoline Ruling[31]

This case involved a challenge by Venezuela and Brazil to certain reformulated gasoline regulations passed under the US Clean Air Act. These regulations, intended to reduce air pollution in the United States, required that gasoline sold in the most polluted cities of the US meet a specific pollution standard (reformulated gasoline standard), and that in the rest of the US, gasoline sold had to meet, at a minimum, the same pollution standards as gasoline sold in 1990 (conventional gasoline standard). One of the objectives of the conventional gasoline standard was to prevent producers from blending pollutants removed from the reformulated gasoline into conventional gasoline. To achieve this objective for producers who were not in operation in 1990 (and for importers), a statutory baseline was established in place of the producer-specific 1990 baseline. At the heart of the dispute was the fact that foreign refiners were not permitted to establish individual baselines, rather, they had to rely on baselines established by importers based on their 1990 records, or rely on the statutory baseline imposed by the Clean Air Act.

The claim against the US was based on the position that the regulations were inconsistent with Article III of the GATT, and not covered by the exceptions in Article XX. The panel and the Appellate Body both concluded that the regulations were inconsistent with Article III of the GATT, in that it treated importers of gasoline less favourably than domestic producers. The panel concluded that the regulation was not justified under paragraphs (b), (d), and (g) of Article XX.

The Appellate Body found that the baseline rules came under Article XX(g), but were not consistent with the requirements of the chapeau of Article XX[32], and were therefore inconsistent with the GATT rules. In other words, the Appellate Body concluded that the measures related to the conservation of an exhaustible resource (clean air) and that they were made in conjunction with restrictions on domestic production or consumption, but nevertheless constituted an arbitrary or unjustifiable discrimination, and were therefore not exempt under Article XX. The basis for this conclusion was that domestic producers had the choice to establish their own 1990 baselines or rely on the statutory baseline, whereas foreign producers had to use the statutory baseline or baselines established by importers. Foreign refiners therefore did not have the option of establishing facility specific baselines.

In drawing the line between the chapeau and paragraph (g), the Appellate Body appears to have drawn a line between purpose and effect. As long as the measure

[31] WTO AB 20 May 1996, *United States – Standards for Reformulated and Conventional Gasoline*, WT/DS/9, WT/DS2/AB/R. (AB 1996 – 1).

[32] See Section 5, below for a discussion of Article XX and its chapeau.

(gasoline baselines) is primarily aimed at the conservation of clean air, and does not completely single out importers,[33] it meets the requirements of paragraph (g). In order to be exempt under Article XX as a whole, however, the measure also has to meet the test in the chapeau of not arbitrarily or unjustifiably discriminating between domestic and foreign products.[34] The discrimination in this case was found to be unjustifiable because the objective of conserving clean air could have been achieved to the same standard without discriminating against foreign producers.[35] In essence, the US was unable to convince the Appellate Body that it had valid reasons for not either applying a statutory baseline to all producers or allowing all producers to establish individual baselines.

(b) Beef Hormone Ruling[36]

In this case, Canada and the United States claimed that measures by the European Community to prohibit the importation of meat and meat products from cattle treated with certain hormones violated the GATT rules, specifically Articles 3.1, 5.1, and 5.5 of the SPS Agreement. The panel concluded that the import ban violated Article 3, in that it was not based on an international standard. It also agreed with Canada and the United States that the import ban was in contravention of Article 5.1 of the SPS Agreement in that it was not based on a risk assessment. It also held that the ban was in violation of Article 5.5 in that it adopted arbitrary or unjustifiable distinctions in the levels of sanitary protection applied in different situations.

The Appellate Body overturned the panel decision on Article 3.1, holding that countries are entitled to set their own standards higher than international standards. It also overturned the panel's conclusion that the measures were arbitrary and unjustifiable under Article 5.5. It nevertheless found that the EU standard violated

---

[33] *Supra* note 31, at page 19. This was found to mean there had to be an element of even-handedness, but not identical treatment of domestic and foreign producers. In the case of the reformulated gasoline standards, there were restrictions on both domestic and foreign producers, and this was found to be sufficient to meet this part of paragraph (g) of Article XX.

[34] *Ibid.*, at page 20.

[35] *Ibid.*, at page 21. The Appellate Body set out three prohibitions in the chapeau for the application of the exemptions in Article XX: 1) arbitrary discrimination; 2) unjustifiable discrimination; and 3) disguised restriction on international trade. For a detailed critique of this ruling, see R. Quick *et al*, "An Appraisal and Criticism of the Ruling in the WTO Hormone Case" (1999), Journal of International Economic Law 603.

[36] WTO AB 16 January 1998, *European Communities – Measures Affecting Meat and Meat Products*, WT/DS26/R/USA, WT/DS26/AB/R, WT/DS48/R/CAN, WT/DS48/AB/R (AB – 1997 – 4) . For a detailed critique of this ruling, see R. Quick *et al*, "An Appraisal and Criticism of the Ruling in the WTO Hormone Case" 2:3 (1999) J. Int'l Econ. L. 603, K. A. Ambrose, "Science and the WTO" (2000) 31 Law & Pol'y Int'l Bus. 861, and R. Neugebauer, "Fine-Tuning WTO Jurisprudence and the SPS Agreement: Lessons from the Beef Hormone Case" (2000) 31 Law & Pol'y Int'l Bus. 1255.

Article 5.1 in that it was not based on a risk assessment. Specifically, it concluded that the measure was not objectively or rationally connected to the risk identified.

The basis for the decision of the Appellate Body was that the scientific evidence supported a risk only in case of misapplication of a hormone, whereas the risk assessment was conducted only based on appropriate application. It therefore concluded that the risk of inappropriate application of hormones was not subjected to an appropriate risk assessment, and that the measure was furthermore not designed to prevent inappropriate applications but prevented all applications.[37] The measures were therefore found to be in violation of Articles 5.1 and 3.3 of the SPS Agreement. While indicating that the violation of Article 5.1 implied a violation of the more general Article 2.2, the Appellate Body, applying the principle of judicial economy, declined to specifically rule on whether the measure was in violation of Article 2.2. In the process, it has become clear that Article 5.1 sets out a substantive, not just a procedural obligation to justify the measure through risk assessment. While stressing that such an assessment does not have to be based on scientific proof, the Appellate Body in this case did not accept the scientific evidence presented in support of the SPS measure. Through these findings, the Appellate Body has positioned itself as the final arbiter of scientific evidence presented to justify a given SPS measure.[38]

Finally, the Appellate Body concluded that the burden of proof, at least initially, rests on the complaining Party to establish a *prima facie* case of a violation of the SPS Agreement. If the complaining Party makes out a *prima facie* case, the burden then shifts to the Party whose measures are being challenged to demonstrate that they are consistent with applicable WTO rules.

(c) Shrimp Turtle Ruling (No. 1)[39]

Under the US Endangered Species Act, US shrimp trawlers are required to use 'turtle excluder devices' (TEDs) in their shrimp nets when fishing in areas that are likely to be turtle habitat. In 1989, the US passed further laws to restrict the importation of shrimp and shrimp products only from States that have comparable regulations in place or demonstrate that their fishing practices do not pose a threat to turtles. In practice, exporting countries had to demonstrate the use of TEDs in order to be certified to import into the US under this law.

---

[37] *Ibid.*, DS48/AB ruling 16 January, 1998, at para. 113.

[38] For a more detailed discussion of opposing views on the relationship between science, precaution and risk assessment under the SPS Agreement, see J.M. Wagner, *supra* note 18 above, at 857; K.A. Ambrose, *supra* note 36 above, at 861; and L. Hughes, *supra* note 18 above, at 915.

[39] See Shrimp Turtle (No.1), *supra* # 8 above.

India, Malaysia, Pakistan, and Thailand challenged this certification requirement under Articles I, III, and XI of the GATT. The panel concluded that the certification requirement violated Article XI of the GATT, and was not justified under Article XX. The Appellate Body reached the same conclusion on different grounds and in the process followed the Reformulated Gasoline decision in its interpretation of Article XX.[40] It concluded that while the measure to protect turtles was provisionally justified under Article XX(g),[41] it did not meet the conditions set out in the chapeau of Article XX,[42] and was therefore not exempt under Article XX. In so doing, it continued a trend of interpreting provisions (a) to (g) of Article XX broadly, but applying more strictly the test of 'arbitrary or unjustifiable discrimination' where the same conditions apply and act as a 'disguised restriction on international trade'. In the process, a distinction was drawn between the provisional validity of the measure itself, and the 'trade fairness' of its application. In its decision, the Appellate Body went to great length to affirm the rights of Member States to adopt effective measures to protect the environment, and that the decision was about how the turtles were protected, not whether they should or could be protected.[43]

(d) Shrimp Turtle Ruling (No. 2)[44]

At the conclusion of the first round of rulings on the shrimp turtle dispute, the Dispute Settlement Body made certain recommendations on measures to be taken by the United States to come into compliance with the GATT rules as they applied to its regulation of shrimp products. In response, the US did not change the applicable law, but changed its application. Specifically, it developed criteria for the certification of foreign harvesting nations aimed at ensuring that foreign shrimp products were treated in accordance with the GATT rules and the US objective of protecting turtle populations was still met. A second challenge was brought by a number of shrimp exporting countries. They argued that the new requirement amounted to an obligation to demonstrate that measures in an importing country were 'comparable in effect' to the TEDs, rather than to require the use of TEDs.

The Appellate Body in Shrimp Turtle (No. 2) upheld the new rules as consistent with Article XX of the GATT, essentially concluding that the change to 'com-

---

[40] *Ibid.*, AB ruling, para. 118.
[41] *Ibid.*, at paras. 128-145.
[42] *Ibid.*, at paras. 146-186.
[43] *Ibid.*, at paras. 185-186.
[44] WTO AB 12 October 1998, *United States – Import Prohibition of Certain Shrimp and Shrimp Products, Recourse to Article 21.5 of the DSU by Malaysia*, WT/DS58/AB/RW (AB – 2001 – 4) (hereinafter 'Shrimp Turtle (No. 2)'). As discussed below, this ruling is limited to the implementation of the Shrimp Turtle (No. 1) ruling.

parable in effect' removed the violation of the chapeau of Article XX.[45] In the process, the Appellate Body considered the obligation on the US to pursue international cooperation in the protection of turtles, and the level of flexibility required for the revised rules to meet the requirements set out in the chapeau of Article XX. On the first issue, the Appellate Body confirmed that negotiation in good faith is sufficient, and that there is no obligation to conclude an agreement. What the US had to do and ultimately did in this case was to provide all exporting countries with similar opportunities to conclude an agreement. With respect to the flexibility of the new rules, the Appellate Body concluded that the test of 'comparable effectiveness' provided sufficient flexibility to take into account special circumstances in the exporting country, while providing the necessary assurance to the importing country that its environmental objective could still be met.[46]

(e) Australian Salmon Ruling[47]

This dispute involved a challenge by Canada regarding salmon import prohibitions imposed by Australia. The regulations related to certain treatment requirements for the import of fresh salmon products.[48] In order to prevent the introduction of any infectious or contagious diseases, the Director of Quarantine had to approve of the importation of any fresh fish products before they were permitted. In response to a request to allow the importation of fresh salmon, the Director exercised this discretion to prohibit the importation.

Both the panel and the Appellate Body concluded that the requirements were in violation of Articles 2 and 5 of the SPS Agreement, as Australia had failed to justify the requirements on the basis of available scientific evidence. The Appellate Body further concluded that the measure was not based on a risk assessment within the meaning of Article 5.1 of the SPS Agreement.[49] Consistent with the Beef Hormone dispute, the focus was on Article 5.1, and the violation of Article 2.2 was seen to arise from the violation of the more specific obligation in Article 5.1. The Appellate Body confirmed that countries can determine their own ac-

---

[45] *Ibid.* at paras. 111-152.

[46] *Ibid.* at para. 146. The revised guidelines specifically reference that any demonstrable differences will be taken into account.

[47] WTO AB 20 October 1998, *Australia – Measures Affecting Importation of Salmon*, WT/DS18/R, WT/DS18/AB/R (AB – 1998 – 5).

[48] *Ibid.* The issue was essentially one of whether the treatment of fresh salmon with heat amounts to an import prohibition of fresh salmon, or whether it amounts to a treatment requirement. Given that the requirement for heat treatment changed the salmon from fresh to cooked salmon, the Appellate Body concluded that the requirement amounted to an import prohibition. This import prohibition was found not to be based on any risk assessment and therefore in violation of the SPS Agreement.

[49] *Ibid.*, see para. 121 for criteria for risk assessment under Article 5.1.

ceptable level of protection, including a level of protection based on zero risk, and that the appropriateness of the measure is to be considered in light of the level of protection chosen. This, however, does not absolve a Party from basing its measures to achieve this level of protection on a risk assessment.[50] Finally, the Appellate Body confirmed and applied the burden of proof requirements set out in the Beef Hormone Rulings discussed above.

(f) Japan Agriculture Ruling[51]

In this case, the United States challenged regulations passed by Japan to prohibit the importation of certain agricultural products that failed to comply with Japan's quarantine measures. The panel concluded that Japan's measures were inconsistent with Article 2.2 and 5.6 of the SPS Agreement. The Appellate Body upheld the panel's finding based on the conclusion that the quarantine requirement was maintained without sufficient scientific evidence within the meaning of Article 2.2 of the SPS Agreement, and was furthermore inconsistent with Article 5.7 as a temporary measure. The ruling in this dispute failed to follow the precedent set in the Beef Hormone and Australian Salmon disputes by deciding, for the first time, that a measure is in violation of Article 2.2 on its own, without relying on the violation of Article 5.1 as the basis for this conclusion. It concluded that while Article 5.1 may be seen as a specific application of the basic obligation set out in Article 2.2, this did not mean that the two provisions had to be considered together in all cases.[52]

(g) Asbestos Ruling[53]

This dispute arose out of certain measures imposed by France to prohibit the importation and sale of certain types of asbestos and asbestos products in France. Canada, an exporter of asbestos, challenged these measures under the TBT Agreement, and under Article III of the GATT. The Panel concluded that the regulations in question did not constitute technical regulations. It therefore found that the TBT Agreement did not apply. The Panel further concluded that asbestos and cement-based products were 'like products' under Article III and that France violated Article III of the GATT in prohibiting the importation of asbestos. However, the Panel found that the measures in question were saved under Article XX(b) of the GATT as necessary to protect human health or life. On appeal, the

---

[50] *Ibid.*, see para. 125.

[51] WTO AB 22 February 1999, *Japan – Measures Affecting Agricultural Products*, WT/DS76/R, WT/DS76/AB/R (AB- 1998 – 8) (hereinafter 'Agricultural Products Case').

[52] *Ibid.*, at para. 82.

[53] WTO AB 12 March 2001, *European Communities – Measures Affecting Asbestos and Asbestos-Containing Products*, WT/DS135/R, WT/DS135/AB/R (AB – 2000 – 11) (hereinafter 'Asbestos Case').

Appellate Body held that the TBT Agreement applied, as the measures in question were technical regulations.

The Appellate Body applied the following test for determining whether two products are 'like products':

1. The physical properties of the products;
2. The extent to which the products are capable of serving the same or similar end-uses;
3. The extent to which consumers perceive and treat the products as alternative means of performing particular functions in order to satisfy a particular want or demand; and
4. The international classification of the product for tariff purposes.[54]

It disagreed with the Panel in its interpretation of Article III of GATT and concluded that in considering likeness of products, France was entitled to take into account the health risk associated with products containing asbestos, and was not limited to considering the nature and quality of the product itself.

The Appellate Body concluded that all relevant evidence of likeness or differences between products must be considered. In this case, it concluded that the health risk associated with the asbestos-based product was relevant. This lead to the conclusion that Canada had failed to show that the products were like products, in that it failed to respond to evidence that the health risk associated with the asbestos product resulted in it having different physical properties and different consumer tastes and habits. The Appellate Body left open the possibility that the four criteria or characteristics can be added to, and made no final determination that differences considered under Article III must fall under these categories.[55]

(h) Sardines Ruling[56]

Peru, supported by a number of other countries, challenged certain EC regulations dealing with the marketing of certain fish products, specifically products to be marketed as 'preserved sardines'. The EC regulations accepted only one local species of sardines for purposes of its labeling regulations, whereas the

---

[54] *Ibid.*, AB ruling, at para. 101.

[55] *Ibid.*, AB ruling, at para. 117. In the context of climate change, there are at least two issues to consider. Can the environmental impact of producing the product become an additional factor to be considered? Would consumer preference for products that are climate change friendly in their production qualify as differences in whether consumers are 'willing to use these products to perform these functions'?

[56] WTO AB 26 September 2002, *European Communities – Trade Description of Sardines*, WT/DS231/R, WT/DS231/AB/R (AB – 2002 – 3) (hereinafter 'Sardines Case').

Codex Alimentarius covers 21 fish species.[57] The Panel concluded that the labeling restriction violated Article 2.4 of the TBT Agreement on the basis that it was an ineffective or inappropriate means of fulfilling the legitimate objectives of the EC regulations.

The Appellate Body considered a number of significant issues in its review of the panel ruling. As a starting point, it confirmed its three-point test in its reasons in the Asbestos Ruling[58] of the definition of 'technical regulation' in the TBT Agreement:

1. The document must apply to an identifiable product or group of products;
2. The document must lay down one or more characteristics of the product; and
3. Compliance with the product characteristic must be mandatory.[59]

Furthermore, the Appellate Body ruled that an international standard does not have to be adopted by consensus in order to meet the definition of a standard under Annex 1.2 of the TBT Agreement. In other words, a standard can be used for the purposes of the TBT Agreement even though the Party, against which it is being applied, objected to the standard when it was developed.[60]

Finally, the Appellate Body concluded that Peru had met its burden of demonstrating that the international standard was an effective and appropriate means of fulfilling the legitimate objective of the EC regulation. On this basis, it concluded that the European Community was not entitled to develop its own standard under the provisions of the TBT Agreement.

5. THE WTO RULES AND CLIMATE CHANGE MEASURES

The TBT and SPS agreements and the above trade rulings with environmental implications serve to provide the context for the assessment of key GATT provisions with respect to States' ability to mitigate climate change and motivate others to follow suit. In this Section, Articles I, III and XX of the GATT are assessed in the context of the questions posed at the start of the Chapter.

---

[57] *Ibid.*, AB ruling, at paras. 5 and 6.

[58] See Asbestos Case, *supra* note 53 above, AB ruling, at para. 66-70.

[59] See Sardines Case, *supra* note 56, AB ruling, at para. 176.

[60] *Ibid.*, at para. 227. This principle is potentially important in the context of whether any standards developed under the climate change regime, such as emission reduction commitments under Kyoto, have relevance under the trade rules of the WTO. Note that in the context of the TBT itself, the question of the applicability of the international standard is linked to the further test of whether the standard is an effective or appropriate means for the fulfillment of a legitimate objective pursued, allowing Parties in circumstances where the international standard is not sufficient, to go beyond the international standard.

(a) Are Products With Differing Climate Change Impacts 'Like Products' Under the GATT Rules? (Articles I and III)

Article I of the GATT (often referred to as the 'Most Favoured Nation' clause) is one of the core provisions of the GATT. It essentially requires Member States of the WTO to treat products from any Member State in an equivalent manner to the best treatment afforded 'like products' when imported from any other State, or to treat every Member State as it treats a 'Most Favoured Nation' on a product by product basis. Article III further applies this concept by requiring States to treat like products from Member States no less favourably than domestic products (often referred to as the 'National Treatment' clause).[61]

In the context of climate change, this raises the question of what constitutes like products, given that the requirements of Articles I and III only apply to like products. In other words, if two products, which a State wants to treat differently for climate change policy reasons are not considered like products, Articles I and III do not limit a State's right to treat the products differently. If the two products are considered to be 'like', Articles I and III apply to limit the right of a State with respect to the relative treatment of the two products.

In considering the issue of like products, it may be useful to consider several different scenarios relevant to the implementation of climate change policy and measures by way of example:[62]

1. Two products have the same physical characteristics and same applications once produced, but have different levels of GHG emissions associated with production because of differences in energy efficiency of production methods (for example, two cars, one built with all virgin material, the other with 50% recycled content, or some other more energy efficient production method);
2. Two products have the same physical characteristics and same applications once produced, but have different levels of GHG emissions associated with production because of differences in energy sources available to the producer (i.e. one production facility uses electricity from coal, the other from wind and solar);

[61] For a general overview of the application of the GATT rules in an environmental context, see C. Wofford, "A Greener Future at the WTO: The Refinement of WTO Jurisprudence on Environmental Exceptions to GATT" (2000) 24 Harv. Envtl. L. Rev. 563.

[62] In terms of existing cases, the Shrimp Turtle and Asbestos Rulings probably come closest to dealing with similar facts situations, but the five scenarios proposed here offer more subtle differences than those explored in any of the cases to date. While the Asbestos Ruling, for example, deals generally with the distinction between fitness for the intended purpose and the question of likeness of products, and introduces the idea that differences in health risks can be a factor in determining that products are not alike, the decision does not draw the distinctions between impacts at the production, consumption and disposal stages of the lifecycle of the product.

3. Two products serve the same application once produced, but the physical characteristics are sufficiently different resulting in different GHG emissions from disposal of the product after use (i.e. one car manufactured with 100% recyclable or reusable content, the other essentially requires complete disposal);
4. Two products serve the same purpose, and have the same general characteristics, but have different GHG emissions associated with their use (i.e. fuel efficiency differences for cars, or differences between gasoline, hybrid and fuel-cell cars); and
5. Two products serve the same ultimate purpose, but achieve that purpose in a different manner. (i.e. cars versus buses, bikes, and trains).

In each of these scenarios, there is one product that has relatively low GHG emissions associated with its life cycle, and another with higher GHG emissions. The two products are either offered for import from different exporting States (Article I), or the low GHG emission product is a product produced domestically and the high GHG emission product is an import. In either case, if the two products serve the same ultimate purpose, a State trying to implement effective policies and measures to address climate change will have to consider whether Article I imposes any restrictions on its ability to treat the two products differently so as to allow it to require or encourage the import, production and use of the low GHG emission product within its territory. The question for each scenario is therefore whether a State can take measures to favour the low GHG emission product over the high GHG emission product without violating Articles I and III.

Under Scenario 1, the difference between the two products for purposes of this analysis is the energy efficiency of the production of the product. Beyond this, the two products are identical, including in their physical characteristics and function (PPMs). This means the difference relates to something that arises from the process and production of the product, but is not reflected in the final product in any tangible way.

The climate change concerns are twofold. First, given the global nature of the problem, a State cannot prevent climate change within its own territory in isolation. This means a State concerned about climate change has a legitimate stake in the GHG emissions from the production in another State of a product that is imported into its territory. Second, if the lower GHG emissions are achieved at an additional cost, treating the two products the same will result in a competitive advantage for the product with the higher GHG emissions, resulting in a disincentive for producers to achieve lower GHG emissions in producing the product, regardless of whether the producer operates domestically within a State or in an exporting State.

On the surface, the two pre-WTO Tuna-Dolphin cases[63] seem to close the door on considering PPMs under Articles I and III in the context of like products. These cases specifically reject the argument that PPMs can form the basis for a conclusion that two products are not like products for purposes of Article III. The pre-WTO Canada-US dispute on Salmon and Herring[64] considered PPMs to be valid for the purposes of Article XX exceptions, but did not consider the threshold question of whether differences in process and production methods can exclude products from the application of Articles I and III in the first place. In Shrimp Turtle, the measure to protect turtles was found to violate Article XI. As a result, the Appellate Body declined to consider whether the shrimp caught with different impact on turtles would be considered to be like products under Article I and III. With the general recognition that PPMs received in the Shrimp Turtle rulings, the role of PPMs in assessing whether products are like products must now be considered an open question. To an extent, Shrimp Turtle marks a clear departure from the approach in Tuna-Dolphin with respect to PPMs in general and in the context of Article XX in particular. Conversely, there is no ruling to date that explicitly accepts a PPM as a basis for a finding that two products are not like products under Articles I or III.

There is one other WTO ruling that sheds some light on how this issue may be addressed in the future: the ruling of the Appellate Body in the Asbestos dispute.[65] As a starting point, the Appellate Body applied a number of criteria in considering the likeness of two products, including consumer tastes and habits (or perceptions and behaviour) with respect to the product.[66] The test in that case was then applied to asbestos products to conclude that otherwise like products, one containing asbestos, the other not, were not considered like products for purposes of Article III, based on the health risk associated with the product. Depending on the level of consumer interest in addressing climate change on one side, and the link between the PPM and environmental and human health impacts of climate change on the other, it would seem reasonable to conclude that a similar argument could be made for products that differ in their GHG emissions from production.

Scenario 2 is essentially identical to Scenario 1, in that it identifies a non-product related PPM. The only difference is that its focus is on sources of energy with different levels of GHG emissions, rather than on energy efficiency. There is no basis in the GATT provisions or WTO rulings to date to conclude that Scenario 2 would be treated differently under Articles I or III than under Scenario 1.

---

[63] See Tuna-Dolphin Case, *supra* note 28.

[64] See Herring and Salmon Case, *supra* note 26.

[65] See Asbestos Case, *supra* note 53.

[66] *Ibid.* at para. 101.

Under Scenario 3, the difference is again in the PPMs, but this time the difference is reflected in the product. In this context, the difference, while not affecting the application and use of the product, does affect the GHG emissions from disposal. This type of difference in production method is referred to as product related PPMs. While the distinction may be relevant from a climate change policies and measures perspective, and may have some influence over the treatment of PPMs in the context of the issue of like products in Articles I and III, in the absence of any rulings on point it is difficult to predict how the distinctions among the first three scenarios may impact on future WTO tribunal rulings. It seems reasonable, however, to conclude that there is likely to be more comfort with a submission that two products are not like products on the basis of product related PPMs than in the case of non-product related PPMs.

Under Scenario 4, the difference is not in the process and production method, but in the GHG emissions from the use of the product. The products are not functionally different, in that both provide the same means of transportation, but they have different GHG emissions associated with their use. Under this scenario, the reasoning in the Asbestos decision would appear to apply directly, resulting in a finding that the two products are not like products, given their difference in energy efficiency, the operational costs associated with the difference in efficiency, and the established difference in consumer tastes and habits with respect to the energy efficiency of vehicles. In considering the application of the Asbestos Appellate Body ruling, it is important to keep in mind that the ruling stresses the need to examine all the evidence related to all the criteria established for the likeness test, and to reach a balanced overall conclusion based on this examination.[67]

Under Scenario 5, the products are functionally different. These products would clearly not be considered like products under Article 1. Member States are therefore not limited by Article 1 of the GATT in their treatment of these products relative to each other, in that the requirement to treat like products alike would not apply in this scenario.

Based on the provisions of Articles I and III of the GATT and relevant interpretations in recent rulings, it is fair to conclude that while PPMs have to date not been recognized as a basis for distinguishing between goods for the purposes of Articles I and III, under at least some of the scenarios outlined; a finding that products are not like products based on their different climate change impacts would appear to be a reasonable application of the principles advanced in recent WTO rulings. In other words, depending on the circumstances, countries may very well be able to justify differential treatment of goods purely based on the climate change impact from process and production methods, in some circum-

[67] *Ibid.*, at para. 102.

stances even on the basis that the products are not alike due to their different climate change impacts.

(b) Are Measures to Require Products to Meet a Specific GHG or Energy Standard Technical Barriers to Trade in Violation of Articles I and III of the GATT?

At this stage of the analysis, it is assumed that two products are considered like products for the purposes of Articles I and III. The question then becomes whether two products are being treated differently by an importing Member State. In the case of Article I, the question is whether two products from two different countries are being treated differently by the importing country. Under Article III, the issue is whether the importing country treats imports differently than domestic products.[68]

As a starting point, it seems likely that measures that States might consider to reduce GHG emissions from the production, transportation, use or disposal of a product would be considered to be a technical regulation[69] or standard under the TBT Agreement.[70] Assuming, therefore, that the TBT Agreement applies, it is important to note that under the agreement, there is a presumption of consistency with Articles I and III of the GATT in cases where a State relies on an international standard in developing its technical regulation.[71]

This raises the question of the role of Kyoto and the UNFCCC in providing an international standard to create this presumption.[72] In short, it is difficult to see how the requirements under the UNFCCC or Kyoto can be used directly to support any specific measure to reduce the GHG emissions associated with the production, transportation, use or disposal of a product. Neither agreement directs States at a sufficient level of detail on how Parties are to meet the overall GHG emission reduction targets to enable a State to make a convincing case that its technical

---

[68] Note that there are other restrictions on a Member State's ability to pass technical regulations, such as Article XI on quantitative restrictions. For the purposes of this analysis, however, the focus will be on Articles I and III as they provide the foundation for the GATT and are the provisions most likely to be applied in a climate change context.

[69] Technical regulation is defined in Annex I of the TBT Agreement as follows: 'Document which lays down product characteristics or their related processes and production methods, including the applicable administrative provisions, with which compliance is mandatory. It may also include or deal exclusively with terminology, symbols, packaging, marking or labeling requirements as they apply to a product, process or production method'. See TBT Agreement, *supra* note 4.

[70] *Ibid.*, Article 1.

[71] *Ibid.*, Article 2.5.

[72] For a good general discussion of the role of international environmental standards in trade disputes, see W. M. Donahue, 'Equivalence: Not Quite Close Enough for the International Harmonization of Environmental Standards', 30:2 *Envtl. L.* (2000), 363.

regulation is 'in accordance with relevant international standards' as required for the presumption in Article 2.5 of the TBT Agreement. For one thing, it will be impossible to determine in isolation whether a specific measure on a specific product is implemented to comply with Kyoto, especially given the many ways countries can meet their obligations, including making use of the Protocol's flexible mechanisms in deciding how much each sector of their societies is to contribute to a reduction target.[73]

The next issue to be considered here is the existence of specific links between the WTO and the climate change regime. As indicated above, Kyoto requires that Parties implement it in such a manner as to minimize effects on international trade.[74] A likely and reasonable interpretation of these provisions would be that they generally propose a requirement for measures to be consistent with the trade rules of the WTO, given that the WTO is the international body tasked with minimizing barriers to international trade. It is therefore not likely that the provisions of the UNFCCC and the Kyoto Protocol will have a significant influence over whether GHG emission reduction measures will be upheld under the TBT Agreement and Articles I and III of the GATT.

It is difficult to predict how the general requirement under Articles I and III to treat foreign products from different countries in the same or equivalent manner to each other and to domestic products, will be applied to GHG emission reduction measures, without knowing the specifics of the measures considered. We can consider the parameters within which the WTO is likely to operate in considering measure specific cases. One interesting issue in this regard is whether a GHG emissions limit or an energy intensity limit would be more likely to be upheld under Articles I and III and the TBT Agreement. Clearly, a limit on GHG emissions is more directly linked to the objective of preventing climate change. On the other hand, a target based on GHG emissions in the production of a given product would treat products differently simply based on the energy sources available in that country.

The problem this might create is that some low-emitting sources of energy, such as geothermal, solar, wind, tidal, and hydropower, may simply not be available in some countries, whereas they may be abundant in others. It would then in effect treat products differently based on the energy sources of the State producing the product. An energy intensity limit, on the other hand, would focus on energy efficiency, but would not deal with the need to switch to less GHG emitting energy sources, thus providing a partial solution only to the climate change problem.

---

[73] The UNFCCC creates similar challenges, in that the objectives of the Convention are so broad and general, that it is difficult to say what the standard is other than the general commitment to reduce global emissions to safe levels.

[74] Kyoto Article 2(3), see also Section 1 above.

One hint of how the WTO Appellate Body might deal with this issue can be found in its Shrimp Turtle Ruling (No. 2). In that case, the Appellate Body noted the importance of the US taking into account the specific conditions of Malaysian shrimp production in the way it chose to implement the requirement to protect turtles from the adverse impacts of the shrimp fishery.[75] This would suggest that the availability of energy sources may become a factor in deciding whether a technical regulation that sets a GHG emission standard is upheld under Article XX.[76]

Beyond this general issue of GHG versus energy intensity requirements on products, the provisions of the TBT Agreement offer the following substantive direction on how to design technical regulations in accordance with the GATT and TBT Agreement rules:

- Member States have a general obligation to ensure that technical regulations do not have the effect of treating products imported from one country less favourably than like products from another country or domestic products.[77] This is essentially a restatement of Articles I and III of the GATT.
- Technical regulations shall not have the purpose or effect of creating unnecessary obstacles to international trade.[78] It would appear that technical regulations can be challenged both in terms of purpose and effect. The subjective nature of any challenge based on the purpose of a technical regulation that has the effect of meeting a legitimate objective makes it difficult to see how a challenge of the purpose of the technical regulation can be a constructive part of the analysis. The effects-based analysis should therefore be considered in most cases to be the more appropriate test in deciding on the legitimacy of technical regulations under Article 2.2.
- Technical regulations shall not be more trade restrictive than necessary to fulfill a legitimate objective. Human health and safety, animal or plant life or health, and the environment are considered legitimate objectives in this regard.[79] What will be interesting in this context is whether the WTO will look at the legitimate objective on a product–by-product basis, or more generally. If done on a product-by-product basis, the WTO would accept the Member State's sovereignty to decide to reduce GHG emissions through the production, transportation, use and disposal of that particular product, and would focus on whether the technical regulation in question is the least trade restrictive way of reducing those emissions. The alternative would require

---

[75] See Shrimp Turtle (No. 2), *supra* note 44, at para. 146.

[76] The process of determining whether an exception under Article XX applies is discussed in the next Section below.

[77] See TBT Agreement, *supra* note 4 above, Article 2.1.

[78] *Ibid.*, Article 2.2.

[79] *Ibid.*, these legitimate objectives are directly connected to the exceptions in Article XX of the GATT and are discussed below in the next Section.

the WTO to consider whether reducing GHG emissions in the life cycle of the particular product in question is the most effective and least trade restrictive way for the Member State to reduce its GHG emissions, or whether the emission reduction should take place in another sector, where the impact on international trade would be less. Any attempt to take the latter approach would severely limit a State's ability to achieve GHG emission reductions, and would likely be seen as interference with the domestic policies of that State.[80]

- There is reference in Article 2.2 to the need to consider the risk involved in not meeting the stated objective, suggesting a risk assessment approach. Specifically, Article 2.2 proposes that the risk be assessed considering available scientific and technical information, related processing technology or intended end-uses of products. It is not as clear as in the SPS Agreement that a member can determine its own level of risk, but the language of the TBT Agreement does support a risk assessment approach.[81] In the context of climate change, it is difficult to see how a risk assessment approach would be applied, given that the risk is one that is based on global emissions over years and decades. Perhaps the best practical approach to this in the context of climate change would be that countries are free to determine their own GHG emission or energy targets on a product-by-product basis, and that the analysis under the TBT Agreement should focus on whether that product specific target is implemented in the least trade restrictive manner.
- There is strong language to require Member States to rely on international standards wherever they can be used to meet the legitimate objective of the Member State, and to accept the standards of other Member States whenever those standards adequately fulfill the objectives of the Member States' own regulations.[82] The issue of international standards was most recently considered in the Sardines Ruling.[83] The standard, in that case, was a product-based standard, where the issues related to what level of acceptance for the standard was required, and whether the standard applied to the particular product in question. In the context of climate change, there are no clear product-based

---

[80] A good indication of this reluctance to have to account for decisions at this level were the negotiations on compliance under Kyoto, where there was strong resistance by many States to international direction on policies and measures needed to bring countries who failed to meet their targets back into compliance. At COP 6 in The Hague, for example, a proposal that was put forward by some countries to require a compliance action plan to be approved by the enforcement branch of the compliance body was rejected as an invasion of State sovereignty.

[81] See TBT Agreement, *supra* note 4 above, Article 2.2. Legitimate objectives listed include human health, animal or plant life or health, and the environment.

[82] *Ibid.*, Articles 2.4, 2.5, 2.6, and 2.7.

[83] See Sardines Case, *supra* note 56 above, para. 110, where the Appellate Body concluded on the issue of when a measure is based on an international standard that 'there must be a very strong and very close relationship between two things in order to be able to say that one is the basis for the other'. It does not appear that anything in the current climate change regime would meet this test in the context of a product specific measure.

standards. As they are developed, it will be interesting to see whether States will be successful in taking the position that these standards are insufficient to allow them to meet their objectives in terms of overall GHG emission reductions in their countries. In the meantime, neither the UNFCCC nor the Kyoto Protocol includes anything that could be considered a product-based standard. It would therefore be reasonable to conclude that whether a product-based measure is linked to compliance with Kyoto may have little or no impact on its treatment under the WTO regime.

- Member States are encouraged to develop performance-based standards rather than design specific standards.[84] In the context of transportation scenarios, this may become a question of whether the technical regulation requires cars to be hybrids or fuel cell vehicles, or whether the standard is a fuel-efficiency standard. Article 2.8 clearly suggests that fuel-efficiency standards might be easier to justify under the TBT Agreement; however, if a gasoline driven vehicle and a fuel-cell vehicle are not considered like products under Articles I and III of the GATT, there would be no requirement under the TBT Agreement or the GATT to treat the two products the same, and, therefore, the issue of compliance with Article 2.8 of the TBT Agreement would not arise.

(c) Are Technical Regulations That Require Products to Meet a Certain GHG or Energy Standard Saved by Article XX?

In this Section, a *prima facie* violation of Articles I or III is assumed, and the application of the exceptions under Article XX is considered in this context. Article XX provides for certain circumstances where technical regulations that treat like products from different countries differently[85] may nevertheless be permitted under the GATT. In theory, Article XX has to be considered in connection with the TBT Agreement, because it also lists legitimate objectives of technical regulations.[86] From a practical perspective, Article XX would appear to be more specific, and cases dealing with the question of legitimate objective or exception under Article XX have made little reference to the TBT Agreement. It would appear, therefore, that the provisions of Article XX are likely to be determinative when faced with the question whether technical regulations, which are

---

[84] See TBT Agreement, *supra* note 4 above, Article 2.8.

[85] This refers to either products from two or more different importing countries or imported versus domestic products, see Articles I and III of GATT.

[86] See TBT Agreement, *supra* note 4 above, Article 2.2. Human health and safety, animal or plant life or health, and the environment are considered legitimate objectives in this regard.

prima facie in violation of Articles I and III because they treat like products differently, are saved by the legitimate objective they serve.[87]

Article XX consists of a chapeau and a list of specific exceptions to the requirement to treat like products alike. The specific exceptions of interest in the context of climate change are (b) and (g), referring to measures 'to protect human, animal or plant life or heath' and measures 'relating to the conservation of exhaustible natural resources if such measures are made effective in conjunction with restrictions on domestic production or consumption'.[88] The application of these two exceptions is limited in the chapeau by requiring that the measures not be applied so as to constitute 'a means of arbitrary or unjustifiable discrimination between countries where the same conditions prevail' or constituted 'a disguised restriction on international trade'.[89]

Through the rulings of the Appellate Body in the Reformulated Gasoline and Shrimp Turtle cases, a two-step test has been developed for the application of Article XX. First, the measure needs to be provisionally justified under one of the exceptions of Article XX. If it is not, the analysis ends here and the measure is not saved under Article XX. If the measure is provisionally justified under one of the exceptions, it then has to pass the tests of not being 'arbitrary discrimination', 'unjustifiable discrimination' or a 'disguised restriction on international trade' in order to be saved by Article XX. The rulings of the Appellate Body in the Reformulated Gasoline and Shrimp Turtle rulings suggest a fairly generous application of (b) and (g) of Article XX, and a focus on the chapeau in weighing the utility of the measure against its trade restrictiveness.[90]

In the context of the Reformulated Gasoline rulings, the Appellate Body concluded that the measure to reduce pollution from gasoline was provisionally consistent with Article XX(g). In the process, it extended the concept of 'exhaustible natural resource' to 'air'. By not allowing foreign producers to establish

---

[87] It is worth noting that it is less than clear from the TBT Agreement or environmental rulings to date whether the TBT Agreement's focus is on Articles I and III, or whether it was intended to elaborate also on the application of Article XX. From the environmental rulings reviewed here, it would appear that its application does not extend to Article XX, and that the latter therefore stands on its own.

[88] It is important to note that in the Reformulated Gasoline Ruling, the Appellate Body, on page 19, concluded that the requirement of 'in conjunction with restrictions on domestic production or consumption' was met as long as there was some restriction on domestic products. Identical treatment is not required here, just some restriction of a similar nature. See Reformulated Gasoline Case, *supra* note 18 above.

[89] GATT, Article XX.

[90] See also WTO DS 17 June 1987, *United States-Taxes on Petroleum and Certain Imported Substances*,L/6175 – 34S/136, at para. 156, where the Panel concluded as follows: "a balance must be struck between the right of a Member to invoke and exception under Article XX and the duty of that same Member to respect the treaty rights of the other Members".

baselines in an equivalent manner as domestic producers, the Appellate Body then concluded that the measure amounted to 'unjustifiable discrimination' and was therefore not consistent with the chapeau of Article XX.

Similarly, in Shrimp Turtle (No. 1), the Appellate Body ruled that the measure to protect turtles was provisionally consistent with Article XX(g). In the process, the Appellate Body accepted the objective of preventing an extra-territorial environmental impact[91] as a legitimate basis for an exception under Article XX(g). However, by requiring importing countries to protect turtles in the same manner as required in the US, the Appellate Body concluded, the measure was applied in a manner that violated the chapeau. The basis for this conclusion was that the requirement to meet the objective in the 'same' manner rather than in a 'comparable' manner was unjustifiable.[92]

Applying this analysis to climate change, given the threat to human health, forests, agriculture, and biodiversity more generally, it must be considered likely that any measure to reduce GHG emissions will be found to be provisionally consistent with Article XX (b) and (g) of the GATT. The only real issue under Article XX should therefore be whether the measure is designed and implemented in such a manner that it does not discriminate arbitrarily or unjustifiably, and is not a disguised restriction on international trade. This could become a matter of comparing the measure to alternatives that could have achieved the same objective. It is important to note, however, that the Appellate Body has clearly moved away from the test of whether the measure in question is the least trade restrictive manner to achieve the legitimate objective.[93] The implications of considering this

---

91 The protection of turtles in Malaysian waters from the adverse impact of the shrimp fishery in those waters, which lie clearly outside the territory or jurisdiction of the US, is an example. It did so by finding a sufficient connection between the protection of turtles and the interests of the US. It is easy to see how a similar reasoning would lead to the conclusion that measures designed to reduce the impact of production of certain products on climate change, even if the production took place outside a given country, should similarly be of sufficient interest to that country, given that the impacts of climate change are felt globally.

92 Of note is that the Appellate Body went to great length to point out different circumstances in importing countries that would justify allowing importing countries to meet the valid objective of protecting turtles in a different manner. See also *United States – Taxes on Petroleum and Certain Imported Substances*, *supra* note 90, at para. 164. Absent from the analysis, however, is why those differences were not considered in the context of the chapeau of Article XX, given that the chapeau limits the non-discrimination requirement only to situations 'where the same conditions apply'. In the Shrimp Turtle (No. 1), the reverse of what seems to be contemplated in the chapeau of Article XX was found. Shrimp Turtle is an example of identical treatment applied to different conditions rather than different treatment applied where the conditions are the same.

93 Rather, the question of whether there are less trade restrictive ways of achieving the legitimate objective will be a factor to consider in deciding whether the measure is consistent with the chapeau of Article XX. See Reformulated Gasoline, *supra* note 31 and Shrimp Turtle (No. 1), *supra* note 8.

issue in the specific context of a given product versus the more general context of a State's emission reduction target are discussed above.

The wording of the chapeau, referring to a disguised restriction on international trade, again raises the possibility of the WTO dispute resolution process having to consider Parties' motives or the predominant purpose, rather than just the effect. Outside the issue of whether the legitimate objective could have been met in a less trade restrictive manner, it is unclear how this will contribute constructively to the analysis. It may put the Appellate Body in the very difficult position of being asked to interpret the intentions of the Parties rather than focus on the effect of the measures to see if the legitimate objectives could have been met in a less trade restrictive manner.

Different climate change mitigation measures carry with them very different collateral costs and benefits. It seems clear that if a State chooses a relatively more trade restrictive measure because of collateral benefits such as reduced air pollution, or some other social or environmental benefit, it will have to be careful to clearly identify the collateral benefits and bring all benefits under Article XX. Otherwise, a less trade restrictive measure that does not have these collateral benefits associated with it may very well form the basis for a conclusion that the measure chosen is more restrictive than necessary to meet the objective, which is a finding that is still likely to have significant influence over the test in the chapeau of Article XX on whether the measure is arbitrary and unjustifiable in its impact on international trade. In other words, States need to be clear when the objective of a measure is to address more than one legitimate objective under Article XX, in order to prevent a comparison with measures that do not achieve the multiple intended objectives.

The overall conclusion from the rulings to date appears to be that measures to address climate change will be acceptable under GATT if the measures treat like products alike. If not, the different treatment will have to be justified under Article XX. Most likely to succeed under Article XX will be measures with clear environmental objectives and that have as much flexibility as possible on how to meet those objectives. Measures that require the application of specific technologies or measures to meet the stated environmental objective are less likely to be saved under Article XX, unless there is no alternative way of achieving the Article XX (b) or (g) objective. Finally, if there are collateral environmental or social benefits other than the effects of climate change mitigation that justify imposing a requirement for a specific production method or the use of a specific technology over another, careful analysis will be needed to justify the choice of measures, including an analysis of whether the collateral benefits result in the products being different products, or whether the collateral benefits help justify any difference in treatment of like products under the chapeau of Article XX.

(d) Would Requirements to Dislose GHG Emissions from Production on Imported Products Be Upheld Under the WTO Rules?

The definition of technical regulations in Annex I of the TBT Agreement[94] includes labeling, terminology, symbols, packaging and marking requirements. A requirement to provide information about the GHG emissions from, or energy intensity of, the production process, transportation, use, or disposal of a product is therefore likely to be subject to the same tests under Articles I, III and XX of the GATT and relevant provisions of the TBT Agreement as discussed above.

To date, there have been no cases on this point, but one would expect the labeling requirement, when weighed against a legitimate objective under Article XX (b) or (g), to be considered a relatively minor restriction on international trade. Much will likely depend on the particular circumstances involved. The issue of labeling is likely to be first considered by the WTO in the context of genetically modified food products, given the labeling requirements in Europe and the resistance in many food-exporting countries to accept these requirements.[95]

(e) Is the Failure to Internalize Climate Change Impacts Into the Costs of Products a Subsidy?

The issue here is whether a WTO Member State could take the position that a product that is imported at a price that does not internalize the climate change costs from the production, transportation, use and/or disposal of the product is being subsidized and therefore can be the subject of a countervailing measure. The starting point for any assessment of this issue is the Agreement on Subsidies and Countervailing Measures.[96] The Agreement generally categorizes subsidies into three categories: those that are prohibited and subject to an accelerated dispute settlement procedure; those that are non-actionable and therefore permitted;[97] and

---

94 See TBT Agreement, *supra* note 4, above, Annex.

95 The discussions on this issue by the Trade and Environment Committee set up under the WTO are very much at the initial stages, with few indications on whether and how this issue might be resolved through negotiations as opposed to through interpretation of existing provisions of the GATT and the TBT Agreement.

96 Agreement on Subsidies and Countervailing Measures, in GATT *supra* note 3, at 264-314 ('Subsidy Agreement').

97 This exception to actionable subsidies was in place for an initial period, which lapsed at the end of December 1999. Efforts to extend the application of these provisions have failed to date, so the provisions are currently not applicable. See Committee on Subsidies and Countervailing Measures, Minutes of the special meeting held on 20 December, 1999 (G/SCM/M/22). This review took place pursuant to Article 31 of the Agreement on Subsidies and Countervailing Measures, *Ibid.* The consensus required to extend the application of Articles 6.1, 8, and 9 of the Agreement on Subsidies and Countervailing Measures was not achieved at that meeting.

those that are open to challenge under the dispute settlement process or subject to countervailing action. More specifically, the Agreement defines a subsidy as a financial contribution by a government or a public body in a Member State that confers a benefit, such as:

- a direct transfer of funds or liabilities;
- revenues owed that are not collected or are forgiven;
- goods or services other than general infrastructure that are provided; or
- any of these actions that are taken indirectly by a government or public body through a private body.

In addition, the Subsidy Agreement incorporates Article XVI of the GATT into the definition of a subsidy. Article XVI, however, does not provide a clear definition. The wording of Article XVI suggests an approach comparable to the one set out above, but Article XVI leaves considerable room for interpretation.[98]

In the absence of any case law directly on point,[99] it is difficult to come to any firm conclusion on this issue, other than to state that there is no indication to date that the WTO is ready to consider a failure to internalize the environmental cost of producing a product to be a subsidy. It seems clear that any such recognition would have to either come under Article XVI, or would require an amendment of the definition of subsidy under the Subsidy Agreement. An adjustment to the definition of subsidy to explicitly recognize the obligation of Member States to internalize environmental costs incurred during the life cycle of the products it

---

[98] For a good general overview of the application of the Agreement on Subsidies and Countervailing Measures by the Appellate Body of the WTO, see WTO AB 9 December 2002, *United States – Countervailing Measures Concerning Certain Products from the European Communities*, WT/DS212/AB/R, (AB-2002-5), paras. 77 – 83 (hereinafter 'US Counterveiling Measures').

[99] There are some cases, such as the various US-Canada Softwood Lumber Disputes, which have raised some related subsidy issues without addressing the specific issue of whether failure to internalize environmental costs can be considered a subsidy. See, for example, WTO DS 27 September 2002, *United States – Preliminary Determinations With Respect To Certain Softwood Lumber From Canada, Report of the Panel*, WT/DS236/R, where the Panel concluded that selling crown-owned lumber at a lower stumpage fee than private woodlot stumpage fees is not a subsidy under the GATT rules. It should be noted that the Panel in this case did find that a low stumpage fee can constitute a subsidy (see para. 7.30 of the decision). The US lost its argument that the low stumpage fees amounted to a subsidy based on a rejection of the baseline the US thought to establish for determining whether Canada was subsidizing its lumber exports. The US relied on stumpage fees in the US rather than, for example, on stumpage fees in Canada for domestic use. This decision therefore leaves the door open to an argument that the baseline should be the cost to the environment of removing the lumber from the forest, and that any stumpage fee below this cost amounts to a subsidy. To date, however, such an approach to establishing the baseline for an assessment of whether a product is subsidized has not been argued before a WTO tribunal. The traditional approach has been to establish a baseline price based on existing prices of the product in comparable circumstances in the absence of government involvement to lower the price (see for example US Countervailing Measures Ruling, *Ibid*).

trades internationally would be an adjustment that would go a long way to addressing some of the serious challenges in the relationship between international trade and environmental law. In the meantime, given the current definition used in the Subsidy Agreement, it is not clear whether an argument that such failure amounts to a subsidy would succeed.[100]

A related question is whether a measure to protect a good, which has to meet a stricter GHG emissions standard than competing products on the international market, can receive government assistance without fear of retaliation under the Subsidy Agreement. In this context, the definition appears to be clear in including such government assistance as a subsidy. Given the failure to extend the application of Article 8 of the Agreement of Subsidies and Countervailing Measures, which had previously allowed for some un-actionable subsidies, there currently appears to be nothing to prevent countervailing measures, if a State tries to protect products that have to meet a more stringent GHG emission standard through tariffs and duties. This leaves the acceptance of the failure to internalize environmental costs as a form of subsidy as the only promising avenue to encourage positive action on climate change through the use of the Agreement on Subsidies and Countervailing Measures.

(f) Does the Export of Products That Do Not Internalize the Costs of Climate Change Amount to 'Dumping'[101]?

'Dumping' refers to actions of a producer to support its exports, as opposed to subsidies, which refers to actions of the government to support a domestic producer. Because of the focus of dumping, which is on actions of the producer rather than on government action, it is therefore difficult to see how the link between dumping and the failure to internalize the costs of climate change can be made. The failure to internalize the cost of climate change is most likely to be found to be a government failure, not a producer's failure. Given that dumping is defined based on the domestic price as the baseline, unless the product is not sold domestically, the only way these provisions could have application is if a producer chooses to internalize the cost for the domestic market, but not for export purposes. If the producer's actions are as a result of government laws or policies, the action would fall under subsidies rather than dumping.

---

[100] Other existing subsidies to goods that contribute to climate change, however, can still be challenged under this agreement if the subsidies take one of the forms indicated above. Examples might include existing subsidies to fossil fuel producing industries, such as coal and oil. See, for example, Submission of Saudi Arabia, *Energy Taxation, Subsidies and Incentives in OECD Countries*, WTO Committee on Trade and Environment, WT/CTE/W/215, TN/TE/W/9 (23 September 2002).

[101] See Agreement on Implementation of Article VI of the General Agreement on Tariffs and Trade 1994, in GATT, *supra* note 2 above, at 168-196.

(g) Are WTO Members Otherwise Permitted to Take Action Against Other States to Level the 'Playing Field'?

The issue here is whether there are other ways under the WTO rules that a Member State can take measures to prevent products (that are produced without having to reduce or eliminate their GHG emissions) from obtaining a competitive advantage over climate-friendly products (that do have GHG emission reduction requirements associated with all or part of their life cycle) through the imposition of tariffs or duties. The general answer is that Parties to the GATT and members of the WTO have agreed to reduce tariffs and duties in an effort to support the international trade of goods. Any effort to increase tariffs or duties based on a concern that products produced in a manner that contributed to global GHG emissions without internalizing the cost to the climate of those emissions would have to be specifically permitted under the GATT or another WTO agreement. Two such agreements, dealing with subsidies and dumping as a justification for the imposition of tariffs or duties, were considered above in the previous two Sections.

One other justification for imposing new tariffs and duties is the application of Article XIX. It provides for an increase in tariffs and duties, if there is a threat to a domestic industry as a result of an increased quantity of imported product into the domestic market. In circumstances where a Member State that does not impose any GHG emissions or energy intensity limit on a given product, perhaps because it has decided not to ratify Kyoto, develops a significant competitive advantage over other Member States and thereby significantly increases its export of the product into these other States, Article XIX would allow the importing country to take emergency action to prevent the increased quantity of imports.[102]

Outside of these three specific exceptions, in response to a subsidy, dumping or a threat under Article XIX, the imposition of a tariff or duty to address competitiveness implications of climate change measures would seem to be inconsistent with State obligations under the WTO regime.

---

[102] Article XIX was the subject of an appellate body ruling in WTO AB 10 November 2003, *European Communities - Provisional Safeguard Measures On Imports Of Certain Steel Products*; WT/DS248/AB/R, WT/DS249/AB/R , WT/DS251/AB/R , WT/DS252/AB/R , WT/DS2530/AB/R, WT/DS254/AB/R, WT/DS258/AB/R, WT/DS259/AB/R (03-5966). On the facts before it, the Appellate Body concluded that the United States had failed to provide a reasoned and adequate explanation demonstrating that 'unforeseen developments' had resulted in increased imports causing serious injury to the relevant domestic producers.

6. CONCLUSION: WHAT ACTIONS COULD BE TAKEN ON CLIMATE CHANGE IN ACCORDANCE WITH THE RULES OF THE WTO?

Having gone through the relevant provision of the various WTO agreements and rulings dealing with environmental and related issues, it is now time to return to the questions posed at the beginning of the Chapter:

1. Can a State take measures to protect industries that are adversely affected by the State's efforts to implement the Kyoto Protocol (Against Parties to the Kyoto Protocol)?

Product specific measures can be justified in a number of ways. First, the measure to protect industries that are required to meet GHG emission standards cannot be challenged under the WTO rules, unless the products in question are considered to be like products. Differential treatment on the basis that the products are different is more likely to be accepted if the GHG emission standard applies to the use or disposal of the product than if it applies to the production of the product. The Shrimp Turtle and Asbestos Rulings have, however, opened the door to consideration of production methods and resulting environmental impacts in the context of whether two products are like products.

Secondly, measures to protect low-emission GHG products can still be acceptable under Article XX of the GATT. It would seem likely that the objective of reducing climate change would be considered to be a legitimate objective under Article XX. The main question will be whether the measure in question appropriately strikes a balance between the legitimate objective and its impact on international trade. This issue will be debated in the context of the chapeau of Article XX, and will likely draw the WTO into an assessment of the effectiveness of various possible GHG reduction measures. As long as this analysis is conducted with a focus on effect rather than purpose, and based on an acceptance of the right of States to decide which products to target for GHG emission reductions, it is not likely that the WTO and Climate Change regimes will clash significantly on this issue.

Thirdly, in certain circumstances, there may be opportunities to justify the imposition of tariffs or duties. The most promising context is the Agreement on Subsidies and Countervailing Measures. Recognition of failure to address environmental costs as a form of subsidy would actually set a precedent that could lead to a much more progressive relationship between trade and environment, by giving countries that are willing to take the lead in addressing an environmental issue a tool to do so without being economically disadvantaged by having to compete with countries who fail to internalize environmental costs in the price of the goods they produce.[103]

2. Can a State take measures to protect industries that are adversely affected by efforts to address climate change, independently of Kyoto (i.e. what if a State goes beyond Kyoto, whether or not it ratifies Kyoto)? There are no signs in the WTO rules and rulings to date to indicate that the Kyoto targets will have much bearing on future WTO rulings with respect to project-based climate change measures. It will likely not be practical to draw a distinction between measures that are taken to comply with Kyoto and those that do not, other than a distinction on a State-by-State basis, based simply on whether or not a Party has ratified the Kyoto Protocol. Because of this, it is likely that measures will not be treated differently whether or not a Party seeks to justify them based on compliance with Kyoto or based on a more general policy to reduce GHG emissions.[104]

3. Can a State take measures to influence the climate change impact of products imported into that State? The answer to this question may depend on whether the climate change impact results from the production, transportation, use or disposal of the product in question.[105] Secondly, it would depend on whether the measure in question was a technical requirement imposed on the imported product or a countervailing duty. In the former case, if the technical requirements are comparable to the requirements imposed on like domestic products, the measure should be acceptable under the WTO regime. In the latter case, the justification would have to come from a finding that the imported products were subsidies, or a finding that they posed a treat under Article XIX.

4. Can a State take measures to prevent a competitive disadvantage resulting from efforts to address climate change in case of exports? This issue raises the reverse of the previous question in the context of the Agreement on Subsidies and Countervailing Measures. Specifically, it raises the question of whether government assistance to reduce the cost of an exported product by the amount it cost to meet a more stringent domestic GHG emission standard would be considered as a subsidy. Such government assistance would clearly meet the definition of subsidy. In the absence of specific provisions allowing such a subsidy or making it non-actionable, States affected by such a subsidy, upon following the process set out in the Subsidy

[103] This would also be more consistent with the polluter pays principle and with the concept of 'sustainable development' as used in the Rio Declaration. For a detailed analysis of these principles, see de Sadeleer, *supra* note 13, at 21.

[104] This conclusion is supported by the very general objectives of the UNFCCC, which 188 States have ratified, and which can form the justification for almost any effort to reduce life-cycle GHG emissions for a given product.

[105] See the five scenarios set out above.

Agreement, will therefore likely be permitted to impose countervailing duties or other measures to counteract the effect of the subsidy.[106] Financial assistance to domestic industries to protect them from competition against higher GHG emitting products must therefore be considered a risky practice under the WTO regime. Technical regulations to impose GHG emission standards or energy efficiency standards are more likely to be found to be consistent with WTO rules.

To conclude, one open question is to what extent the WTO will consider climate change impacts from product and production methods and transportation in deciding whether two products are like products. Exactly how and to what extent these issues will be considered under Article XX in deciding whether a measure meets the legitimate objective of reducing climate change in a manner consistent with Article XX is also unresolved. Technical regulations, however, designed as much as possible to treat products alike except for their climate change impact, should be upheld under Article XX, if not under Articles I and III. Finally, the extent to which subsidies will evolve through WTO rulings to include the concept of failing to require industry to internalize the climate change cost associated with a given product also remains to be seen.

The bottom line on the relationship between the WTO and the climate change regime would appear to be that as long as the WTO dispute settlement bodies continue to make decisions based on legal principles and precedent, there will be opportunities to develop climate change measures in a way to protect domestic industries from the impact of having to meet more stringent GHG emission reduction requirements; motivate other States to take action; and protect those that do against competition from those that do not.

What remains to be seen in the context of the potential influence of the WTO on compliance with Kyoto is whether States value GHG emission reductions sufficiently to test the WTO regime on this front, and whether the WTO will make principled decisions on these issues, or political ones. This in turn will determine largely whether the WTO will directly influence the evolution of the climate change regime post-2012. These choices by Member States on whether to test the WTO rules with respect to climate change and the WTO Appellate Body on how it responds, may very well determine whether the future evolution of the WTO

---

[106] As previously discussed, Article 8 of the Agreement on Subsidies and Countervailing Measures, which made subsidies to help industries to meet new environmental requirements non-actionable under some limited circumstances, expired on 31 December 1999. See Agreement on Subsidies and Countervailing Measures, *supra* note 96.

regime, through its rulings and through future negotiations, will bring environmental and trade principles closer together or further apart.[107]

It is important to note that the general relationship between trade and environment has been the subject of considerable debate over the past decade. In the context of the WTO, for example, the Committee on Trade and Environment (CTE) was established with a mandate to explore many of the issues raised in this Chapter. To date, it is far from clear how these issues will be resolved in the context of the WTO, let alone whether the two regimes will conflict over time or find a common way forward.[108] Utilizing the rules of the WTO to help achieve the objectives of the UNFCCC may be an effective way to test the boundaries of harmonization in the regimes as they exist today. In the long-term, further efforts are clearly needed, such as the recognition of the failure to internalize environmental costs as a subsidy.

The outcome of these efforts to bring together environmental protection and globalization efforts in turn may answer the question about the normative influence of the WTO on the climate change regime and vice versa. Can the two regimes evolve so as to support each others' objectives? The answer to this question may depend on their respective membership in the long-term, and the relative importance States assign to trade and climate change over the course of the next few decades.

---

[107] For a possible indication of the Canadian Government's view on the role of the WTO in addressing environmental issues, see K. A. McCaskill, "Dangerous Liasons: The World Trade Organization and the Environmental Agenda" Department of Foreign Affairs and International Trade, Policy Staff Paper No. 94/14 (June 1994).

[108] See Howard Mann, *et al*, *The State of Trade and Environmental Law 2003: Implications for Doha and Beyond* (Winnipeg: International Institute For Sustainable Development, 2003), at page 3. See also Magnus Lodefalk, *Climate and Trade Rules – Harmony or Conflict?* (Stockholm: Kommerskollegium, National Board of Trade, 2004).

6

Dispute Settlement Under UNCLOS, Compliance, and the Future of the Climate Change Regime

1. INTRODUCTION

The scientific link between climate change and marine environmental protection has been identified and highlighted, in part, through the work of the Intergovernmental Panel on Climate Change (IPCC).[1] Considering the slow progress on climate change mitigation under the climate change regime to date,[2] linkages between climate change and State obligations under other existing regimes could become more and more important in shaping the global response to climate change. In this context, the United Nations Law of the Sea Convention (UNCLOS) was chosen mainly because of the scientific link between climate change and the marine environment, the inclusion of obligations to protect the marine environment in UNCLOS, and the inclusion of a binding dispute settlement process. Not surprisingly, therefore, the focus of this Chapter is on the potential for enforcement, though normative influences from UNCLOS onto the climate change regime are also possible.

In this Chapter, two central questions are posed. Firstly, can a Party be held to be in violation of its obligations under UNCLOS[3] for failing to mitigate climate change? This issue is considered in the context of Parties' obligations under UNCLOS with respect to the prevention of marine pollution and the protection and preservation of the marine environment. Secondly, can the binding dispute settlement process under UNCLOS be used to require Parties to take appropriate action to reduce their GHG emissions?

These issues are considered in Sections two to six. Section two briefly reviews the state of science on the links between climate change and the marine environ-

---

[1] The IPCC was set up jointly by the WMO and UNEP to assess science, technology, and socio-economic information relevant to climate change. In particular, it is to enhance the understanding of climate change impacts, and adaptation and mitigation options. See Chapter 2 above for more on the role of the IPCC. See also <http://www.ipcc.ch/>.

[2] See assessment of the climate change regime in Chapter 2 above.

[3] *United Nations Convention on the Law of the Sea* (UNCLOS), 10 December 1982, 21 I.L.M. 1261.

ment. Section three proceeds to consider the marine environmental protection provisions of Part XII of UNCLOS from a climate change perspective. Section four provides an overview of the binding dispute settlement process under Part XV of UNCLOS. Section five considers the implications of recent rulings by various international tribunals on the ability to bring a claim under Part XV of UNCLOS. Finally, Section six offers some concluding thoughts on the likely treatment of climate change under UNCLOS should a claim be brought.

2. CLIMATE CHANGE AND THE OCEANS

A comprehensive assessment of our scientific understanding of past and predicted future impacts of climate change, and specifically its impacts on the marine environment is not necessary here, as much has been written on both.[4] A general overview of the state of knowledge on this issue is, however, provided as scientific evidence would be the foundation for any claim of failure to mitigate climate change brought under the UNCLOS dispute settlement procedures. What follows is an overview of the conclusions reached by the IPCC in its Third Assessment Report[5] with respect to climate change impacts on the marine environment.

Ocean-related impacts of human-induced climate change identified by the IPCC include the following:[6]

- global average sea-level rises of 0.09 – 0.88 meters by 2100;[7]
- reductions in sea-ice cover;[8]
- elevated average sea surface temperatures (SST);[9]
- increased storm floods worldwide;[10]

---

[4] For a general overview of the state of the science on climate change, see Chapter 2. For more detail, see Robert T. Watson *et al*, eds., *Climate Change 2001: Synthesis Report: A Contribution of Working Groups I, II and III to the Third Assessment Report of the Intergovernmental Panel on Climate Change* (Cambridge University Press, 2002); John T. Houghton *et al*, eds., *Climate Change 2001: The Scientific Basis: Contribution of Working Group I to the Third Assessment Report of the Intergovernmental Panel on Climate Change (IPCC)* (Cambridge University Press, 2002); James J. McCarthy *et al*, eds., *Climate Change 2001: Impacts, Adaptation and Vulnerability: Contribution of Working Group II to the Third Assessment Report of the Intergovernmental Panel on Climate Change (IPCC)* (Cambridge University Press, 2002); Bert Metz *et al*, eds., *Climate Change 2001: Mitigation: Contribution of Working Group III to the Third Assessment Report of the Intergovernmental Panel on Climate Change (IPCC)* (Cambridge University Press, 2002).

[5] *Ibid.*, Synthesis Report, at 252; and *Ibid.*, Impacts, Adaptation and Vulnerability Report, at 343.

[6] *Ibid.*, Impacts, Adaptation and Vulnerability Report, at 345-379.

[7] *Ibid.*, at 348.

[8] *Ibid.*, at 349. See also at 801 with respect to changes in Polar Regions. One of many species predicted to be affected by these changes is the polar bear, whose hunting season is already being affected by the increasing ice melt.

[9] *Ibid.*, at 348 and 356-362.

[10] *Ibid.*

- accelerated levels of coastal erosion;[11]
- increased seawater intrusions into fresh surface and groundwater;[12]
- accelerating adverse impacts on marine fish;[13] and
- impacts on aquaculture.[14]

In summary, the primary physical changes of sea-level rise, increases in sea surface temperature, an increase in extreme weather events, and reductions in ice cover, are predicted to lead to a long list of changes to marine ecosystems. Specific aspects of the marine environment threatened include coral reefs,[15] polar mammals, coastal ecosystems and commercial and non-commercial species of marine life alike. Resulting social and economic impacts range from loss of property (including land mass), loss of access to potable water, and loss of coastal infrastructure, to the potential depletion to a number of commercial fish-stocks.[16]

Based on the IPCC's Third Assessment Report, the message is clear. Climate change poses a serious risk to many marine species, many ecosystems, and to the marine environment as a whole, resulting in secondary social and economic impacts. This leads to the central question posed in this article, whether the provisions of UNCLOS, designed to protect and preserve the marine environment, provide any legal avenues to motivate States to take action to mitigate climate change so as to reduce its impacts on the marine environment. More specifically, does UNCLOS provide a mechanism for States to seek recourse against other States that have not taken or are not taking adequate action to reduce their contribution to climate change?[17]

3. CLIMATE CHANGE UNDER UNCLOS

There is little indication from either the text of UNCLOS or historical accounts of the negotiations that climate change *per se* was on the minds of negotiators at

[11] *Ibid.*

[12] *Ibid.*

[13] *Ibid.*, at 350.

[14] *Ibid.*, at 354.

[15] See M. G. Davidson, "Protecting Coral Reefs: The Principal National and International Legal Instruments" (2002) 26 Harv. Envtl. L. Rev. 499.

[16] See Synthesis Report, *supra* note 4, at 245 to 253.

[17] The question of what level of climate change mitigation action is adequate for a given country will be one of the challenging evidentiary issues in any such dispute. For a more detailed discussion of the tension between the environmental objectives and the short-term economic cost of meeting these environmental objectives, see J.L. Hafetz, "Fostering Protection of the Marine Environment and Economic Development: Article 121(3) of the Third Law of the Sea Convention" (2000) 15 Am. U. Int'l L. Rev. 583.

the time this Convention was developed.[18] The protection and preservation of the marine environment, however, has a prominent place among the objectives of the Convention. In this Section, the extent to which the objectives of protecting and preserving marine ecosystems were translated into firm obligations and commitments will be considered. Next, the link between these obligations and responsibilities and climate change mitigation will be explored.

(a) General Provisions of UNCLOS

In its Preamble, UNCLOS recognizes the importance of establishing a legal order for the oceans that promotes, among other things, the protection and preservation of the marine environment. At the same time, the Preamble recognizes the tension that exists between these and other objectives designed to protect both the common interest in the oceans as well as State sovereignty. The Preamble specifically urges Parties to adopt a holistic approach to ocean issues.

UNCLOS, Article 1 includes the following definition of 'pollution of the marine environment':

> the introduction by man, directly or indirectly, of substances or energy into the marine environment, including estuaries, which results or is likely to result in such deleterious effects as harm to living resources and marine life, hazards to human health, hindrance of marine activities, including fishing and other legitimate uses of the sea, impairment of quality for use of sea water and reduction of amenities.

A traditional interpretation of UNCLOS including the definition of pollution might focus on what the Parties contemplated at the time UNCLOS was negotiated. Using this approach, it could be argued that climate change was not likely on the minds of negotiators, and the term "energy" therefore could not have been intended by the Parties to include increases in ocean temperature resulting from GHG emissions. To be fair, there is no indication from commentators at the time that climate change was specifically the subject of negotiations at UNCLOS.[19]

It is suggested here that such an interpretation is inappropriate for a number of reasons. First, this approach would be an unduly restrictive interpretation of the intention of the Parties at the time. It seems clear from its plain wording that the overriding objective of the definition of pollution was to capture a full range of

---

[18] For a description of the history of the negotiations on the key provisions of UNCLOS dealing with the protection of the marine environment, see Myron H. Nordquist, *United Nations Convention on the Law of the Sea 1982: A Commentary, Volume 1* (Dordrecht & Boston: Martinus Nijhof Publishers, 1985), *Volume 4* (1991), and *Volume 2* (1993).

[19] See, for example, Myron H. Nordquist, *United Nations Convention on the Law of the Sea 1982: A Commentary, Volume 1* (Dordrecht & Boston: Martinus Nijhof Publishers, 1985).

possible threats to the marine environment. There is no indication that the Parties were intending to limit the definition to specific threats that were clearly identified at the time. Secondly, this approach would treat UNCLOS as a contract, frozen in time at the time it was negotiated. Such an approach would relegate many international treaties to irrelevance soon after they are negotiated. For international treaties to serve a constructive role in international law over time, they have to be interpreted in light of changing circumstances over time.[20]

Furthermore, while the specific cause of temperature increase from climate change may not have been on the minds of negotiators, it is clear from the work of the Joint Group of Experts on the Scientific Aspects of Marine Pollution (GESAMP) that the potential impact of temperature changes to marine ecosystems were within the contemplation of negotiators.[21] The sources of energy under consideration at the time may have been more local, such as land-based effluent from industrial facilities. They likely did not include climate change. The threat of temperature change to the marine environment was, however, clearly identified. The inclusion of "energy" in the definition of pollution is a clear indication that negotiators had turned their minds to this threat.

Finally, Article 293 specifically opens the door to a progressive interpretation of UNCLOS obligations by bringing in other sources of international law not inconsistent with UNCLOS.[22] In this case, it is suggested, a broader interpretation of pollution is more consistent with the plain wording of the provision and its purpose at the time it was negotiated. It is also one that commentators have advocated in the context of Part XII. Crylle de Klemm, for example, suggested even before the signing of UNCLOS that the general provisions be read to include an obligation to protect threatened species and ecosystems.[23] At the same time,

---

[20] Support for this approach can be found in the following ICJ judgment case concerning the Gabcíkovo Nagymaros Project: [1997] I.C.J. Rep. 1.

[21] See IMCO/FAO/UNESCO/WMO/WHO/IAEA/UN/UNEP Joint Group of Experts on the Scientific Aspects of Marine Pollution (GESAMP), *Interchange of Pollutants Between the Atmosphere and the Oceans*, No.13 Reports and Studies (Geneva: World Meteorological Organization, 1980), and IMCO/FAO/UNESCO/WMO/WHO/IAEA/UN/UNEP Joint Group of Experts on the Scientific Aspects of Marine Pollution (GESAMP), *Thermal Discharges in the Marine Environment*, No. 24 Reports and Studies (Rome: Food and Agriculture Organization of the United Nations, 1984). For information on how these issues were considered in the context of project based environmental assessments at that time, see IMCO/FAO/UNESCO/WMO/WHO/IAEA/UN/UNEP Joint Group of Experts on the Scientific Aspects of Marine Pollution (GESAMP), *Environmental Capacity: An Approach to Marine Pollution Prevention*, No. 80 Reports and Studies (Nairobi: United Nations Environment Programme, 1986).

[22] N. Klein, *Dispute Settlement in the UN Convention on the Law of the Sea*, (Cambridge: Cambridge University Press, 2005), at 58, 148-152.

[23] Cyrille de Klemm, "Living Resources of the Ocean" in Douglas Johnston ed., *The Environmental Law of the Sea* (Gland, Switzerland, IUCN, 1981), at 71.

others argued that these provisions include an obligation to protect the fauna and flora of the sea-floor from harm.[24]

Assuming it can be shown that GHG emissions lead to an increase in ocean temperature, it would seem that human-induced greenhouse gas emissions fit within the definition of pollution in two respects. One approach is through the reference to energy in the definition, the other through an effect based interpretation of pollution. These two approaches are briefly considered here.

At the heart of the science behind climate change is the greenhouse effect, which predicts an overall increase in energy within the atmosphere resulting from an increase of greenhouse gases. This in turn is predicted to lead to a rise in energy in the oceans. It is this increase in energy that is predicted to have significant harmful effects on living resources and marine life through changes in water temperature, sea-level, ocean currents, and sea ice.[25] In fact, a recent, yet unpublished study has concluded that there is a clear correlation between GHG emissions and ocean temperature over the past 40 years. In other words, based on this study, there is no reasonable explanation for the changes in global ocean temperatures over the past 40 years other than GHG emissions.[26]

Assuming the science is convincing, there is therefore a strong basis for a position that GHG emissions cause marine pollution due to the increase in energy in the oceans resulting from GHG emissions. An alternative approach is to take an 'effects-based' view of marine pollution, to include within marine pollution the release of any substance that causes harm to the marine environment.[27] Either approach leads to a conclusion that GHG emissions result in pollution as defined in UNCLOS.

---

[24] James N. Barnes, "Pollution from Deep Ocean Mining" in Douglas Johnston ed., *The Environmental Law of the Sea* (Gland, Switzerland, IUCN, 1981), at 259.

[25] The most obvious examples here would be the impact of water temperature on marine life. Some marine life, such as sockeye salmon, will have a significantly reduced living space, due to sensitivity to higher ocean temperatures. Others, such as the Atlantic Cod, will have significantly lower growth rates. Another striking example is the polar bear, whose hunting season is being affected by the reduction in sea ice. See Impacts, Adaptation and Vulnerability Report, *supra* note 4, at 348.

[26] Steve Connor "The Final Proof: Global Warming Is a Man-made Disaster" *The Independent* (19 February 2005). Other recent research on the marine impact of GHG emissions includes: J. M. Roessig *et al*, "Effects of Global Climate Change on Marine and Estuarine Fisheries and Fisheries" (2004) 14:2 Reviews in Fish Biology and Fisheries 251, and U. Riebsell, "Effects of CO2 Enrichment on Marine Phytoplankton" (2004) 60 Journal of Oceanography 719.

[27] This approach to marine pollution in UNCLOS has been proposed in the context of aquatic invasive species, see M.L. McConnell, *GloBallast Legislative Review: Final Report* (London: Global Ballast Water Management Programme, 2002), at 19-21.

(b) Part XII, Marine Environmental Protection

UNCLOS, Part XII deals generally with State obligations with respect to the marine environment. As early as 1991, Part XII was characterized by academics as constitutional in character, reflecting in part existing custom but at the same time providing the first comprehensive statement on the protection of the marine environment in international law.[28]

The starting point for Part XII is a general obligation under Article 192 to 'protect and preserve the marine environment',[29] balanced with a reaffirmation of the right of States to exploit their natural resources 'in accordance with their duty to protect and preserve the marine environment'.[30] Under this part of the Convention, States are obligated to take all measures consistent with the Convention necessary 'to prevent, reduce and control pollution of the marine environment *from any source*, using the best practical means'.[31] Article 194 is central to any analysis of State obligations to mitigate climate change. It acts as the foundation for the following specific obligations that provide some further guidance on what a State may be expected to do to protect and preserve the marine environment:

- an obligation for States to act individually or jointly as appropriate;[32]
- an obligation to take all measures necessary to prevent, reduce and control pollution of the marine environment;[33]
- an obligation to use best practical means at their disposal;[34]
- an obligation to act in accordance with their capabilities;[35]
- an obligation to endeavour to harmonize policies with other States;[36]
- an obligation to control activities under their control or jurisdiction so as to not cause damage by pollution to other States and their environment;[37]
- an obligation to prevent pollution from spreading to areas outside of a State's jurisdiction of control;[38] and

---

[28] See M.L. McConnell, *et al*, "The Modern Law of the Sea: Framework for the Protection and Preservation of the Marine Environment?" (1991) 23 Case W. Res. J. Int'l L. 83, at 84. See also J.L. Hafetz,, *supra* note 17, at 597

[29] See UNCLOS, *supra* note 3, Article 192. This article is considered to reflect customary international law, and as such binding on all States, not only Member States. See J.L. Hafetz, *supra* note 17, at 598.

[30] *Ibid.* UNCLOS, Article 193. This article is also considered to reflect customary international law. See J.L. Hafetz, *supra* note 17, at 598.

[31] *Ibid.* UNCLOS, Article 194.

[32] *Ibid.* UNCLOS, Articles 197, 207(4), and 212(3).

[33] Recall that the definition of 'pollution' includes the addition of energy to the marine environment.

[34] UNCLOS, *supra* note 3, Article 194(1).

[35] *Ibid.*

[36] *Ibid.*

[37] *Ibid.* Article 194(2)

[38] *Ibid.*

- a specific obligation for the preservation and protection of rare or fragile ecosystems, and the habitat of species at risk.[39]

Article 195 directs States on measures to prevent, reduce and control pollution of the marine environment. It does so by obliging States to prevent the transfer of harm from one type or area to another. While the exact scope of this provision is not clear, it does, at a minimum, introduce the concept that mitigation measures must be designed so as to not result in other environmental damage, an issue that has been the subject of considerable controversy in the context of climate change.[40] In so doing, UNCLOS may have been ahead of its time, providing a simple, yet potentially very effective tool to require States to take a holistic approach to addressing environmental issues.[41]

Article 212 is another provision of UNCLOS that, while perhaps not drafted with climate change in mind, can now be reasonably interpreted to apply to the issue. It obligates States to adopt laws and regulations and take other necessary measures 'to prevent, reduce and control pollution of the marine environment from or through the atmosphere'.[42] It essentially obligates States to prevent or control pollution from or through any air space over which a State has jurisdiction.

Similarly, Article 207, dealing with pollution from land-based sources, is sufficiently broad to cover GHG emissions. It requires Party States to endeavour to establish regional and global rules to prevent, reduce, and control marine pollution from land-based sources. In determining a Party's contribution to such efforts, economic capacity and need for economic development are to be taken into account. Article 213 does require States to enforce domestic laws passed in accordance with Article 207 and any other international obligation to address land-based sources of marine pollution. Overall, these provisions appear to be weaker than Articles 192 to 195 in that they only require States to endeavour to control pollution, and are, therefore, less likely to play a significant role in determining State obligations to mitigate climate change.

(c) Part XII and the Duty to Mitigate Against Climate Change

On their face, the provisions of UNCLOS, particularly Part XII, are sufficiently broad to allow for a claim that failure to mitigate climate change violates obli-

---

[39] *Ibid.* Article 194(1), (2) and (5).

[40] Especially in the context of the use of mitigation measures such as nuclear power, and carbon capture to offset emissions, see Chapter 2.

[41] See J.I. Charney, "Implementing the United Nations Convention on the Law of the Sea: Impact of the Law of the Sea Convention on the Marine Environment" (1995) 7 Geo. Int'l Envtl. L. Rev. 731, at 732. See also Jonathan I. Charney, "The Marine Environment and the 1982 United Nations Convention on the Law of the Sea" 28 Int'l Law. 879.

[42] UNCLOS, *supra* note 3, Article 212(1) and (2).

gations to preserve and protect the marine environment. Particularly relevant in this regard, as noted earlier, is the definition of pollution to include energy. Obligations under Article 194 to preserve and protect the marine environment through the prevention, reduction and control of pollution, and the obligation to use best practical means in accordance with a State's capabilities may also prove to be key provisions. Finally, the obligation to prevent pollution from spreading outside a State's jurisdiction of control may prove to be a critical provision.

To date, the substance of Part XII does not appear to have been interpreted by any international tribunal. However, one dispute currently going through the UNCLOS dispute settlement process that has raised the issue of marine environmental protection involves Malaysia and Singapore over land reclamation activity by Singapore.[43] The claim makes reference to Articles 192 and 194, which, as noted above, are two of the central provisions of UNCLOS with respect to marine environmental protection. In the context of the ruling on provisional measures, there was no substantive analysis of these provisions, as preliminary measures are not based on a ruling on the substance, similar to domestic injunction procedures in many States. However, the pending arbitration decision on the substance of this claim may very well become the first ruling under the UNCLOS binding dispute settlement procedures to interpret these provisions. As such, it should provide some guidance on the nature and scope of Party obligations with respect to the protection and preservation of the marine environment.[44]

One issue that will be interesting to follow as these provisions are interpreted by international tribunals is the connection to other international agreements. In particular, for Parties that are also Parties to the Convention on Biological Diversity (CBD),[45] the CBD may provide a whole new context for understanding marine pollution and the effort required to meet the obligations under UNCLOS to protect and preserve the marine environment. Given the wide acceptance of both conventions, the influence of the CBD on the interpretation of the marine environment provisions of UNCLOS may turn out to be substantial. In the context of the interpretation of Parties' obligations under UNCLOS, an interesting connection

---

[43] See International Tribunal for the Law of the Sea, Case concerning Land Reclamation by Singapore in and around the Straits of Johor (*Malaysia v. Singapore*), Request for the Prescription of Provisional Measures under Article 290, Paragraph 5, of the UN Convention on the Law of the Sea (10 September, 2003).

[44] Similarly, UNCLOS Articles 123, 192, 193 194, 197, 206, 207, 211, 212, and 213 have been raised by Ireland in a dispute with the United Kingdom. So far, there only has been a ruling on provisional measures. See International Tribunal for the Law of the Sea, The MOX Plant Case (Ireland v. United Kingdom), Request for the Prescription of Provisional Measures under Article 290, Paragraph 5, of the UN Convention on the Law of the Sea (13 November 2001), reprinted in 41 ILM (2002) 405. The arbitral tribunal procedure of the main dispute in the MOX Plant Case has been suspended. See Klein, *supra* note 22, at 45-52.

[45] *Convention on Biological Diversity of the United Nations Conference on the Environment and Development*, June 5, 1992, U.N. Doc. DPI/1307, reprinted in 31 I.L.M. 818 [CBD].

is the general recognition that climate change is a significant threat to biological diversity, and the further recognition that biological diversity is crucial to good ecosystem health.[46] By providing this context to the connection between climate change and obligations to protect and preserve the marine environment under UNCLOS, the CBD may play a significant role in interpreting obligations of Parties to mitigate climate change. As indicated above, Article 293 invites the application of the CBD as an interpretative tool to the extent that it is not incompatible with UNCLOS.[47] The application of the CBD as an interpretive tool is of course limited to disputes involving Parties who are bound by both treaties.[48]

(d) Making Out a Claim, Causation and Related Issues

Assuming that the obligation of States to protect and preserve the marine environment under UNCLOS extends to mitigating climate change, and assuming that the IPCC's Third Assessment Report is sufficient to establish a connection between GHG emissions and current and predicted future climate change, this leaves the question of causation.[49] The main challenge in any claim made against a State that its failure to mitigate climate change violates its obligations under UNCLOS is the ability of the claimant to establish the link between the failure of a particular State to reduce GHG emissions on one hand and the impacts of climate change on the marine environment on the other.

One way to approach the causation analysis is to consider the standard against which a State will be measured as a way to decide on its contribution to the problem. One possibility would be that the standard is simply whether the State cooperated with the international community, perhaps through the UNFCCC[50] or the Kyoto Protocol.[51] The UNFCCC does not impose any binding obligations on States. The Kyoto Protocol is inadequate in its collective obligations on States in addressing the climate change problem, it is therefore also an inadequate standard

---

[46] For a recent assessment of links between climate change and biodiversity, see Intergovernmental Panel on Climate Change, *Climate Change and Biodiversity* (Technical Paper, IPCC Working Group II, Technical Support Unit) (IPCC, April 2002).

[47] See Klein, *supra* note 22, at 58.

[48] Unless or until the substance of the CBD is recognized as customary international law.

[49] For a more detailed consideration of issues of state responsibility and causation, see Richard S. J. Tol, Roda Verheyen, "State Responsibility and Compensation for Climate Change Damages: A Legal and Economic Assessment" (2004) 32 Energy Policy 1109.

[50] *United Nations Framework Convention on Climate Change*, Intergovernmental Negotiating Committee for a Framework Convention on Climate Change OR, 5th Sess., Annex, UN Doc. A/AC.237/18 (PartII)/Add.1 (1992), 31 I.L.M. 849, online: UNFCCC <http://unfccc.int/resource/docs/a/18p2a01.pdf> [UNFCCC or The Framework Convention].

[51] *Conference of the Parties to the Framework Convention on Climate Change: Kyoto Protocol*, 10 December 1997, U.N. Doc. FCCC/CP/1997/L.7/add. 1, 37 I.L.M. 22 (1998), [hereinafter the Kyoto Protocol].

against which to measure whether a given State has complied with the UNCLOS obligation to protect and preserve the marine environment. Alternatively, the standard could be an allocation of GHG emission rights based on factors such as *per capita* plus/minus depending on historical contribution, capacity, and national circumstances.[52] Finally, one author[53] has considered the standard question with respect to climate change, but not in the context of the UNCLOS obligations. He suggests a reasonable care standard that starts with the country's emissions in the year 1990, the year taken as the time international awareness and understanding was sufficiently high to expect States to act. He then proposes that the reasonable care standard be applied to compare from that point forward what the State reasonably could have been expected to do relative to what was done to mitigate climate change. The gap between the two is what a State would be responsible for.[54]

It may be useful to further consider the causation issue in the context of a specific hypothetical claim. The claim under UNCLOS could reasonably be brought by a number of small island States whose existence is threatened by sea-level rise. Small island States are an obvious claimant for two reasons: they have generally contributed little to the problem as their GHG emissions are minimal; and the threat of climate change to these States is immediate and serious.[55] It would be reasonable to assume that the claim would be brought against a State with a high economic capacity to address the problem and high historical *per capita* contributions to GHG concentrations in the atmosphere above natural levels. Further criteria might be that the State has refused to ratify the Kyoto Protocol, that its *per capita* emissions are among the highest in the world, and that it is a Party to UNCLOS, the CBD, and the UNFCCC.

A State that meets all these criteria is Australia.[56] A claim against Australia could therefore be brought on the basis that it has contributed more than other States to human-induced climate change on a *per capita* basis, that it has among

---

[52] See for example: See Kevin A. Baumert *et al*, eds., *Building on the Kyoto Protocol: Options for Protecting the Climate* (Washington, D.C.: World Resources Institute, 2002), online: WRI <http://pdf.wri.org/opc_full.pdf>, and Hermann E. Ott *et al*, "South-North Dialogue on Equity in the Greenhouse" May 2004, Wuppertal Institute, Germany, online: Wuppertal Institute <www.wupperinst.org/download/1085_proposal.pdf>.

[53] P. Barton, "State Responsibility and Climate Change: Could Canada Be Liable to Small Island States?" (2002) 11 Dal. J. Leg. Stud. 65.

[54] *Ibid.*, Barton, Figure 3.

[55] In fact, Jonathan Charney has argued that any state with a concern about marine protection may initiate dispute resolution, not just directly affected states. See Charney, *supra* note 40, at 737. See also Klein, *supra* note 22, at 161.

[56] For a good overview of current and predicted future total and per capita emissions from various developed countries including Australia, see Farhana Yamin, *The International Climate Change Regime: A Guide to Rules, Institutions and Procedures* (Sussex: Cambridge University Press, 2005), Country Fact Sheet Annex.

the highest economic capacity and responsibility to mitigate climate change, and that it has done less than most other nations to address climate change, either in the context of international cooperation or domestic action.[57]

An obvious response to such a claim is that while Australia's *per capita* contribution is high, and has been high historically, that contribution is nevertheless insignificant in the global context, and that even if it was possible to eliminate Australia's GHG emissions completely, the impact on climate change would be minimal. Furthermore, any claim purely based on a standard of international cooperation, i.e. Australia's decision not to ratify the Kyoto Protocol could be countered on the basis that the difference between business-as-usual and compliance with the first commitment period targets set out in the Kyoto Protocol would be minimal. As suggested above, the Kyoto Protocol, while perhaps a factor in an overall determination of whether a State has taken adequate measures to mitigate against climate change impacts on the marine environment, is clearly not an adequate standard to hold States to in the context of UNCLOS.[58]

A detailed analysis of the causation issue is not possible here. However, given that this issue is likely to be central to a claim under UNCLOS, a few further thoughts on how this issue might be addressed are warranted here. One way of avoiding the defence that the impact would have occurred even if Australia had reduced its emissions would be to bring a claim collectively against a sufficient number of States to overcome the causation problem. Accepting that Kyoto ratification is an insufficient answer to responsibility under UNCLOS, other States that could be part of such a claim include Canada, Japan, and European States that are members of UNCLOS. Collectively, these States would find it much harder to take the position that their impact on climate change is insignificant, and that the claimants have failed to establish causation. Such an approach could only be effective, however, if a tribunal was prepared to consider causation in the context of a collective claim.

Market share liability has been applied successfully to establish causation domestically in the United States, and would appear to be the most reasonable

---

[57] For an Australian perspective on international efforts to mitigate climate change, see V. Cusack, "Perceived Costs Versus Benefits of Meeting the Kyoto Target for Greenhouse Gas Emissions: The Australian Perspective" (1999) 16 *Environmental and Planning Law Journal* 53. See also F. Yamin, *Ibid*.

[58] See Klein, *supra* note 22, 148 – 152. One possible standard might be the commitment in Article 2 of the UNFCCC to prevent dangerous human interference with the climate system, assuming that commitment can be translated into individual State responsibility.

principle to deal with causation in this context.[59] Applied to the issue of climate change, principles for the distribution of responsibility could include *per capita* emissions, historical contribution, economic capacity to mitigate, or some combination of such factors. The starting point for allocation of responsibility based on such factors can be found in the factors that influenced the allocation of the first commitment period targets negotiated in Kyoto. While the targets were pledge based, and not determined on clear principles of responsibility, such principles clearly influenced the negotiations, and have since been the subject of much debate in the context of discussions about allocation of responsibility for mitigating climate change beyond 2012.[60]

One further issue that could arise in the context of causation and liability is whether the concept of balancing environmental protection with economic development would provide a legitimate reason for Parties not to do more about climate change. In other words, could Australia legitimately take the position that it is not required to do more about climate change, because the economic cost of doing more is too high, and it is entitled to make decisions about how to balance its obligation to protect and preserve the marine environment against its right to economic development? Given the repeated reference to the concept that Parties shall act according to their economic capability, it is difficult to see how such a position is supportable for the most developed countries, including Australia.

Economies in transition and developing countries are more likely to be able to rely on these provisions to defend inaction on climate change. How successful such a position might be is far from clear. The concept of common but differentiated responsibility is not reflected in UNCLOS to the extent that it has been reflected in more recent MEAs.[61] There are, however, some signs in Article 119(1)(a) that the special needs of developing countries are to be taken into account in establishing conservation measures. In addition, Article 194 obligates Parties to act in accordance with their capabilities. There is little else, however, to indicate how economic capacity might influence obligations for environmental protection including climate change mitigation.[62]

---

[59] For a more detailed discussion of market share liability as a way to establish causation in the context of a claim for damages related to the impacts of climate change, see P. Barton, *supra* note 53. The article also considers the separate but closely related issue of liability. It considers a range of options from absolute liability to negligence, with strict liability as perhaps a reasonable middle ground.

[60] See Chapter 2.

[61] See for example, UNFCCC, *supra* note 50, Article 3.1.

[62] See J.L. Hafetz, *supra* note 17, at 600 on balancing of economic and environmental objectives under UNCLOS. See also J.L. Hafetz, *Ibid.* at 621 on the economic value of the environment. The latter supports the concept that appears more and more apparent on the climate change issue, that effective climate change mitigation is a precondition for long-term global economic health, not a competing interest to be balanced.

It is important to note in considering this question that the wealthier countries have contributed and continue to contribute more to the problem, they have accepted more of an obligation under the United Nations Framework Convention on Climate Change (UNFCCC)[63] to address it, and their economies are strong enough to favour environmental protection over economic development. In fact, for there to be an overall balance, recognizing the relative inability of developing countries to give priority to environmental protection, there has to be a heavy burden on developed countries to give priority to this issue.[64]

Therefore, in the context of the obligations under Part XII, an attempt to justify inaction on climate change with an economic development imperative argument should carry little weight for developed countries. The weight given for less developed countries would depend upon the level of responsibility for climate change, the current level of contribution to the problem, and the capacity to address the problem. For developed countries generally, and Canada, Australia, and the United States particularly, these factors all point toward an obligation to give priority to environmental protection generally and climate change mitigation specifically.

Finally, there have been suggestions since before UNCLOS actually came into force that much of its substance may already be customary international law.[65] If this extends to Part XII of UNCLOS, and the requirement to protect and preserve the marine environment becomes recognized as customary international law, the duty to take effective climate change mitigation measures could extend beyond UNCLOS Parties to all States. If this transition takes place, all States will have an obligation to mitigate climate change, but only Parties to UNCLOS will have access to the binding dispute settlement process to resolve disputes about the adequacy of such efforts for the protection and preservation of the marine environment. The potential role of this dispute settlement process is considered in the following Section.

---

[63] See UNFCCC, *supra* note 50, Article 3.1.

[64] See Baumert and Ott, both at *supra* note 52. See also Chapter 9.

[65] See, for example, John King Gamble Jr., *et al*, "The 1982 Convention and Customary Law of the Sea: Observations, a Framework, and a Warning" (1984) 21 San Diego L. Rev. 491.While there are no indications that the marine environmental protection provisions specifically are on the brink of being recognized as customary law, there are signs generally that the provisions of UNCLOS are broadly accepted and may have become customary international law. See also, M.L. McConnell, *supra* note 26, at 84. For recognition of UNCLOS rules dealing with the EEZ, the continental Shelf and boundary delimitation, see, for example the International Court of Justice Ruling on the *Delimitation of the Maritime Boundary in the Gulf of Maine Area (Canada/United States of America)* (1981-1984) (October 21, 1984, Ruling note 67) reported in 23:6 ILM 1197. In fact, through state practice, many of the rules established in the UNCLOS had been accepted by state practice well before UNCLOS came into force in 1994.

4. THE DISPUTE SETTLEMENT PROCESS UNDER UNCLOS

The UNCLOS dispute settlement procedures have been described as rivalling the process set up under the World Trade Organization in terms of potential for resolving Party-to-Party disputes and power conferred on dispute resolution tribunals,[66] although they are limited in scope to narrowly defined subject matters set out in UNCLOS.[67] The process is set out in three parts. The first provides an opportunity for the Parties to agree on a dispute settlement tool.[68] The second sets rules for initiating a dispute settlement process in the absence of agreement.[69] The third Section provides some limited exceptions to the binding dispute settlement process.[70] Two features of the UNCLOS process that have been particularly noted by commentators are the ability to initiate the process without having to agree on a procedure, on a case-by-case basis, and the binding nature of the outcomes.[71]

This Section will consider how the dispute settlement process under UNCLOS would respond to a claim that a State's failure to take climate change mitigation seriously violates its obligations to protect the marine environment. The focus is on aspects of the UNCLOS dispute settlement process that are likely to be relevant to a dispute over the adequacy of a Party's climate change mitigation measures.

As with many international dispute settlement procedures, the overriding obligation on Parties is to resolve disputes through peaceful means.[72] Parties are encouraged to seek agreement on how to resolve disputes, and the binding settlement process set out in Part XV of UNCLOS is intended as a safeguard for cases where disputes cannot be resolved by Parties on their own, and where they are unable to agree on a process for resolving the dispute peacefully.[73] Consensus on how to resolve a dispute will often take the form of a specific agreement reached at some point after the dispute arises. Agreement can also arise from State obligations enshrined in another treaty that the disputing countries are Parties to, if that treaty sets out a process for resolving disputes under UNCLOS.[74] Whether

---

[66] See J.L. Hafetz, *supra* note 17, at 597 and 632. See also D. Brack, *International Environmental Disputes*, (Royal Institute of International Affairs, 2001), at 11.

[67] See Y. Shany, *The Competing Jurisdictions of International Courts and Tribunals* (Oxford: Oxford University Press, 2003), at 5. See also Klein, *supra* note 21, at 29-124.

[68] See UNCLOS, *supra* note 3, Part XV, Articles 279 to 285.

[69] *Ibid.*, Articles 286 to 296.

[70] *Ibid.*, Articles 297 to 299.

[71] See, for example, B. Kwiatkowska, "The Australia and New Zealand v. Japan Southern Bluefin Tuna (Jurisdiction and Admissibility) Award of the First Law of the Sea Convention Annex VII Arbitral Tribunal" (2001) 16 Int'l J. Mar. & Coast. L. 239.

[72] See UNCLOS, *supra* note 3, Article 280.

[73] *Ibid.* Articles 279-281.

[74] *Ibid.* Article 288(2). For a treaty that specifically relies on the UNCLOS binding dispute resolution process, see the 1995 UN Fish Stocks Agreement (adopted August 4, 1995), reproduced at 34 ILM 1542, Article 30.

the UNCLOS dispute settlement process can be initiated therefore depends in general terms on whether a settlement is reached through an alternate process agreed to by the Parties, and whether the alternate process excludes the application of the UNCLOS dispute settlement process.[75]

Subject to these conditions, any Party to UNCLOS can initiate a binding dispute settlement process against another Party to the Convention.[76] The choice of procedure is determined based on a number of factors, including any declarations filed by the Parties under Article 287(1) on which of the following procedures are acceptable to it:

- the International Tribunal for the Law of the Sea (ITLOS) process under Annex VI;[77]
- the International Court of Justice (ICJ);[78]
- an arbitral tribunal established under UNCLOS, Annex VII;[79] or
- a special arbitral tribunal established under UNCLOS Annex VIII.[80]

In the absence of a declaration, the arbitral tribunal procedure under Annex VII shall be deemed to be the procedure selected by a Party.[81] In cases where two Parties to a dispute have not selected any common procedure in their respective declarations, the arbitral tribunal procedure under Annex VII will likewise be the applicable procedure. Regardless of the choice of procedure, the tribunal chosen has general jurisdiction concerning the interpretation and application of UNCLOS, and has the authority to determine its own jurisdiction to hear a particular dispute.[82] In addition to UNCLOS, a tribunal selected to resolve a dispute under these provisions is authorized to consider other rules of international law to the

---

[75] *Ibid.* UNCLOS, Article 281(1). This provision was at the heart of the recent Bluefin Tuna Arbitral Tribunal ruling discussed below.

[76] *Ibid.* Article 286.

[77] For an overview of the ITLOS process and its rulings to date, see A. Rest, "Enhanced Implementation of International Environmental Treaties by Judiciary – Access to Justice in International Environmental Law for Individuals and NGOs: Efficacious Enforcement by the Permanent Court of Arbitration" (2004) 1 Macquarie Journal of International and Comparative Environmental Law 1, at 13. For a survey of recent rulings, see also Robin Churchill, "The International Tribunal for the Law of the Sea: Survey for 2002" (2003) 18 Int'l J. Mar. & Coast. L. 447.

[78] For a discussion of the potential for conflict between the UNCLOS dispute settlement process and the ICJ, see Y. Shany, *supra* note 67, at 32-33.

[79] See UNCLOS, *supra* note 3, Article 287(1)(c).

[80] *Ibid.* Article 287(1)(d).

[81] *Ibid.* Articles 297 to 299 provide opportunities to further limit the options Parties have with respect to binding dispute settlement procedures. These provisions include a number of limitations that are fairly specific and restricted in scope. They do not appear to apply to obligations and responsibilities in UNCLOS that could be relevant to a dispute over climate change mitigation, and are therefore not considered further here.

[82] *Ibid.* Article 288.

extent that they are not incompatible with the rules set out in UNCLOS.[83] A tribunal's findings are final and binding on the Parties to the dispute, but not binding on other Parties to UNCLOS, and, therefore, at least in theory, are not precedent setting for purposes of interpreting the provisions of UNCLOS.[84]

While there are a number of differences between the four procedures set out in Article 287, the most important factors for Parties are likely to be both the level of control over the selection of members of a tribunal and the level of expertise of those members. ITLOS and the ICJ have the advantage of being permanent tribunals and as such are more likely to make predictable rulings, and rulings that take into account the implications of specific rulings for the future of dispute settlement under UNCLOS. The arbitration process under Annex VII has the advantage of providing a Party with more control over the membership of the specific tribunal hearing a particular dispute. In addition, the Annex VIII special arbitration tribunal process has the advantage of the flexibility to be able to ensure special expertise in the subject matter under dispute.[85]

Given the consistency in terms of jurisdiction, scope, and outcome of these four processes, a detailed comparison of the four options is not necessary for purposes of determining whether a claim based on failure to mitigate climate change could be brought under UNCLOS. More important for purposes of this analysis of whether a Party can bring a successful claim under UNCLOS is the question of jurisdiction with respect to a claim, as well as substantive issues related to such a claim. On the issue of jurisdiction to force a tribunal ruling under UNCLOS, the most important rulings to date under UNCLOS have been two rulings related to a dispute over Southern Bluefin Tuna. In addition, the recent ruling of the ITLOS in the dispute between Ireland and the United Kindom (The Mox Plant Case) offers some indication that the ITLOS will be reluctant to accept claims that the UNCLOS process is unavailable where the issue under dispute is also relevant under another international agreement with a dispute settlement process.[86]

(a) The Bluefin Tuna Decisions

To date there have been two rulings by two different tribunals under UNCLOS that have dealt with aspects of the dispute between Australia, New Zealand, and Japan over the management and protection of bluefin tuna. These rulings provide

---

[83] *Ibid.* Article 293(1).

[84] *Ibid.* Article 296.

[85] Such as expertise on climate change impacts and mitigation, which is expertise a standing tribunal such as the ITLOS or the ICJ may not always possess to the same extent.

[86] See V. Hallum, "International Tribunal for the Law of the Sea: The Mox Nuclear Plant Case" (2002) 11 R.E.C.I.E.L. 372, at 373.

one of the first indications of the application of the binding dispute settlement procedure under UNCLOS, particularly with respect to the jurisdiction of the UNCLOS process in a case of overlap with dispute settlement processes under other international treaties.[87] These rulings could give considerable insight into the question whether a successful claim could be made under the UNCLOS dispute settlement process, as they may assist in predicting how the problem of overlapping dispute settlement processes under UNCLOS and UNFCCC might be resolved.

The initial ruling in the Southern Bluefin Tuna dispute was a finding by the ITLOS for the prescription of preliminary measures against Japan. In the context of this ruling, the ITLOS concluded that there was *prima facie* jurisdiction under UNCLOS to hear the dispute between the Parties over Japan's actions with respect to the conservation of bluefin tuna.[88]

The issue of jurisdiction was first raised by Japan in response to a request for provisional measures filed by Australia and New Zealand. Japan's position was that the dispute was one under the Convention for the Conservation of Southern Bluefin Tuna, (CCSBT)[89] not under UNCLOS. In the alternative, Japan argued that Australia and New Zealand had failed to attempt 'in good faith to reach a settlement in accordance with the provisions of UNCLOS Part XV, Section 1'.[90] Japan further took the position that the dispute among the Parties was a scientific dispute rather than a legal dispute. Australia and New Zealand argued that the dispute was at least in part a dispute under UNCLOS, that it was partly a legal dispute over the interpretation and application of provisions of UNCLOS,[91] and that both Australia and New Zealand had in good faith attempted to reach a settlement.

---

87 For a good analysis of overlapping jurisdiction among international tribunals, see Y. Shany, *supra* note 67.

88 For a review of the rulings of the ITLOS on provisional measures, see B. Kwiatowska, "The Southern Bluefin Tuna (New Zealand v Japan; Australia v Japan) Cases" (2000) 15 Int'l J. Mar. & Coast. L. 1. See also D. Morgan, "Implications of the Proliferation of International Legal Fora: The Example of the Southern Bluefin Tuna Cases" (2002) 43 Harv. Int'l L.J. 541, and L. Sturtz, "Southern Bluefin Tuna Case: Australia and New Zealand v. Japan" (2001) 28 Ecology L. Q. 455.

89 *Convention for the Conservation of Southern Bluefin Tuna* (Adopted in Canberra, Australia 10 May 1993) 26 Law of the Sea Bulletin 57, online: Commission for the Conservation of the Southern Bluefin Tuna <http://www.ccsbt.org/docs/pdf/about_the_commission/convention.pdf>.

90 See International Tribunal for the Law of the Sea, Southern Bluefin Tuna Cases (New Zealand and Australia v. Japan), Request for the Prescription of Provisional Measures under Article 290, Paragraph 5, of the UN Convention on the Law of the Sea (27 August, 1999) [hereinafter 'ITLOS Ruling'], at paragraph 33.

91 Relying on Article 288 of UNCLOS.

In response, the ITLOS made the following findings. It concluded that the fact that Australia and New Zealand were alleging violations of various provisions of UNCLOS by Japan, including Articles 64, and 116 to 119, and that Japan denied that it was in violation of these provisions, supported the position that there was a legal dispute under UNCLOS.[92] The ITLOS also concluded that the fact that the Parties agreed to the CCSBT did not exclude the rights and obligations under UNCLOS.[93] Furthermore, the ITLOS found that 'the fact that the Convention of 1993 applies between the Parties does not preclude recourse to the procedures in Part XV, Section 2' of UNCLOS.[94] It was on the basis of these conclusions that the ITLOS determined that there was *prima facie* jurisdiction under Part XV of UNCLOS to arbitrate the dispute.[95]

The ITLOS proceeded to award certain provisional measures to be put in place pending the outcome of the Annex VII arbitration process. The Parties then went on to the UNCLOS arbitration, with Japan repeating its preliminary objection based on lack of jurisdiction. The Arbitral Tribunal in accordance with Annex VII proceeded to hear the Parties on the jurisdictional issue in May 2000, and issued its final ruling denying jurisdiction under Part XV of UNCLOS on August 4, 2000.[96] In its ruling, the Arbitral Tribunal went to great length to point out that it did not reject any of the conclusions on jurisdiction reached by the ITLOS tribunal; the latter had simply made a preliminary ruling that was not based on a consideration of all jurisdictional arguments.

In other words, the Arbitral Tribunal agreed with the ITLOS that the issues raised by Australia and New Zealand amounted to a dispute over the application and interpretation of UNCLOS, and that the CCSBT did not replace obligations under UNCLOS. It also confirmed that Australia and New Zealand had made sufficient efforts to settle the matter before invoking the binding dispute settlement process under Part XV of UNCLOS.

Having accepted the ITLOS ruling of *prima facie* jurisdiction to this point, the Arbitral Tribunal concluded that there was one key issue with respect to the issue of jurisdiction that had not been argued before the ITLOS tribunal and had not

---

92 ITLOS Ruling, *supra* note 90, paragraphs 44-45.

93 *Ibid.*, paragraph 51. This is an important finding in the context of this article. It suggests that Parties to the Kyoto Protocol will not be able to absolve themselves of responsibilities to protect the marine environment from climate change on the basis that the Kyoto Protocol is a subsequent agreement on what to do about climate change impacts on the marine environment. The Kyoto Protocol therefore cannot be used by Parties to imply they are permitted harm the marine environment through climate change as long as they comply with Kyoto.

94 *Ibid.*, paragraph 55.

95 *Ibid.*, paragraph 62.

96 See UNCLOS Arbitral Tribunal: Southern Bluefin Tuna Case (August 4, 2000), 39 ILM 1359 [hereinafter 'Arbitral Tribunal Ruling'].

been specifically discussed in the ITLOS ruling. That issue was the link between Article 16 of the CCSBT and Article 281 of UNCLOS, and it was this issue that led to the ruling that there was no jurisdiction to apply the binding dispute settlement process under UNCLOS.[97]

The essence of the Arbitral Tribunal's ruling is that the Parties to the dispute agreed through Article 16 of the CCSBT to exclude any dispute settlement procedure that can be initiated by a single Party (or otherwise without the consent of all Parties involved) and replace it with various settlement options, the selection of which all Parties have to agree to. In other words, Article 16 directs Parties to the CCSBT to resolve disputes following a process selected 'with the consent in each case of all Parties to the dispute'. In doing so, the Arbitral Tribunal found, the Parties to the CCSBT excluded 'any further procedure', which in turn, under Article 281(1), prevents a Party from initiating the binding dispute settlement procedure under UNCLOS.[98]

The real difficulty with the ruling is that the CCSBT process is not a binding settlement process, in that a Party cannot initiate the process under the CCSBT without consent of all other Parties to the dispute. That means a dispute under the CCSBT can only be settled with the cooperation of all Parties to the dispute. This raises the question whether Article 16 of the CCSBT amounts to an agreement 'to seek settlement of the dispute by a peaceful means of their own choice' or whether the Parties to the CCSBT actually agreed to exclude any further procedure. Notably, there is no express agreement in Article 16 of the CCSBT to exclude 'any further procedure'. The Arbitral Tribunal rather drew an inference from the provision in Article 16 that requires the consent of all Parties to refer a dispute to the ICJ or to arbitration.

The Arbitral Tribunal concluded that the intention of the Parties under Article 16 was to exclude other dispute settlement procedures as a result of subsection 2, which simply indicates that Parties have an obligation to continue to try to resolve the dispute by peaceful means referred to in subsection 1 of Article 16, even where they cannot agree on a process for settling the dispute. A reasonable interpretation of this provision would have been that in the absence of any other procedure, the Parties agreed to continue their efforts to resolve matters in accordance with Article 16.

Instead, the Arbitral Tribunal concluded that the intention of the Parties to the CCSBT was to exclude the one process that could resolve a dispute over the procedure for settling a dispute. In the process, the Arbitral Tribunal allowed the Parties to the CCSBT to replace the binding dispute settlement process under

---

[97] *Ibid.*, at 1390, paragraph 59.
[98] *Ibid.*, at 1389, paragraph 57.

UNCLOS with a non-binding process under the CCSBT without express wording in the CCSBT to exclude the UNCLOS binding dispute settlement process.[99]

The ruling is not binding on future tribunals tasked with determining the jurisdiction to initiate the binding dispute resolution process under UNCLOS. It is, however, the first such ruling, and as such will undoubtedly influence future rulings on jurisdiction. In terms of the impact of this ruling, it is important to note that it was decided in the context of a provision in a regional fisheries agreement specifically contemplated under UNCLOS as a preferred mechanism for the implementation of the general obligations under UNCLOS.

This certainly made it more difficult for the applicants to argue that there was a dispute under UNCLOS that went beyond the issues in dispute under the CCSBT. While upholding the ITLOS finding that the CCSBT did not preclude a dispute under UNCLOS, the Arbitral Tribunal was clearly influenced by the fact that the Parties to the dispute had negotiated the CCSBT to resolve issues related to bluefin tuna conservation, the very issue that was the subject of the dispute. It is therefore unlikely that this aspect of the ruling will have much influence over questions of jurisdiction in case of conflicts with dispute settlement procedures under a completely separate regime, such as the UNFCCC.

Of note is also the fact that UNCLOS actually came into force *after* the CCSBT, even though the latter was clearly developed as an implementing convention under UNCLOS. This leaves open the argument that the Parties to the CCSBT did not contemplate the application of the UNCLOS dispute settlement process as a justification for not requiring specific language excluding the binding dispute settlement process. This would then suggest that, for any treaty negotiated after the coming into force of UNCLOS, more explicit language excluding the UNCLOS dispute settlement process would be required.

In the end, it is disappointing to see how willing the Arbitral Tribunal was to replace the binding dispute settlement process under UNCLOS with a non-binding process under CCSBT, one that can be permanently stalled by one Party's refusal to agree on a process for resolving the dispute. If nothing else, one would hope that future rulings will clarify that Article 281 of UNCLOS only allows Parties to exclude the UNCLOS dispute settlement process in cases of agreement on a binding dispute settlement process to take its place; or in the case of an actual agreement on the process for resolving a particular dispute. At a minimum, explicit language should be required to exclude a binding dispute settlement process and

---

[99] For a discussion of these implications of the ruling, see Y. Shany, *supra* note 66, at 203 and 235. For a critique of the ruling, see also M.D. Evans, "Decisions of International Tribunals: The Southern Bluefin Tuna Arbitration" (2001) 50 I.C.L.Q. 447. See also Klein, *supra* note 22, at 35-43.

replace it with a non-binding process. Otherwise, a more appropriate interpretation of the intention of the Parties would be that the non-binding process needs to be explored before there can be recourse to the binding process rather than that it completely blocks access to a process the Parties agreed to in the context of UNCLOS.[100]

(b) Implications For Future Role of UNCLOS Dispute Settlement Process

In considering the implications of these rulings for the future of the UNCLOS process, it is important to keep in mind that these are only the first rulings on the jurisdiction under Part XV of UNCLOS, and that these rulings are not binding on future tribunals. Furthermore, it is unusual for international arbitral tribunals to decline to exercise jurisdiction.[101] On most of the points raised by Japan in its submissions before both tribunals the two tribunals actually agreed, and agreed with Australia and New Zealand in rejecting Japan's arguments against jurisdiction.

On the argument made by Japan that prevailed in the end, only the Arbitral Tribunal concluded that the CCSBT amounted to an agreement to exclude the binding dispute settlement process.[102] While the Arbitral Tribunal suggested that this issue may not have occurred to the ITLOS tribunal, this must be considered unlikely. More reasonably, the ITLOS simply dismissed the idea because there is no indication in Article 16 of the CCSBT that the Parties intended to exclude the UNCLOS process. This leaves two conflicting rulings on this issue, and leaves the door open for future tribunals to clarify how specific Parties have to be in declaring their intention to exclude the UNCLOS process.[103]

(c) Implications For Dispute Settlement Over Climate Change

This brings us back to the relationship between UNCLOS and climate change, and the UNFCCC[104] more specifically. Assuming all relevant Parties have ratified both conventions, a response to a claim under the UNCLOS process might be that the Parties agreed to settle their disputes with respect to climate change

---

[100] For a critical look at the Bluefin Tuna decisions, see Tim Stephens, "The Limits of International Adjudication in International Environmental Law: Another Perspective on the Southern Bluefin Tuna Case" (2004) 19 Int'l J. of Mar. &Coast. L. 177.

[101] See B. Kwiatkowska, *supra* note 71, at 240.

[102] It is important to note that there is no indication in the ITLOS ruling that Japan had either raised this specific point in its initial pleadings or had raised this argument before the ITLOS.

[103] For a detailed analysis of these and other possible conflicts between international tribunals, see Y. Shany, *supra* note 67, particularly at 203.

[104] See UNFCCC, *supra* note 50.

mitigation under the dispute settlement process in the UNFCCC.[105] This is, however, a difficult argument to make, because there is no binding obligation under the UNFCCC on individual States to take action to prevent impacts of greenhouse gas (GHG) emissions originating from their territories from harming the marine environment. Similarly, the Kyoto Protocol, which relies on the same dispute settlement process as the UNFCCC, does not impose obligations on Parties to prevent harm to the marine environment, it simply imposes obligations on certain Parties to reduce GHG emissions by 2012.[106] A dispute over whether the GHG emissions originating from a given State causes harm to the marine environment and violates the provisions of UNCLOS is therefore unlikely to be considered a dispute under the UNFCCC or the Kyoto Protocol.

Article 14 of the UNFCCC would apply assuming a tribunal found that the dispute was, in fact, a dispute under the UNFCCC or the Kyoto Protocol. It sets out the dispute settlement process under the UNFCCC and the Kyoto Protocol. The starting point is negotiations between the Parties or other mutually agreed means. The Parties also have the option to agree on a binding dispute resolution process. In the absence of a declaration to that effect, after 12 months of negotiation, any Party to the dispute can refer the matter to conciliation. Under the conciliation phase, a commission is appointed by the Parties. It makes a recommendatory award with respect to the dispute that is to be considered by the Parties in good faith.

An important difference between the UNFCCC process and the CCSBT process is that the latter essentially required the Parties to continue their efforts to reach agreement on a mutually acceptable process, whereas the UNFCCC process has a definite end point, but one that may not resolve the dispute. The Commission's recommendatory award is the end point in the UNFCCC process; however, it is conceivable that the Parties will not accept the recommendations made. In that case, there would appear to be no legal impediment to a Party initiating the UNCLOS dispute settlement process, even if the dispute was considered to be one under UNFCCC, and the UNFCCC was considered to take precedent, both of which must be considered unlikely. In the final analysis, therefore, the Bluefin Tuna rulings do not appear to impose any legal impediments to a claim for failure to implement effective climate change mitigation.

5. CONCLUSION

The work of the IPCC would suggest that a convincing case can be made that failure to mitigate climate change results in pollution of the marine environment

---

[105] For a general discussion of potential conflicts between two different specialized tribunals, see Y. Shany, *supra* note 67, at 47.

[106] Kyoto Protocol, *supra* note 51, Article 19.

as defined under UNCLOS. Failure to prevent pollution can be considered a violation of Parties' obligations to protect and preserve the marine environment. Success will largely depend on the issue of causation; on the extent to which the contribution to climate change by a particular Party or number of Parties can be isolated; how much higher the contribution of that country is relative to other countries; how relevant capacity to reduce emissions is considered; and the effect of the historical contribution to the problem in determining whether a Party has failed to take sufficient action to mitigate its climate change impact on the marine environment.[107]

The conclusion that a claim is technically possible and that it could be brought under the UNCLOS binding dispute settlement process raises more questions than it answers. Who can bring a claim, and against what countries can a claim be brought? What is the likelihood of such a claim? What would the implications of such a claim be for the climate change regime and international relations more generally? What would be the exact standard to which a Party would be held? Finally, there are unanswered questions about possible remedies. Would remedies be limited to a finding that the Party was in violation of its obligations, or would they extend to an order to reduce GHG emissions, either generally or by a specific amount? Furthermore, could remedies include an award of damages or perhaps even an order to assist other Parties in adapting to climate change?[108]

This type of claim could be brought by any State Party to UNCLOS against any other Party to UNCLOS. While there may be opportunities for non-state actors to make submissions before a tribunal struck to deal with such a claim, the procedure itself could only be initiated by a State. Most likely the Claimant State would be a developing country with low GHG emissions, high vulnerability to climate change, and high reliance on the marine environment by its population. The defending Party would most likely be a developed state, with high *per capita* or total GHG emissions. The higher the overall contribution to global emissions

---

[107] Other factors would include the composition of the tribunal hearing the case, and the willingness of the members to progressively interpret its provisions. As will be discussed in Chapter 7 in the context of human rights, courts have been more willing to progressively interpret international obligations in some areas than others. One question this raises is what is needed for environmental protection to achieve a status sufficient to encourage a progressive approach to the interpretation of treaty obligations.

[108] For a more detailed discussion of state liability issues, see, for example, A. E. Boyle, "Globalising Environmental Liability: The Interplay of National and International Law" (2005) 17 J. Envtl. L. 3; J. Brunnee, "Of sense and Sensibility: Reflections on International Liability Regimes as Tools for Environmental Protection" (2004) 53 Int'l. & Comp. L. Q. 354; J. Peel, New State Responsibility Rules and Compliance with Multilateral Environmental Obligations: Some Case Studies of How the New Rules Might Apply in the International Environmental Context" (2001) 10 R.E.C.I.E.L 82, and A. Daniel, "Civil Liability Regimes as a Complement to Multilateral Environmental Agreements: Sound International Policy or False Comfort?" (2003) 12 R. E. C. I. E. L. 225.

historically and presently, the better the chance of success. Whether or not a State is Party to the UNFCCC or the Kyoto Protocol is not likely to be determinative, although compliance with current and future emission reduction targets may influence a tribunal's finding of whether mitigation measures have been adequate to reduce or eliminate liability.

The US has not ratified UNCLOS, and is therefore not at risk unless the relevant provisions of UNCLOS are declared to be customary law. Even then, it will not be subject to a binding ruling from the ICJ unless it once again accepts the compulsory jurisdiction of the ICJ.[109]Canada and Australia, the two other nations with the highest *per capita* GHG emissions, would be the next most obvious targets. The main difference between them is that Canada has ratified the Kyoto Protocol.[110] This leaves Australia as the most likely target. Even if such a claim is never brought, the above analysis demonstrates the linkages between Parties' obligations under UNCLOS and UNFCCC. A better understanding of such linkages may provide additional motivation for States to take climate change seriously and increase their efforts to seek agreement internationally and implement those agreements effectively at home.

[109] The U.S. withdrew its acceptance under Article 36(2) of the ICJ Statute in 1985. See U.S. Department of State's statement reprinted in (1985), 24 I.L.M. 1743.

[110] For status of ratifications, see UNFCCC website at <http://unfccc.int/resource/kpthermo.html>. With the ratification by Russia in November, 2004, the Protocol came into force on February 16, 2005.

7

International Human Rights and the Future of the Climate Change Regime

1. INTRODUCTION

According to the Intergovernmental Panel of Climate Change (IPCC)[1], unabated climate change is likely to cause some Small Island States to disappear as a result of sea level rise. Flooding in parts of the world is expected to displace millions of people.[2] Other regions of the earth, particularly Polar Regions, are predicted to lose whole ecosystems, and in the process, threaten the future of communities that depend upon them for food, shelter and culture. Changes in precipitation patterns, wind currents, ocean currents, temperature, and increases in extreme weather events are expected to displace millions of people around the globe, and create new challenges for meeting the basic needs of many more.

In spite of these predictions and the surprising level of consensus that has developed on them over time,[3] action at international and national levels to date has been slow. From the time international negotiations started in the late 1980s,

---

[1] For a general overview of the most recent findings of the IPCC, see Robert T. Watson *et al* (eds), *Climate Change 2001: Synthesis Report: A Contribution of Working Groups I, II and III to the Third Assessment Report of the Intergovernmental Panel on Climate Change (IPPC)* (2002); John T. Houghton *et al* (eds), *Climate Change 2001: The Scientific Basis: Contribution of Working Group I to the Third Assessment Report of the Intergovernmental Panel on Climate Change (IPPC)* (2002); James J. McCarthy *et al*(eds), *Climate Change 2001: Impacts, Adaptation and Vulnerability: Contribution of Working Group II to the Third Assessment Report of the Intergovernmental Panel on Climate Change (IPCC)* (2002); Bert Metz *et al* (eds), *Climate Change 2001: Mitigation: Contribution of Working Group III to the Third Assessment Report of the Intergovernmental Panel on Climate Change (IPCC)* (2002).

[2] See, for example, *Ibid.* McCarthy, at 14, 17.

[3] For a comparison of the science between 1997 and 2001, compare the second and third assessment reports of the IPPC. For the second IPPC assessment reports, see John T. Houghton *et al*(eds), *Climate Change 1995: The Science of Climate Change: Contribution of Working Group I to the Second Assessment of the Intergovernmental Panel on Climate Change (IPPC)* (1996); Robert T. Watson, Marufu C. Zinyowera and Richard H. Moss (eds), *Climate Change 1995: Impacts, Adaptations and Mitigation of Climate Change: Scientific-Technical Analyses: Contribution of Working Group II to the Second Assessment of the Intergovernmental Panel on Climate Change (IPPC)*(1996); James P. Bruce, Hoesung Lee and Erik F. Haites, (eds), *Climate Change 1995: Economic and Social Dimensions of Climate Change: Contribution of Working Group III to the Second Assessment of the Intergovernmental Panel on Climate Change (IPPC)* (1996).

global emissions have continued to increase significantly. More specifically, emissions in developed countries with the highest *per capita* emissions have continued to rise, often more dramatically than those in developing countries with lower emissions to start with.[4] Three countries stand at the top of the list; Canada, the United States, and Australia. Per capita emissions in these countries are almost double the average *per capita* emissions in other developed countries, and 10 to 30 times the emissions of many developing countries. What is worse, emissions in these countries have increased a further 10 to 15% since 1990. Of the three, only Canada has ratified the Kyoto Protocol,[5] a very tentative step toward addressing the challenge of climate change. The other two countries, the United States and Australia, have refused to ratify the protocol citing economic hardship and a concern that developing countries are not obligated to make reductions in the first commitment period of 2008 to 2012 under the Kyoto Protocol.

Underlying the position that economic hardship and the refusal of developing countries to commit to action justifies not participating in this international effort to address climate change is an assumption that the governments of the United States and Australia have an obligation to protect the economic well being of the current generation of their citizens ahead of any obligation to prevent climate change for the benefit of current and future generations of their own citizens.[6] Implicit in the response of these countries to climate change is furthermore an assumption that there are no obligations beyond the borders of these countries with respect to climate change that override obligations to maximise short-term economic prosperity for their own citizens.

---

4 For an overview of recent emission trends, see Metz *et al*, *supra* note 1, 27-28. It shows recent trends for energy use and CO2 emissions in various regions of the world from 1971 to 1998. The most significant increases have occurred in industrialized countries and in Asia Pacific developing States.

5 *Conference of the Parties to the Framework Convention on Climate Change: Kyoto Protocol*, 10 December 1997, U.N. Doc. FCCC/CP/1997/L.7/add. 1, 37 I.L.M. 22 (1998), [hereinafter the Kyoto Protocol]. The Kyoto Protocol is the first international agreement with binding obligations to reduce GHG emissions.

6 In spite of the rhetoric that economic wellbeing of countries like the US is a precondition for addressing climate change globally, it is clear from the level of Official Development Assistance (ODA) in these countries that climate change is not likely to be solved through an economic growth first approach. In fact, it seems clear that this is the approach which has led to the current situation. For a discussion of levels of ODA and their role in achieving sustainable development in developing States, see Bruch, Carl *et al*, "The Road from Johannesburg: Type II Partnerships, International Law, and the Commons" (2003) 15 Geo. Int'l Envtl. L. Rev. 855; Dernbach, John C., "Targets, Timetables and Effective Implementation Mechanisms: Necessary Building Blocks for Sustainable Development" (2002) 27 Wm. & Mary Envtl. L. & Pol'y Rev. 79; and McFarlane, Amy, "Between Empire and Community: The United States and Multilateralism 2001 – 2003: A Mid-Term Assessment: Development: In the Business of Development: Development Policy in the First Two Years of the Bush Administration" (2003) 21 Berkley J. Int'l. L. 521.

This Chapter seeks to test these assumptions in one specific area, the area of international human rights. Given that two of the three countries with the highest *per capita* greenhouse gas (GHG) emissions are in North America, this Chapter will focus on the Inter-American Human Rights Regime (IAHR) as the regime within which this issue is most likely to be raised.

First, however, some general context for this analysis is warranted in two areas, the science of climate change and international human rights. To this end, the impacts of climate change as predicted by the IPCC in its third assessment report will first be briefly summarized for two areas of the globe that are and will continue to feel the impacts of climate change first and most dramatically, Polar Regions and Small Island States. This is followed with a brief introduction to the concept of human rights in international law. Given the tremendous influence of the Universal Declaration on Human Rights on the evolution of regional human rights regimes such as the IAHR regime, and given the ongoing influence of the United Nations on the substantive development of international human rights, a brief reference to the UN human rights regime is also included.[7]

According to the IPCC 2001 Synthesis Report, Polar Regions have experienced increases in temperature of up to five degrees Celsius as well as significant changes in precipitation in the past century. Sea ice has been decreasing by 2.9% per decade. In addition, permafrost areas have been significantly affected. The IPCC concludes that "the Arctic is extremely vulnerable to climate change, and major physical, ecological, and economic impacts are expected to appear rapidly."[8] The report further predicts "severe disruption for communities of people who lead traditional lifestyles," and "severe damage to buildings and transportation infrastructure." Finally, the report concludes that "for indigenous communities who lead traditional lifestyles, opportunities for adaptation to climate change are limited. Changes in sea ice, seasonality of snow, habitat, and diversity of food species will affect hunting and gathering practices and could threaten longstanding traditions and ways of life."[9]

---

[7] For a more detailed assessment of the UN Human Rights System as it relates to the environment, see C. Dommen, "Claiming Environmental Rights: Some Possibilities Offered by the United Nations Human Rights Mechanisms" (1998) 11 Geo. Int'l Envtl. L. Rev. 1, and L. Malone, "Exercising Environmental Human Rights and Remedies in the United Nations System" (2002) 27 Wm. & Mary Envtl. L. & Pol'y Rev. 365.

[8] See Watson *et al Climate Change 2001: Synthesis Report: A Contribution of Working Groups I, II and III to the Third Assessment Report of the Intergovernmental Panel on Climate Change (IPPC)*, *supra* note 1, at 276.

[9] *Ibid.* at 277.

These general findings have been confirmed more recently in the Arctic Climate Impact Assessment.[10] It reached the following key conclusions about the climate change impact on the Arctic:

- Artic climate is now warming rapidly and much larger changes are predicted (almost twice the rate of the rest of the world, with further increases of 4 - 7 degrees C expected in the next 100 years).
- Arctic warming and its consequences have worldwide implications (such as the impact of ice melting on sea-level rise and ocean circulation).
- Arctic vegetation zones are very likely to shift, causing wide ranging impacts.
- Animal species' diversity, ranges, and distribution will change (polar bears, seals, seabirds, caribou, and reindeer are among the species expected to be most affected).
- Many coastal communities and facilities face increasing exposure to storms.
- Reduced sea ice is very likely to increase marine transport and access to resources.
- Thawing ground will disrupt transportation and other infrastructure.
- Indigenous communities are facing major economic and cultural impacts.
- Elevated ultraviolet radiation levels will affect people, plants and animals.
- Multiple influences interact to cause impacts to people and ecosystems.[11]

Small Island States can be found in the Pacific, Atlantic, and Indian Ocean regions, in addition to areas within the Caribbean and Mediterranean Seas. These islands are particularly vulnerable to climate change due to their small size, low elevation, limited resources, isolation, small economies, high population densities, and limited capacity to deal with natural disasters such as severe weather events and sea-level rise.[12] Islands in these regions are already experiencing a number of changes that cause significant challenges to their inhabitants. They include changes to rainfall regimes, soil moisture budgets, prevailing winds, changes in local and regional sea-level, and patterns of wave action.[13] These challenges are expected to increase as temperature and sea-level changes in affected areas continue to accelerate. It is clear from the IPCC Third Assessment Report that Small Island States will face significant loss of landmass as a result

---

[10] Susan Joy Hassol *et al.*, *Arctic Climate Impact Assessment* (Cambridge, Cambridge University Press, 2004).

[11] *Ibid.* at 10 and 11.

[12] See McCarthy *et al Climate Change 2001: Impacts, Adaptation and Vulnerability: Contribution of Working Group II to the Third Assessment Report of the Intergovernmental Panel on Climate Change (IPCC), supra* note 1, 847.

[13] See Watson *et al Climate Change 2001: Synthesis Report: A Contribution of Working Groups II III and I to the Third Assessment Report of the Intergovernmental Panel on Climate Change (IPPC), supra* note 1, 277.

of a combination of sea-level rise, storm surges, and the lack of capacity to adapt.[14] Other impacts identified in the IPCC report include impacts on biodiversity, water resources, tourism, food security, settlement infrastructure, and human health.

These current and predicted impacts of climate change, when considered from a human rights perspective, raise a number of issues that require consideration and are the focus of this study. One consideration is how far human rights have evolved to specifically recognize the need for a clean and healthy environment as a precondition for the right to life and other established human rights. Furthermore, how far has international law evolved to recognize the right to a clean and healthy environment as a separate human right? A third issue is the state of recognition and connection between other evolving human rights and the right to a healthy environment.

Given the complexity of the climate change issue, and the interconnectedness of various human rights to the numerous expected impacts of climate change, the list of possible human rights violations is long. While a complete assessment of various impacts of climate change is impossible at this stage, scientific impacts predicted for areas such as Small Island States and Polar Regions include deterioration of human health and well being, displacement of people and communities, as well as threats to culture and development as likely outcomes of human-induced climate change. Based on this overview of the expected impacts of climate change, in an effort to focus the discussion in this paper, the substantive analysis will be limited to the following questions relating to the human rights and the impacts of human-induced climate change:

(1) Under what conditions would the failure to reduce GHG emissions be considered a violation of the human right to life, liberty and security of the person?
(2) Under what conditions would the failure to reduce GHG emissions be considered a violation of the right to development?
(3) Under what conditions would indigenous human rights be violated by the failure to reduce GHG emissions?
(4) What is the status of a human right to a clean environment, and under what conditions would the failure to reduce GHG emissions be considered a violation of the right to a clean environment?

These four issues, while by no means exhaustive, are collectively the most obvious battleground for individuals and communities whose way of life is threatened by human-induced climate change. Before considering these issues in detail to assess whether and under what conditions human rights violations might be made out, it is important to have some understanding of the context within which

[14] See McCarthy *et al Climate Change 2001: Impacts, Adaptation and Vulnerability: Contribution of Working Group II to the Third Assessment Report of the Intergovernmental Panel on Climate Change (IPCC)*, *supra* note 1, 855-870.

these issues are likely to be debated. This context is provided in the form of the Inter-American Human Rights Regime as the regime most likely to deal with these issues in the context of climate change.

This study is therefore undertaken in the following sequence. First, a general overview of the United Nations Human Rights Regime is offered. This is followed with a more detailed review of the Inter-American Human Rights Regime, given that it appears to have progressed further in recognizing links between environmental protection and human rights than the UN regime. Furthermore, the IAHR regime is the most likely battleground for assessing human rights and climate change, given that Canada and the United States, two of the three highest *per capita* contributors to the problem are from this region. Finally, both of these countries include regions and citizens that are particularly vulnerable to climate change. Following the review of the IAHR regime, a claim of a human rights violation will be considered in the procedural and substantive context of the IAHR regime. It is this part of the study that the paper will return to the four questions posed here, and will consider specifically whether and under what conditions a human rights claim under the IAHR regime is likely to succeed. The prognosis for success as well as possible implications of a successful claim is considered in the final Section.

2. HUMAN RIGHTS AND THE UNITED NATIONS

Any discussion of the link between climate change and human rights must consider the scope and meaning of human rights. While there is no universally accepted definition of human rights, there are a number of broadly accepted concepts which assist in defining the scope of human rights: They include the following:

- Regardless of their ultimate origin or justification, they represent individual and group demands for shaping and sharing of power, wealth, enlightenment and other important values in the community process that limit State power.
- Human rights refer to a wide continuum of value claims ranging from the most justiciable to the most aspirational. They represent both the "is" and the "ought" of social interaction.
- A human right is general or universal in character, equally possessed by all human beings everywhere.
- Most, but not all, are qualified by the limitation that the rights of any particular group or individual are restricted as much as is necessary to secure the comparable rights of others and the aggregate of common interest.

- Human rights are commonly assumed to refer, in some vague sense, to "fundamental" as distinct from "non-essential" claims.[15]

An attempt at defining a human right is provided by Maurice Cranston as:

> A universal moral right, something which all men [sic] everywhere, at all times ought to have, something of which no one may be deprived without a grave affront to justice, something which is owing to every human being simply because he [sic] is human.[16]

Human rights are considered by some to have evolved in three generations of rights. The first generation, civil and political rights, consists of such rights as the right to life, liberty and security of the person, freedom from slavery and torture, and the right to own property. The second generation, made up of economic, social and cultural rights, includes rights such as the right to health and well being, the right to social security, and the right to education. The third generation, commonly known as solidarity rights, includes the right to self-determination, the right to economic and social development, the right to peace, and the right to a healthy environment. The third generation is the most controversial in that its focus is on collective rather than individual rights. There is some debate about whether this is overall a constructive categorization of human rights.[17] It does, however, quite usefully illustrate varying levels of uncertainty about the status of rights, suggesting that some, such as the right to life, are universally recognized and binding while the status of others, such as the right to sustainable development or the right to a healthy environment is less certain.

Central to any consideration of the state of global recognition is the United Nations (UN) Human Rights Regime; a regime that consists of a large number of international human rights norms. These norms, which are in turn reflected in regional and national human rights initiatives, are set out in a complex system of UN instruments that establish the substance of these norms as well as the procedures and mechanisms necessary for their implementation. Within the United Nations system, the recognition of human rights is contained in the United Nations Charter.[18] Article 55 of the Charter obligates UN Member States to encourage solutions to economic, social, cultural, health and other related issues, and to promote universal respect for human rights.

The most recognized substantive source of human rights is generally considered to be the Universal Declaration of Human Rights adopted by the General

---

[15] B. H. Weston, "Human Rights" (1986) 6 Hum. Rts. Q. 257, at 262.

[16] Maurice Cranston, *What Are Human Rights?* (New York: Taplinger Publishing Co.,1973) 36.

[17] See, for example, Practicia Birnie *et al*, *International Law & the Environment*, 2nd ed. (Oxford: Oxford Univeristy Press, 2002), at 253.

[18] Reprinted in Kindred *et al*, *International Law: Cheifly as Interpreted and Applied in Canada*, 6th ed. (Toronto: Emond Montgomery Publications Ltd., 2000).

Assembly of the United Nations on December 10, 1948.[19] Most of the human rights set out in this declaration are recognized as first generation human rights, which are reflected in numerous regional and national human rights statements. Included in the Universal Declaration are such basic rights as the right to life, liberty and security of the person, the right to freedom of religion, association and expression, and the prohibition of torture and slavery.[20] The Declaration also includes a right to property and the right not to be deprived of such property arbitrarily as well as a number of other second-generation rights.[21]

The Declaration was supplemented with two other key UN instruments, the International Covenant on Civil and Political Rights and the International Covenant on Economic, Social and Cultural Rights, both developed by the United Nations Commission on Human Rights.[22] These covenants embrace much of the Declaration, but they each include additional rights that are generally considered to be second generation human rights, such as social and cultural rights. Other binding UN human rights instruments, particularly those dealing with the elimination of discrimination and with the rights of children[23] also contribute to the global recognition of the link between environmental health and human rights.[24]

Through these instruments six UN treaty bodies are created to implement human rights recognized under the UN regime. Four of these bodies are considered to have particular relevance in the environmental context. They are the Human Rights Committee, the Committee on Economic, Social and Cultural Rights, the Committee on the Rights of the Child, and the Committee on the Elimination of Racial Discrimination. The primary monitoring mechanism for these committees is the review of reports submitted by obligated States. Based on the review of filed reports and follow-up communication with the State in question, the committees then report on the status of a State's compliance with human rights.

---

[19] *Universal Declaration of Human Rights* GA Res 217 (III), UN GAOR, 3d Sess., Supp. No. 13, UN Doc. A/810 (1948) 71, reprinted in B. H. Weston *et al*, *Basic Documents in International Law and World Order,* 2nd ed, (St. Paul, MN: West Pub. Co., 1990) 298.

[20] *Ibid.* arts. 3, 4, 5, 19.

[21] *Ibid.* art 17.

[22] *International Covenant on Civil and Political Rights*, 16 December 1966, 6 I.L.M. 368 (entered into force 23 March 1976), *International Covenant on Economic, Social and Cultural Rights*, 16 December 1966, 6 I.L.M. 360 (entered into force 3 January 1976).

[23] See *International Convention on the Elimination of All Forms of Racial Discrimination*, 21 December 1965, 5 I.L.M. 350 (entered into force 4 January 1969). See also the *Covenant on the Rights of the Child*, 20 November 1989, 28 I.L.M. 1448 (entered into force 2 September 1990). Rights recognized in this Convention that have particular relevance to environmental issues include the right to be free from discrimination (art. 2), the right to life (art. 6), the right to health (art. 24), and the right for children of minorities and indigenous populations to enjoy their own culture (art. 30).

[24] For a more detailed discussion of the recognition of the link between racial discrimination and environmental health and between the rights of children and the right to a clean environment, see Dommen, *supra* note 7.

Some committees, including the Human Rights Committee, can communicate with other States in the course of their examination and, perhaps most importantly for purposes of this study, can receive individual complaints. Under Article 1 of the Optional Protocol to the International Covenant on Civil and Political Rights 1966,[25] the Human Rights Commission actively investigates and rules on individual complaints. This procedure is the oldest, most utilized, and considered to be the most authoritative within the UN regime. Otherwise these committees are limited in their ability to conduct active independent investigations by a combination of limited power, limited resources, and limited capacity.[26]

A number of non-binding and regional international instruments have also contributed to the development of human rights in relation to a right to a healthy environment. Their status is much less clear than that of the Universal Declaration and the Covenants on Civil and Political Rights and on Economic, Social and Cultural Rights. On one hand, there are three major soft law initiatives; the Stockholm Declaration on the Human Environment, the World Charter for Nature, and the Rio Declaration.[27] On the other hand, there have been a number of global and regional efforts to codify a human right to a clean and healthy environment.[28] These efforts are considered in more detail below.

As indicated above, reference to human rights generally includes both recognized rights such as those in the Universal Declaration and rights that are claimed but not universally recognized. The state of recognition of a given human right in a given jurisdiction can depend on a number of factors, including domestic

---

[25] Reprinted in C. A. R. Robb *et al*, eds., *Human Rights and Environment, International Environmental Law Reports*, vol. 3 (Cambridge: Cambridge University Press, 2001), at 898.

[26] See Dommen, *supra* note 7, at 7-10 and 22.

[27] Report on the United Nations Conference on the Human Environment at Stockholm: Final Documents, 16 June 1972, 11 I.L.M. 1416, Declaration of the United Nations Conference on the Human Environment, 16 June 1972, UN GAOR, U.N. Doc. A/CONF.48/14/Rev.1 (1973), 11 I.L.M. 1416, online: United Nations Environment Programme <http://www.unep.org/Documents/Default.asp?DocumentID=97&ArticleID=1503>, [hereinafter Stockholm Declaration], World Charter for Nature, G.A. Res. 37/7, UN GAOR 37th Sess., U.N. Doc. A/RES/37/7 (1982), online: United Nations <http://www.un.org/documents/ga/res/37/a37r007.htm>, Report of the United Nations Conference on Environment and Development, (Rio de Janeiro, 3-14 June 1992) ACONF 151/26 vol. 1, online: United Nations <http://www.un.org/documents/ga/conf151/aconf15126-1annex1.htm>, Rio Declaration on Environment and Development, UN CEDOR, Annex, Agenda Item 21, UN Doc. A/CONF.151/26/Rev.1 (1992) [hereinafter Rio Declaration].

[28] Such efforts range from the Draft Declaration of Principles on Human Rights and the Environment, U.N. Doc. E/CN.4/Sub.2/1994/9 Annex. 1 (1994) to regional efforts such as the San Salvador Protocol in the Additional Protocol to the American Convention on Human Rights, 17 November 1988, O.A.S.T.S. 69 (1988), 28 I.L.M. 156 reprinted in Basic Documents Pertaining to Human Rights in the Inter-American System, OEA/Ser.L.V/II.82 doc.6 rev.1 at 67 (1992) (entered into force 16 November 1999) [hereinafter San Salvador Protocol], art. 11. For a more comprehensive account and discussion of these efforts see M. Dejeant-Pons & M. Pallemaerts, Human Rights and the Environment (Strasbourg: Council of Europe Publishing, 2002), at 118.

law, acceptance by the jurisdiction of applicable international law,[29] and the development of customary international law. The inclusion of "claimed rights" is based on the concept of human rights as an inherent right that exists whether or not that right is recognized by a given jurisdiction. Hence the accused State's acceptance of human rights may have little effect on the existence of a human right. However, at the same time, domestic factors in an accused State will clearly affect its recognition of international rulings on violations and the ability of the international community to impose consequences.[30]

It is important to understand that the concept of human rights as inalienable is in direct conflict with the concept of State sovereignty and the resulting foundation for international law that Sates can only be bound to obligations they freely choose to accept. This creates a dilemma. The concept of human rights entails recognizing rights that derive simply from being human, and it is generally accepted that such rights should not depend on where you live, or what form of government you live under. Nevertheless the concept of State sovereignty works against the application of international law until a State has in some form accepted the obligation as binding.[31]

The solution to this dilemma suggested here is to consider human rights in two stages; stage one being the international recognition that all human beings are entitled to a given right, and the second stage, the process of implementing or developing an acceptance of that human right in States around the world.[32] The challenge for international human rights law is how to determine the substance and scope of these inalienable rights to which all human beings should be entitled, in the absence of agreement by nation States either through treaties or conduct.

It seems undisputed that the human rights that are recognized by international agreements or through customary international laws by conduct should be considered the foundation for all human rights. It has furthermore become clear in the context of international human rights that finding a strategy to move past a consensual approach among nation States is essential if human rights are to have

---

[29] Examples include agreements developed under regional human rights regimes such as the *The American Convention on Human Rights*, 22 November 1969, 1144 U.N.T.S. 123, 9 I.L.M. 673 (entered into force July 18 1978), reprinted in Inter-American Commission on Human Rights, Handbook of Existing Rules Pertaining to Human Rights, OAS Off. Rec., OEA/Ser.L/V/II.50 (1980) 27.

[30] See E. M. Kornicker Uhlmann, "State Community Interests, *Jus Cogens* and Protection of the Global Environment: Developing Criteria for Peremptory Norms" (1998) 11 Geo. Int'l Envtl. L. Rev. 101.

[31] See P. E. Taylor, "From Environmental to Ecological Human Rights: A New Dynamic in International Law?" (1998) 10 Geo. Int'l Envtl. L. Rev. 309, and J. Lee, "The Underlying Legal Theory to Support a Well-Defined Human Right to a Healthy Environment as a Principle of Customary International Law" (2000) 25 Colum. J. Envtl. L. 283.

[32] See Jack Donnelly, *International Human Rights*, 2nd ed. (Boulder: Westview Press, 1998).

meaning as inalienable rights.[33] What is suggested here is that there needs to be more focus on credible independent sources to consider whether a human right claimed should be recognized as falling under the category of rights to which every human being should be entitled.[34] Such an approach supports the evolution of existing human rights to recognize connections such as the need for a healthy environment as a precondition for the right to life, liberty and security of the person. Alternatively, it can also underpin a completely new and separate human right, such as a right to culture, community-based human rights for indigenous peoples, and an independent right to a clean environment.

Credible sources for the recognition of such new and emerging rights would be pronouncements from global and regional human rights tribunals, and other authoritative bodies such as the International Court of Justice. Rulings of international tribunals with the authority to make declarations on the state of international human rights have a fairly well-defined role in this process of establishing the state of evolution of international human rights. In addition, efforts by credible international organizations such as the UN High Commissioner for Human Rights, the UN General Assembly, and the IUCN can be an indication that our understanding of the scope of human rights is evolving.[35]

The role of other international efforts is less clear, and they are therefore considered here in a more limited sense. They are considered to the extent that they assist in determining the current state of thinking among international human rights experts. They are also considered to the extent that they are likely to influence international tribunals in their thinking on the evolution of human rights. In the context of this study, the focus will therefore be on the influence any such pronouncement on the recognition of links between human rights and a healthy environment has or is likely to have on the IAHR Commission, the international monitoring body at issue here. The question of what the role of such efforts should be in the evolution of international human rights law remains, at least for purposes of this study, to be decided by international tribunals which are given the task of rendering opinions on the state of human rights. The definition of human rights

---

33 Traditional sources of international law as recognized by the International Court of Justice include conventions, customs, general principles of law recognized by civilized nations, and judicial decisions and academic literature as subsidiary means of determining the rules of law; see *Statute of the International Court of Justice*, June 26, 1945, 59 Stat. 1055, 33 UNTS 993, (1945) arts. 38(1)(a), (b), (c), (d).

34 See Luis E. Rodriguez-Rivera, "Is the Human Right to Environment Recognized under International Law? It Depends on the Source" (2001) 12 Colo. J. Int'l Envtl. L. & Pol'y 1, at 2-3.

35 A good example are the efforts of the IUCN and other organizations on the 1995 Earth Charter and the ongoing work to further develop and gain international recognition for the Charter. For an overview of its evolution and content, see Lynn, William S., "Situating the Earth Charter: An Introduction, in Global Ethics and the Earth Charter" (2004) 8 Worldviews 1; and N. A. Robinson, "The IUCN Academy of Environmental Law: Seeking Legal Underpinnings for Sustainable Development" (2004) 21 Pace Envtl. L. Rev. 325.

offered by Cranston above provides some moral if not legal guidance for evaluating whether the concept of human rights has evolved in a particular way.[36]

With this general context of international human rights, particularly in the context of the United Nations Human Rights Regime, the study will now turn to the Inter-American Human Rights Regime. As indicated, it is the regional regime that will most likely be confronted with the questions posed in this study, whether and under what circumstances a State's failure to effectively address climate change may constitute a human rights violation. This regime includes Canada and the United States, two of the three countries[37] that have in the past and are at present contributing the most greenhouse gases on a *per capita* basis. Yet these States are among the few nations in the developed world that have the greatest capacity to address climate change. Ironically, these States have shown the least commitment to cooperate internationally to address the climate change issue.

3. CLIMATE CHANGE AND THE INTER-AMERICAN HUMAN RIGHTS REGIME

In this Section, the relationship between the climate change regime and the Inter-American Human Rights (IAHR) regime is assessed. The Section starts with a general overview of the IAHR Regime. Procedural issues are examined, such as who can bring a claim, and what conditions must be met before a claim is considered. The respective roles of the various IAHR bodies are identified, as is the relevance of the various substantive instruments covered under the regime. The four specific questions raised in the introduction are then considered in the context of the IAHR regime.

Structurally, the IAHR regime consists of the Inter-American Commission on Human Rights,[38] and the Inter-American Court of Human Rights, both of which report to the Organization of American States (OAS). Following a brief overview of these institutions, the specific issue explored in this Section is the potential for a ruling or opinion under the IAHR regime to demonstrate that failure by a Member State to address climate change amounts to a violation of human rights against individuals affected by climate change now, and in the future.

---

[36] Cranston, *supra* note 16.

[37] The third is Australia.

[38] Resolution establishing the *Inter-American Commission on Human Rights;* Res. VIII, Fifth Meeting of Consultation of Ministers of Foreign Affairs, OASOR, 12-18 August 1959, Final Act, at 10-11, OEA/ser. C./II.5 (1960). In 1970, an OAS Charter amendment referred to as the "Protocol of Buenos Aires" opened for signature on 27 February 1967, 21 U.S.T. 607, (entered into force February 27 1970) changed the status of the Commission to that of an official organ of the OAS. The Commission now serves a dual role. It is an organ of the OAS and the body responsible for the implementation of the Convention.

The Inter-American Human Rights regime was chosen over the UN Human Rights regime,[39] in part because it has dealt with a number of recent claims brought by aboriginal communities that provide insight into how the Commission and the Court might respond to a claim that a State's failure to address climate change constitutes a violation of human rights. In addition, the IAHR regime was chosen because, given the current United States opposition to the Kyoto Protocol, it is perhaps the most obvious respondent to any claim brought forth.

The IAHR regime has evolved under the Organization of American States (OAS) since the formation of the OAS in 1948. The approval of the American Declaration of the Rights and Duties of Man (the Declaration) coincided with the adoption of the OAS Charter bringing into existence the Organisation of American States.[40] The Declaration was initially approved as soft law, but has since been found to be a source of binding international obligations for Member States.[41]

In 1959, the implementation of the Declaration was supported through the creation of the Inter-American Commission on Human Rights.[42] This was followed up with a negotiation process that led to the adoption of the American Convention on Human Rights in 1969,[43] and the eventual creation of the Inter-American Court of Human Rights in 1979, the year the Convention came into force. Since then, there have been a variety of efforts to refine and further develop human rights under the IAHR regime, in the form of protocols and draft declarations. Structurally, the regime now consists of the IAHR Commission and the IAHR Court. Substantively, the following agreements relevant to the possible connection between climate change and human rights are currently in force or developing under the IAHR regime:

- The Charter of the Organization of American States (OAS Charter);

---

[39] For an overview of the UN Human Rights system as it applies to a human right to a clean or healthy environment, see Dommen, *supra* note 7.

[40] The The *American Declaration of the Rights and Duties of Man*, adopted at the 9th International Conference of American States, Bogotá, Colombia, 1948, online: The Inter-American Commission on Human Rights <http://www.cidh.org/Basicos/basic2.htm>, reprinted in Inter-American Commission on Human Rights, *Handbook of Existing Rules Pertaining to Human Rights*, OAS Off. Rec., OEA/Ser.L/V/II.50 (1980) 17. See also Weston, *supra* note 15, at 293.

[41] For an interpretation of the *American Declaration of the Rights and Duties of Man* within the framework of Article 64 of the *American Convention on Human Rights* see IACHR *Advisory Opinion*, OC-10/89, July 14, 1989, IACHR (Ser. A) No. 10 (1989); *James Terry Roach and Jay Pinkerton v. United States*, Case 9647, Res. 3/87, September 22, 1987 Annual Report 1986-1987, [46-49]; *Rafael Ferrer-Mazorra et al v. United States*, Report No. 51/01, Case 9903, April 4, 2001. There is some controversy over the exact status of the Declaration due to some ambiguity in the language of the advisory opinion.

[42] *Inter-American Commission on Human Rights, supra note 38*, at 4-11.

[43] *The American Convention on Human Rights*, *supra* note 29. The Court was created under Chapter VII, Part II of the *American Convention on Human Rights*.

- The American Declaration of the Rights and Duties of Man (Human Rights Declaration);
- The American Convention on Human Rights (Human Rights Convention);
- The Protocol of San Salvador on Human Rights in the Area of Economic, Social and Cultural Rights (San Salvador Protocol);[44] and
- The Proposed American Declaration on the Rights of Indigenous Peoples (Indigenous Rights Declaration).[45]

The status of these instruments ranges from binding on all to not binding on any Member States. The OAS Charter and the Human Rights Declaration are binding on all Parties.[46] The Human Rights Convention and the San Salvador Protocol are only binding on Member States that have specifically ratified the respective agreement. The Indigenous Rights Declaration has not yet been adopted. The substance of the human rights applicable to a claim will therefore, to some extent, depend upon the status of ratification in the Member State against which the claim is being made.[47] In the case of the United States, for example, a claim would have to be based on the OAS Charter and the Human Rights Declaration, as neither the Human Rights Convention nor the San Salvador Protocol has been ratified by the US.[48] Currently, Canada would be in the same position, also not having ratified either the Human Rights Convention or the San Salvador Protocol.

Substantively, the OAS Charter sets the general context for relations among Member States, including reference to the mutual respect of State sovereignty, and a commitment to the peaceful settlement of disputes.[49] Of particular interest here is Article 35, which commits Member States to "refrain from practicing policies and adopting actions or measures that have serious adverse effects on the development of other Member States".[50] With respect to human rights, Article 45

---

[44] See San Salvador Protocol, *supra note28*, art. 11.

[45] Draft Inter-American Declaration on the Rights of Indigenous Peoples, approved by the Inter-American Commission on Human Rights on February 26, 1997. Reprinted in *Basic Documents Pertaining to Human Rights in the Inter-American System*, OEA/Ser/L/V/II, 90 Doc. 9 rev. 1.

[46] Member States include all states of North, Central, and South America including Cuba. Cuba has however been prevented by resolution of Member States of the OAS from participating in the organization.

[47] Subject to an argument that regardless of the status ratification, these agreements reflect an evolution in the understanding of existing human rights and their relationship to human dependence on a clean and healthy environment.

[48] The United States signed the Convention on June 1, 1977, but the Senate did not ratify it. See 'Current State of Conventions and Protocols on Human Rights', in *Inter-American Yearbook on Human Rights* (1994) vol 1, 108.

[49] See Preamble and article 2 of the *Charter of the Organization of American States,* 119 U.N.T.S. 3, (entered into force December 13 1951) reprinted in B. H. Weston *et al*, *Basic Documents in International Law and World Order,* 2nd ed, (St. Paul, MN: West Pub. Co., 1990) *50.*

[50] *Ibid.* art. 35.

provides that "all human beings, without distinction as to race, sex, nationality, creed, or social condition, have a right to material well being and to their spiritual development, under circumstances of liberty, dignity, equality of opportunity, and economic security".[51]

The Human Rights Declaration is the only source of substantive human rights in the context of the OAS that is accepted and now considered binding on all Member States.[52] Rights recognized in the Human Rights Declaration that are particularly relevant to the link between climate change and human rights include the right to life, liberty and security of the person,[53] the right to protection against abusive attacks on private and family,[54] the right to inviolability of the home,[55] the right to the preservation of health and to well being,[56] and the right to property.[57]

The Human Rights Convention was signed in 1969 and entered into force in 1978 upon ratification by 11 Member States of the Organisation of American States. Neither the US nor Canada have ratified the Human Rights Convention or accepted the jurisdiction of the IAHR Court established by the Convention. Overall however, 25 of the 35 Member States have ratified the Convention, and 22 of those States have formally accepted the jurisdiction of the IAHR Court. Substantively, while there are clearly differences between the Declaration and the Convention, in practice they have often not resulted in different standards for human rights.[58]

The San Salvador Protocol was adopted in 1988, and came into force in 1999. To date, twelve Member States have ratified the Protocol. Under Articles 10 and 11, the Protocol provides that everyone has a right to health and a right to a healthy environment. Perhaps of most interest is the obligation in Article 10 for States to "promote the protection, preservation, and improvement of the environment." Furthermore, Article two of the Protocol obligates States to pass domestic legislation to protect these guaranteed rights thereby making these rights a reality in their countries. While it is not clear from the Protocol whether the right to health

---

[51] *Ibid.* art. 45(a).

[52] See referred cases in above n 39. See also *Christian B. White and Gary K. Potter v The United States of America and the Commonwealth of Massachusetts* Inter-American Court of Human Rights 25, 38, O.E.A./ser.L./V./II.54, docv.9 rev. 1 (1980-81).

[53] The American Declaration of the Rights and Duties of Man, *supra* note 40. See also Weston, *supra* note 15, at 293.

[54] *Ibid.* art. V.

[55] *Ibid.* art. IX.

[56] *Ibid.* art. XI.

[57] *Ibid.* art. XXIII.

[58] See Thomas Buergenthal, *International Human Rights*, 2nd ed. (St. Paul, MN: West Pub. Co., 1995) 176.

and a healthy environment is seen as an extension or evolution of existing rights or alternatively as a new human right, there is a clear trend within the IAHR regime to recognize the right to a healthy environment as a human right. Other related rights recognised in the Protocol include the right to food, and the right to the benefits of culture.

As a further step in the evolution of the IAHR regime, the draft Indigenous Rights Declaration was approved by the Inter-American Commission on Human Rights (IACHR) in 1997.[59] As a starting point, the draft declaration makes reference to similar initiatives in other forums, including the United Nations and specifically the International Labour Organization's Convention 169.[60] The declaration seeks to apply general human rights to an indigenous context, and to supplement these general rights to address the unique circumstances of indigenous peoples. The declaration addresses issues ranging from the right to belong to indigenous peoples, protection from forced assimilation, to cultural rights such as the right to cultural integrity.[61] With respect to environmental rights, the declaration progresses further than the San Salvador Protocol in that it recognizes the special relationship between indigenous peoples and the environment and their cultural, social and economic dependence on the environment.[62] A focus of the right to environmental protection is the recognition of the special importance of collective rights in indigenous communities.

It is difficult to predict whether and when the draft declaration is likely to be adopted by Member States. In the meantime, given its approval of the declaration, the Commission will likely be influenced by it in interpreting more generic human rights provisions in the Declaration, the Convention and the San Salvador Protocol.

Procedurally, the following documents are relevant to understand the respective roles of the IAHR Commission and the IAHR Court in overseeing the implementation of the various substantive instruments under the IAHR regime:

- Statute of the Inter-American Commission on Human Rights;[63]

---

[59] *Draft Inter-American Declaration on the Rights of Indigenous Peoples*, *supra* note 45.

[60] *International Labour Organization Convention (no. 169) Concerning Indigenous and Tribal Peoples in Independent Countries*, 27 June 1989, 72 I.L.O. Official Bulletin 59, (entered into force 5 September 1991), online: University of Minnesota <http://www1.umn.edu/humanrts/instree/r1citp.htm>, reprinted in B. H. Weston *et al*, *Basic Documents in International Law and World Order,* 2nd ed, (St. Paul, MN: West Pub. Co., 1990) 489.

[61] *Draft Inter-American Declaration on the Rights of Indigenous Peoples*, *supra* note 45, arts. 7 and 10.

[62] *Ibid.* art. 13.

[63] *Statute of the Inter-American Commission on Human Rights*, OAS Res No. 447, 9th Sess., (entered into force October 1979), online: University of Minnesota <http://www1.umn.edu/humanrts/oasinstr/zoas6cts.htm>.

- Rules of Procedure of the Inter-American Commission on Human Rights;[64]
- Statute of the Inter-American Court of Human Rights;[65] and
- Rules of Procedure of the Inter-American Court of Human Rights.[66]

These instruments set out the procedure before the Commission and the Court. To the extent that procedural matters are likely to have an influence over the substance of the claim considered here, these procedural requirements are examined below.

(a) Who Can Bring a Claim?

For purposes of this analysis, the focus is on the Commission, as all complaints must initially be brought before the Commission, and can only subsequently be referred to the Court by the Commission under appropriate circumstances. According to the current rules of procedure of the Commission,[67] any person, group of persons or non-governmental entity may submit a petition, as long as the petition is with respect to an alleged violation of a human right recognized under the IAHR regime. A petition can be brought by an individual on his or her own behalf, or by a person or organization on behalf of an individual whose rights have allegedly been violated.[68] The petitioner also has the right to designate a person to represent him or her before the Commission. From a practical point of view, a climate change claim would most likely be brought by, or on behalf of residents of a Small Island State or on behalf of communities from Polar Regions. In fact, in December 2003, the Inuit Circumpolar Conference (ICC) announced

---

[64] *Rules of Procedure of the Inter-American Commission on Human Rights*, approved by the Commission at its 109th Sess., December 2000, (entered into force 1 May 2001). Available online at Inter-American Commission on Human Rights <http://www.cidh.org/basicos/basic16.htm> (at 23 August, 2004).

[65] *Statute of the Inter-American Commission on Human Rights*, OAS Res No. 447, 9th Sess., (entered into force October 1979) available online at University of Minnesota < http://www1.umn.edu/humanrts/oasinstr/zoas6cts.htm > (at 31 August 2003).

[66] *Rules of Procedure of the Inter-American Commission on Human Rights*, approved by the Commission at its 109th Sess., December 2000, (entered into force 1 May 2001). Available online at Inter-American Commission on Human Rights <http://www.cidh.org/basicos/basic16.htm> (at 23 August, 2004). Collectively, the rules set out in these four instruments guide the overall process. As the US has not accepted the jurisdiction of the IAHR Court, and the focus of this study is on a potential claim against the United States, the emphasis here is on the procedures before the Commission.

[67] *Rules of Procedure of the Inter-American Commission on Human Rights*, approved by the Commission at its 109th Sess., December 2000, (entered into force 1 May, 2001). Available online at Inter-American Commission on Human Rights <http://www.cidh.org/basicos/basic16.htm> (at 23 August, 2004), art. 23.

[68] See Prudence E. Taylor, "From Environmental to Ecological Human Rights: A New Dynamic in International Law?" (1998) 10 Geo. Int'l Envtl. L. Rev. 309, at 359.

its intention to commence a claim against the United States under the IAHR regime.[69]

One issue that was not raised in the context of a claim by the Inuit against the United States, but would arise in cases with similar claims by an individual or communities in a Small Island State is whether claimants have to be residents or citizens of the State against which they are making a claim. In one regard human rights appear to be primarily directed at State action against their own citizens.[70] Conversely, the origin of this focus is the struggle in international law between State sovereignty and individual rights. This would suggest that there is no reason to limit the application of international human rights law only to violations by a State against its own citizens, especially if there are no avenues for the citizen's own State to protect its citizens from the harm, or the State does not pursue avenues to protect its citizens. It may simply be that global environmental problems, including climate change, ozone depletion, and acid rain, among others, make a very convincing case for the need to extend these rights in recognition of the ever increasing trans-boundary impact of environmental degradation.[71]

(b) Who Can a Claim Be Brought Against?

The various instruments that make up the substance of the IAHR regime impose obligations on its Member States. A claim brought before the Commission therefore must be brought against a Member State bound by the substantive obligation under the IAHR regime that the claimant alleges has been violated.

The jurisdiction of the Commission to hear the complaint therefore depends upon the claimant having standing as a person whose rights are alleged to have been violated by a Member State, upon the Member State being bound to comply with the substantive obligation alleged to have been breached. The jurisdiction of the Commission is further limited to consider petitions that have been filed in accordance with the requirements set out in the respective instrument which is alleged to have been violated.

---

[69] Announced at a side event organized by the Center for International Environmental Law at the 9th Conference of the Parties to the UNFCCC in Milan Italy on December 10, 2003. The ICC intends to argue that the US failure to take serious steps to reduce its GHG emissions amounts to a violation of the human rights guaranteed to the Inuit under the IAHR regime.

[70] See Jack Donnelly, *supra* note 32, at 1. The author makes the point that human rights are generally intended to protect individuals from actions of their own governments.

[71] It should be noted that the European Court of Human Rights has come to the conclusion in the criminal law context that states may owe obligations to protect the human rights of individuals who are not their own citizens. For a more detailed discussion, see R. Currie, "Human Rights and International Mutual Legal Assistance: Resolving the Tension" (2000) 11 Crim. L.F. 143, at 151-152.

(c) What is the Procedure For Bringing a Claim?[72]

A petition must contain certain standard information as set out in Article 28 of the Rules of Procedure. Included in this basic information is a statement of facts surrounding the claim, the State the petitioner considers to be responsible, information regarding the exhaustion of domestic remedies, confirmation that there is no duplication with another international settlement process under way, and any request to withhold the identity of the petitioner.[73]

Once the petition is received by the Commission, it is registered, reviewed to assess compliance with applicable requirements, and a decision is made on whether the petition contains more than one distinct claim. If so, the claims may be divided into multiple separate claims. Similarly, multiple petitions surrounding the same fact situation may be joined into a single claim.

Before any decision is made on the admissibility of the petition, the relevant parts of the petition are forwarded to the alleged State in question for comment. Ordinarily, the State will have two months to respond. The Parties may be provided with additional opportunities to comment on the claim, after which the Commission makes a ruling on the admissibility of the claim.[74] The ruling takes the form of a formal decision of the Commission that is made public and included in its annual reports.[75] If a claim is ruled to be admissible, a formal case is opened to consider the merits of the case. Other than a determination as to whether or not there is a basis for the allegation, there is no consideration of the merits of the case until after the admissibility of a claim is decided.

One crucial step in the process of considering the admissibility of a claim is a finding that the petitioner has exhausted domestic remedies. The rules governing this issue are generally favourable to the petitioner.[76] For example, the exhaustion of domestic remedies requirements do not apply where the State does not afford

---

[72] For a detailed overview of the IAHR process, see Inara K. Scott, "The Inter-American System of Human Rights: An Effective Means of Environmental Protection?" (2000) 19 Va. Envtl. L.J. 197, at 203. See also J. Taillant, "Environmental Advocacy and the Inter-American Human Rights System" in R. Picolotti *et al*, eds., *Linking Human Rights and the Environment* (Tucson: University of Arizona Press, 2003) 118.

[73] *Rules of Procedure of the Inter-American Commission on Human Rights*, approved by the Commission at its 109th Sess., December 2000, (entered into force 1 May, 2001). Available online at Inter-American Commission on Human Rights <http://www.cidh.org/basicos/basic16.htm> (at 23 August, 2004), arts. 8, 32 and 33.

[74] *Ibid.* art. 30.

[75] *Ibid.* art. 37.

[76] *Ibid.* art. 31.

due process of law, where there is no access to effective remedies, and where there has been unwarranted delay.[77]

Once a claim is ruled to be admissible, the Commission is required to afford the petitioner a period of two months to submit additional information or arguments on the merits of the claim. Relevant portions will then be submitted to the accused State for a response within another two month period. Following this, there is the option of pursuing a friendly settlement process, the possibility of a hearing, on-site investigations, and finally a decision on the merits.

The report of the decision on the merits shall identify whether or not there have been violations by a Member State. The decision is to be based both on the information provided and any other information that is a matter of public knowledge.[78] If violations have been identified, the Commission shall prepare a preliminary report with proposals on how to address the violations. This preliminary report is presented to the violating State for a response. The State has the opportunity to report on efforts to comply with the recommendations before a final report of the Commission is issued. If the violating State has accepted the jurisdiction of the Court, the Commission will provide the petitioner an opportunity at this stage of the process to consider the response of the violating State to the recommendations of the Commission and to comment on whether the case should be referred to the Court. In the absence of a referral to the Court, the Commission is free to publish its final report within three months of completing its preliminary report.[79] The Commission may then adopt a follow-up program to monitor the implementation of its recommendations or otherwise take measures to monitor whether the violation continues.

In addition, Article 49 provides that the Commission has jurisdiction to receive and review petitions with respect to alleged violations by States who are not Parties to the American Convention on Human Rights. These petitions will be considered in the context of the American Declaration of the Rights and Duties of Man. The same rules of procedure apply to these petitions.[80]

---

[77] *The American Convention on Human Rights*, *supra* note 29, art. 25(1) which provides that "Everyone has the right to simple and prompt recourse, or any other effective recourse, to a competent court or tribunal for protection against acts that violate his fundamental rights recognized by the constitution or laws of the state concerned."

[78] *Rules of Procedure of the Inter-American Commission on Human Rights*, approved by the Commission at its 109th Sess., December 2000, (entered into force 1 May, 2001). Available online at Inter-American Commission on Human Rights <http://www.cidh.org/basicos/basic16.htm> (at 23 August, 2004), arts. 42(1) and 43.

[79] *The American Convention on Human Rights*, *supra* note 29, art. 45.

[80] *Rules of Procedure of the Inter-American Commission on Human Rights*, approved by the Commission at its 109th Sess., December 2000, (entered into force 1 May, 2001). Available online at Inter-American Commission on Human Rights <http://www.cidh.org/basicos/basic16.htm> (at 23 August, 2004), art. 50.

The Standard of Proof [81] is comparatively less formal in international legal proceedings than domestic proceedings. There is specific recognition of the limited ability of claimants to get access to evidence to prove human rights violations in the context of the IAHR regime.[82] Under Article 39 of the Rules of Procedure, for example, there is a presumption that information submitted by a claimant is accurate, as long as it is not inconsistent with other information provided by the claimant. This presumption can be rebutted by the State in question through the submission of information that challenges the claimant's information.

(d) Substance of a Claim

Human Rights with links to some form of environmental protection were described by Philippe Sands as follows in 1995:

> The right to life; prohibition against cruel, inhuman or degrading treatment; the right to equal protection against discrimination; the right to an effective remedy by competent national tribunals for acts violating fundamental rights; the right to receive information; the right to a fair and public hearing by an independent and impartial tribunal in the determination of rights and obligations; the right to protection against arbitrary interference with privacy and home, prohibition against arbitrary deprivation of property; and the right to take part in the conduct of public affairs.[83]

To consider the substance of a claim in the climate change context, the scope of human rights under the IAHR regime must first be assessed to determine to what extent it includes rights with clear links to environmental protection. A good starting point for the substantive scope of the IAHR Regime is the OAS Charter[84] Article 45(a), which holds that all human beings have a right to "material well-being and to their spiritual development, under circumstances of liberty, dignity, equality of opportunity, and economic security".

The American Declaration of the Rights and Duties of Man (1948)[85] includes the following relevant human rights:

---

[81] For a detailed discussion on the standard of proof required in proceedings before the IAHR Court, see *Velasquez Rodriguez Case*, Inter-American Court on Human Rights, (ser. C) No. 4 (1988), reprinted in (1989) ILM 291, paras 128-138.

[82] See *Report on the Situation of Human Rights in Ecuador* (Ecuador Report), (24 April 1997), reprinted in IACHR OEA/ser. L. /V./II. 96, doc. 10 rev. 1, 81-83 and 90.

[83] Philippe Sands, *Principles of International Environmental Law I: Frameworks, Standards and Implementation* (Manchester: Manchester University Press, 1995), at 229. See also general discussions on freedom of association as an additional human right in Dinah Shelton, "What Happened in Rio to Human Rights?" (1992) 3 Y.B. Int'l Env. L. 75.

[84] *Charter of the Organization of American States*, (entered into force 13 December 1951) reprinted in Weston, *supra* note 49, at 50.

[85] The *American Declaration of the Rights and Duties of Man*, *supra* note 40. See also Weston, *supra* note 15, at 293.

- Article I: Right to life, liberty, security of the person
- Article VIII: Right to residence and movement
- Article XI: Right to preservation of health and well-being
- Article XIII: Right to benefits of culture
- Article XXIII: Right to Property

The IAHR Convention (1978)[86] includes the following relevant rights:

- Article 4: Right to Life
- Article 7: Right to Personal Liberty and Security
- Article 21: Right to Property
- Article 22: Right to Freedom of Movement and Residence
- Article 26: Progressive Development in Accordance with OAS Charter

The San Salvador Protocol (1999)[87] adds the following relevant provisions:

- Article 10: Right to Health
- Article 11: Right to a Healthy Environment
- Article 12: Right to Food
- Article 14: Right to the Benefits of Culture

Finally, the Proposed American Declaration on the Rights of Indigenous Peoples[88] includes the following rights that are particularly relevant in the environmental context:

- Article VII: Right to cultural integrity
- Article XII: Right to health and well-being
- Article XIII: Right to environmental protection
- Article XVIII: Right to traditional forms of ownership and cultural survival, and the rights to land, territories and resources
- Article XXI: Right to development

Collectively, these provisions in the OAS Charter, the Declaration, the Convention, the San Salvador Protocol and the Indigenous Rights Declaration are at the heart of any consideration of the state of recognition between the health of the environment and human rights under the IAHR Regime. However, it is important to recognize that each of these instruments carry different weight in determining the status of the environment under the IAHR regime. As previously

[86] *The American Convention on Human Rights*, *supra* note 29.

[87] *Supra* note 28. For a more comprehensive account and discussion of these efforts see M. Dejeant-Pons and M. Pallemaerts, *Human Rights and the Environment* (2002) 118. art. 11.

[88] *Draft Inter-American Declaration on the Rights of Indigenous Peoples*, *supra* note 45.

indicated, the OAS Charter and the Declaration are generally considered to be binding on all OAS States. The Convention and the San Salvador Protocol have come into force, but have not been ratified by all OAS States. The provisions of these two instruments are binding on States that have not ratified them only to the extent that the provisions have attained the status of customary international law. Finally, the Indigenous Declaration has been approved by the Commission, but has not been formally adopted by Member States.

We now return to the four questions posed in the introduction to consider them one by one in the context of the IAHR regime.

(1) Under what conditions would the failure to reduce GHG emissions be considered a violation of the human right to life, liberty and security of the person?

In this Section, the right to life, liberty and security of the person is considered as representative of those human rights listed by Sands which are well established and provide a connection to environmental protection. Reference to the right to life, liberty and security of the person can be found in most if not all global and regional human rights regimes. It is one of the most recognized, fundamental human rights. Within the IAHR Regime, this basic human right is recognized under the OAS Declaration and the IAHR Convention. Other rights identified by Sands as relevant to the environmental context, and some not included in his list, such as the right to equality before the law without discrimination, are also contained in the Declaration. In the IAHR regime, the right to life, liberty and security of the person therefore clearly creates binding obligations on all Member States. This provides the foundation for consideration of the link between these generally recognized first generation human rights, and the right to some level of environmental protection in the IAHR context.

Both the Court and the Commission have generally interpreted the right to life broadly. A good example is the 1985 report on the Status of the Yanomami Indians.[89] This case was initiated through a petition by the Yanomami Indians, an indigenous population in Brazil. At the heart of the claim was the construction of

[89] See: Status of the Yanomami Indians, available online at the American University College of Law <http://www.wcl.american.edu/pub/humright/digest/Inter-American/english/annual/1984_85/res1285.html> (at August 19, 2004). The court clearly recognized in this case that environmental harm amounted to a threat to the claimants' right to life, liberty and security of the person. The Court stopped short, however, of recognizing an independent right to a healthy environment, and instead focussed on the link between the environmental harm caused by the road construction and the resulting threat to the claimants' right to life, liberty and security of the person. On the facts, the threat to the claimants' right to life was so apparent, that no consideration of a separate right to a healthy environment was needed to deal with the claim.

a highway through their traditional lands, the mining activity this attracted, and the resulting impact on the indigenous population. The petition was brought under the right to life, liberty and personal security, the right to residence and movement, and the right to the preservation of health and well-being. The claim was successful, even though much of the actual harm was directly caused by non-state actors, not the Brazilian government.

The specific harm claimed by the Yanomami Indians included the displacement of native villages, the introduction of prostitution into native communities, the introduction of disease, and other physical and psychological threats to their survival. These claims were linked both to the construction of the highway directly and the resulting mining activity once rich mineral deposits were discovered. It is important to note that in this case, the Commission did not find that there was an independent right to a clean environment, nor was any compensation ordered for damage to the environment. The Commission's recommendations were limited to the establishment of protected boundaries and the recognition and protection of the cultural heritage and identity of indigenous peoples.

The 1997 Ecuador Report,[90] a more recent ruling of the Commission, further illustrates the willingness to interpret the right to life broadly. In this report, the Commission dealt with a claim by the Huarorami Indians to the effect that the granting of oil concessions by the government of Ecuador had violated their human rights. Specifically, the claimants alleged that oil exploitation activities in Ecuador were permitted to operate such that they had and would continue to contaminate the water, air and land of local communities to the detriment of the health and lives of their inhabitants.

In response to the petition, the Commission conducted a general investigation into the effect of oil development on indigenous communities in Ecuador. In the process, the Commission established that the environmental impact of oil development and handling of this industry by the State of Ecuador violated the human rights of indigenous communities. In the process, the Commission formally recognized, for the first time in the IAHR regime, the connection between the right to life and the right to a healthy environment.

The Commissions 1997 Report on Ecuador[91] clearly supports the view that a human rights claim under the IAHR regime, can be based on environmental harm. Upon investigation of the claim that oil development was causing serious environmental harm in Ecuador violating the claimants' human rights, the Commission concluded as follows:

---

[90] The Ecuador Report, (24 April 1997), reprinted in IACHR OEA/ser. L. /V./II. 96, doc. 10 rev. 1.
[91] *Ibid.*

> The American Convention on Human Rights is premised on the principle that the rights inhere in the individual simply by virtue of being human. Respect for the inherent dignity of the person is the principle which underlies the fundamental protection of the right to life and to preservation of physical well-being. Conditions of severe environmental pollution, which may cause serious physical illness, impairment and suffering on the part of the local populace, are inconsistent with the right to be respected as a human being.[92]

The Commission proceeded to make the point that this does not mean that human rights discourage or prevent development, rather "they require that development take place under conditions that respect and ensure the human rights of affected individuals."[93] The view that the IAHR Commission in the 1997 Ecuador Report established the link between recognized human rights and environmental protection is well supported in the literature.[94]

Consistent with this approach by the Commission, the IAHR Court has made it clear that it considers the rights under various instruments, from the American Declaration of the Rights and Duties of Man to the San Salvador Protocol, as evolving over time.[95] The Court defined its role in this regard as "determining the legal status of the American Declaration by appropriately looking at the Inter-American system of today in light of the evolution it has undergone".[96]

Most recently, the IAHR Court issued a ruling on a petition filed by the indigenous people of Awas Tingni Community in Nicaragua.[97] This case involved an indigenous community that historically had occupied a significant area of rain forest along the Atlantic coast of Nicaragua. The Community holds any rights it has collectively, and had attempted to have its claims recognized domestically for some time. In 1995, the government of Nicaragua granted a 30-year concession

---

[92] *Ibid.*

[93] *Ibid.*

[94] See Scott, *supra* note 72. See also John Alan Cohan, "Environmental Rights of Indigenous Peoples under the Alien Tort Claims Act, the Public Trust Doctrine and Corporate Ethics, and Environmental Dispute Resolution" (2001) 20 ULCA J. Envtl. L. & Pol'y 133, and S. J Anaya, "Indigenous Rights Norms in Contemporary International Law" (1991) 8 Ariz. J. Int'l & Comp. L. 1, at 24.

[95] For an interpretation of the *American Declaration of the Rights and Duties of Man* within the framework of Article 64 of the *American Convention on Human Rights* see Inter-American Court of Human Rights *Advisory Opinion*, *supra* note 41. See also Scott, *supra* note 72, at 222 for a detailed discussion of the Courts willingness to consider the substance of human rights to evolve without the need for amendments to the instruments that make up the substantive base for the IAHR regime.

[96] *Ibid.* Advisory Opinion.

[97] *The Mayagna (Sumo) Awas Tingni Community Case* (*Tingni Community Case*), Case No. 11.557, IACHR 79 (2001) available online at the University of Minnesota <http://www.umn.edu/humanrts/iachr/AwasTingnicase.html> (at 19 August 2004). See generally S. J. Anaya, "The Awas Tingni Petition to the Inter-American Commission on Human Rights: Indigenous Lands, Loggers, and Government Neglect in Nicaragua" (1998) 9 St. Thomas L. Rev. 157.

to log over 62,000 hectares of land that was also subject to this indigenous claim, and which had traditionally been occupied by the petitioners.

The case first went before the Commission, which found in favour of the petitioners and sent the matter on to the Court. The Court confirmed the Commission's finding that human rights of the Awas Tingni communities had been affected by the granting of the licence. In so finding, the Court affirmed the collective rights of indigenous peoples to their traditional lands, resources, and environment.[98]

Another issue of relevance here is the fact that the case dealt at least in part with a claim of future harm, which was accepted by both the Commission in the first instance and by the Court. The present harm identified was the interference with the community's land claim rights. The future harm was the resulting interference by the Korean company with its traditional lands, resources and the environment the community depended on for their culture, their health and to meet their basic needs. The fact that there is a delay between the decisions being made today and any future climate change impacts that result from those decisions will therefore not impede a claim under the IAHR regime, as long as the harm leading to a human rights violations can be supported by sufficient evidence.

Finally, it is important to note that the Commission and the Court clearly have the legal basis through Articles 29 and 30 of the IAHR Convention[99] to recognize human rights as inherent rights that were not dependent upon a State's ratification of a specific instrument, such as the San Salvador Protocol. Furthermore, this is consistent with the Commission's comments in the 1997 Ecuador Report confirming the Commission's willingness to apply the principle of inherent rights in the environmental context. This clearly opens the door for the Commission to rely on instruments not ratified by the accused State to determine whether a claimant's human rights have been violated. Given that the Commission is limited to giving advisory opinions on the allegation of a human rights violation, the exercise of this function would seem appropriate and furthermore essential to the evolution of human rights within the IAHR regime. Otherwise, as discussed previously, it would be difficult to see how human rights could evolve to protect individuals and communities from any State action which the accused State has not already agreed to cease.

The recognition of the possible link between environmental degradation and a violation of a right to life must then be considered to be inevitable given the

---

[98] Jennifer A. Amiott, "Environment, Equality, and Indigenous Peoples' Land Rights in the Inter-American Human Rights System: *Mayagna (Sumo) Indigenous Community of Awas Tingni v. Nicaragua*" (2002) 32 Environmental Law 873, at 877.

[99] *The American Convention on Human Rights*, *supra* note 29.

right factual basis. In other words, it is safe to predict that the recognition of the violation of existing human rights such as those listed by Sands as a consequence of human-induced climate change will primarily be an evidentiary issue, one that is likely to be at the heart of any claim that the right to life has been violated by a State's failure to address climate change.[100]

Assuming the IPCC's predictions in its third assessment report come to pass,[101] the question of whether the evidentiary burden can be met is, at best, a question of timing. The IPCC's conclusions summarized above clearly support a claim that the life, liberty and security of individuals from Small Island States and in Polar Regions are, and will, be threatened as a result of human impact on the climate system. The evidentiary question is whether the level of certainty is sufficient for a current finding of human rights infringement.[102] This then raises the related question of whether potential claimants will have to wait for the climate change impact to take place before a claim can be made, or whether it will be enough that the decisions resulting in climate change and human rights violations have been or are being made.[103]

Even if claimants were limited to impacts that have already occurred, a claim could still succeed. The main remaining major evidentiary issue would then be the debate between human-induced versus natural climate change and the related question of whether the impacts of climate change have already resulted in a threat to life, liberty and security of individuals. If the claim can be made in advance of the impacts being felt, the issue of natural versus human-induced climate change becomes less significant, and instead the focus shifts to whether predictions of future climate change from past, current, and expected future GHG emissions are accurate, and what level of certainty on those predictions is required for the claim to succeed.

---

[100] See Scott, *supra* note 72, at 211. The author comes to the conclusion that "environmental degradation that directly threatens the health of a community should be adjudicated as a violation of the right to life." See also the author's comment on the expansive interpretation of the right to life by the Commission and the Court on the same page.

[101] See *IPCC Third Assessment Reports*, *supra* note 1. For the purposes of this analysis, the 2001 report is an accurate reflection of the state of knowledge about climate change, the human impact on it, and the likely future changes to the global climate resulting from the human influence on the climate system.

[102] See Watson *et al Climate Change 2001: Synthesis Report: A Contribution of Working Groups I, II and III to the Third Assessment Report of the Intergovernmental Panel on Climate Change (IPPC)*, *supra* note 1, at 276 for a summary of IPCC findings with respect to Polar Regions and Small Island States that the life, liberty and security of the person of people living in these regions will be affected by climate change.

[103] Especially given that the choices made today by states regarding the level of effort they are willing to make to reduce GHG emissions will determine the extent of climate change for decades to come. For a discussion of a similar line of reasoning, see the *Tingni Community Case*, *supra* note 97.

Impacts of climate change in Polar Regions provide a good illustration of these issues. If the claim were limited to current impacts, it would be based on changes in ice cover, ice thickness, permafrost areas, and temperature changes and the impact of those changes on settlements, hunting practices, and generally on the lifestyle of resident communities such as the Inuit. In this context the real issue would be how much of the observed change has resulted from human-induced climate change, and how much has been the result of natural variations in the climate of the Polar Regions.

If the concept of future change predicted as a result of decision made at present on GHG emissions is accepted as the basis for a claim, the focus then shifts to future predictions based on emission scenarios and computer modeling. In that scenario, it will be less important to distinguish between natural and human-induced changes observed in Polar Regions. Instead the focus of the debate is likely to be the level of faith in the predictions of future changes. The question then becomes, for example, whether the prediction of 10 to 15 degree Celsius changes in average temperature in the Polar Regions, and the consequential effects for sea ice, permafrost, etcetera, are sufficiently certain to form the basis of a claim.[104] The other factual issue regarding future impacts of climate change would be for affected individuals and communities to establish that these changes have affected or will affect their right to life liberty and security of the person.

The *Tingni Community Case* suggests that both current and likely future impacts are relevant. The main issue will likely be the level of certainty and the nature of the evidence necessary to show the link between GHG emissions, climate change and the threat to life, liberty or security of the person. To the extent that a claim is based on future harm, the question of the reliability and relevance of future predictions will be crucial. The rulings of the Commission and the Court collectively suggest that a human rights violation based on a State's failure to address climate change could be made out if the consequences of that failure is to cause serious illness, impairment and/or suffering on the part of affected populations. The rulings furthermore suggest that a claim would not have to be based on present impacts of climate change, but could be based in part or in whole on impacts that are the likely or inevitable outcomes of past or present decisions made by the government of accused States. Finally, it seems clear that a State would not be able to defend a claim simply on the basis that the emissions were caused by non-state actors within the accused State.

---

[104] See for example Watson *et al*, *Climate Change 2001: Synthesis Report: A Contribution of Working Groups I, II and III to the Third Assessment Report of the Intergovernmental Panel on Climate Change supra* note 1, at 276. Based on the assessment report it appears that the level of certainty is sufficiently high to meet a balance of probability test.

(2) Under what conditions would the failure to reduce GHG emissions be considered a violation of the right to development?

The UN Declaration on the Right to Development defines development to include economic, social, cultural and political development.[105] It further imposes the primary responsibility on States to create national and international conditions favourable to the realization of the right to development. This is further explained to impose a duty on States to cooperate in eliminating obstacles to development.[106] Article 8 requires States to take necessary measures for the realisation of the right to development, without drawing a clear distinction between development within a given State, and measures a State may take to effect the right to development outside its borders, either in a positive or negative manner.

Within the context of the IAHR Regime, the right to development is not recognized in the OAS Human Rights Declaration. There is reference in the Convention to progressive development.[107] The purpose and scope of this provision appears to be similar to the UN Declaration on the Right to Development, in that it imposes a general duty on Member States to work toward the objectives set out in the OAS Charter. It does not appear to extend obligations of States further, such as by imposing a duty to ensure capacity for maintaining minimum levels of development. The San Salvador Protocol does not include any reference to a right to development.

The Draft Indigenous Human Rights Declaration, on the other hand, does make specific reference to a right to development in Article XXI. This provision makes specific reference to development rights that are likely to be threatened by human-induced climate change. Article XXI refers to the rights of indigenous people to choose their own development path, even if it is different from the path chosen by their national governments. This would suggest a right of choice to continue customary traditional practices relying on renewable resources among indigenous communities. Furthermore, Article XXI establishes a right to restitution or compensation when indigenous rights to development have been compromised.

In the 1997 Ecuador Report,[108] the IAHR Commission focussed on the link between the right to development and the right to be protected from harm resulting from pollution. It made the point that the right to development is not unlimited, but that it has to be exercised within the context of other human rights. Here we

---

105 *Declaration on the Right to Development*, GA Res.41/128, UN GAOR, annex 41 UNDocA/41/53 (adopted on 4 December 1986), art. 1.

106 *Ibid.* art 3.

107 *The American Convention on Human Rights, supra* note 29.

108 The Ecuador Report, (24 April 1997), reprinted in IACHR OEA/ser. L. /V./II. 96, doc. 10 rev. 1, 81 – 83 and 90.

are considering the reverse question. Given that development cannot take place without the resources provided by nature, at what stage of interference with nature does harm to nature and natural systems, such as interference with the global climate system constitute a violation of a right to development? It seems clear from the IPCC reports that climate change does and will continue to affect the right to development in Small Island States and in Polar Regions.

Such a human rights obligation to prevent changes to our climate system that seriously interfere with development is consistent with the objectives set out in the UNFCCC.[109] Article 2, for example, sets as one of its objectives the stabilization of GHG emissions at levels that will allow sustainable development to take place. There is considerable evidence both from the IPCC and other sources that somewhere in the range of two degrees Celsius increases in the global average temperature[110] may be the upper limit of what our climate system can absorb without seriously compromising the objectives of Article 2.[111] Given what is already being observed in more sensitive regions, including temperature increases in the range of five degrees in Polar Regions, it would appear likely that the right to development is already being violated by nations responsible for allowing continual accumulation of GHG concentrations. Current policies and actions of States such as the United States, in terms of historical contribution to the problem, current emissions, domestic response, and international role, all point to a significant contribution to violations of the right to development in those regions.

(3) Under what conditions would indigenous human rights be violated by the failure to reduce GHG emissions?[112]

As with the concept of human rights itself, there is no generally accepted definition of indigenous peoples. Various definitions of indigenous peoples have been offered by a variety of sources, including the International Labour Organi-

---

[109] See Intergovernmental Negotiating Committee for a Framework Convention on Climate Change, *United Nations Framework Convention on Climate Change*, available online at United Nations Framework Convention on Climate Change <http://unfccc.int/resource/docs/a/18p2a01.pdf> (at 19 August 2004).

[110] And a corresponding limit on CO2 concentrations in the atmosphere of approximately 550 parts per million (ppm).

[111] See *IPCC Third Assessment Reports*, *supra* note 1, and German Advisory Council on Global Change, *Climate Protection Strategies for the 21st Century, Kyoto and Beyond*, available online at German Advisory Council on Global Change <http://www.wbgu.de/wbgu_sn2003_presse_engl.html > (at 19 August 2004).

[112] For a general discussion on the links between indigenous rights and the environment, see, for example: Lawrence Watters, "Indigenous Peoples and the Environment: Convergence from a Nordic Perspective" (2001) 20 UCLA J. Envtl. L. & Pol'y 237, Cohan, *supra* note 94, and Anaya, *supra* note 94.

zation (ILO),[113] the Independent Commission on International Humanitarian Issues, and the World Bank. Collectively, they include the following factors which offer some guidance for what constitutes indigenous peoples:

1. Pre-existent, or descendent from original populations of a given country;
2. Non-dominance;
3. Close connection to traditional lands and other natural resources;
4. Subsistence oriented;
5. Cultural differences, strong sense of separate identity;
6. Indigenous language; and
7. Self identification as indigenous.[114]

The Proposed American Declaration on the Rights of Indigenous Peoples[115] include the following rights that are particularly relevant in the environmental context:

- Article VII: Right to cultural integrity;
- Article XII: Right to health and well-being;
- Article XIII: Right to environmental protection;
- Article XVIII: Right to traditional forms of ownership and cultural survival, and the right to land, territories and resources; and
- Article XXI: Right to development.

Regardless of the status of the Draft Indigenous Human Rights Declaration, given the willingness of the Court and the Commission to interpret existing rights both broadly and dynamically evolving over time, it is likely that the content of the declaration will have a significant influence over the Court's and the Commission's interpretations of rights set out in the Declaration and the Convention. For example, it would be natural for the Commission to consider the Draft Indigenous Human Rights Declaration in determining the scope of the right to life, liberty and security of the person in an indigenous context. That would mean considering the special relationship and dependence of indigenous communities with the environment and natural resources for food, shelter, and cultural needs. This approach is clear from the ruling of the Court in the *Tingni Community Case*,[116] where the court applies the right to property in the Convention to extend to the protection of indigenous rights in a way that clearly recognizes an evolution

---

[113] *Convention Concerning Indigenous and Tribal Peoples in Independent Countries ILO Convention 169*, (entered into force 5 September 1991), available online at the University of Minnesota <http://www1.umn.edu/humanrts/instree/r1citp.htm> (at 31 August 2004).

[114] See Cohan, *supra* note 94, at 133, 136.

[115] *Draft Inter-American Declaration on the Rights of Indigenous Peoples*, *supra* note 45.

[116] For a detailed assessment and review of the case see Amiott, *supra* note 98.

of our understanding of property rights and how they relate to indigenous communities.[117]

Finally, it is worth noting that the status of an indigenous human right to a clean environment has potential implications beyond indigenous human rights. It is suggested here that the indigenous context might be a good indicator of future directions in the evolution of the link between human rights and the environment. It seems clear that indigenous communities have a special connection to nature and depend culturally, socially, and economically on a healthy environment more than the general population. At the same time, similar claims can and undoubtedly will be made by others who also choose to live in a manner that makes them vulnerable to changes in the condition of the environment in which they depend upon for food, shelter, and cultural enrichment. The fact that courts and tribunals are demonstrating increasing willingness to recognize these factors in the context of indigenous human rights suggests that the door may be open in the future to similar claims by non-indigenous claimants who are able to establish a similar dependence on the environment under threat.

(4) What is the status of a human right to a clean environment, and under what conditions would the failure to reduce GHG emissions be considered a violation of the right to a clean environment?

(i) *History and Current Status*[118]

The first clear global expression of something close to a human right to a healthy environment was provided in the Stockholm Declaration.[119] It did so most notably in principle one, which states in part that humans have the right to live "in an environment of a quality that permits a life of dignity and well being", and that humans bear "a solemn responsibility to protect and improve the environment for present and future generations". The Declaration further explores the responsibility of humans to protect and preserve as well as share the benefits of all natural resources, again, pointing to a human right to a healthy environment and an equitable share of access to and benefits from the resources it provides.

The Stockholm Declaration was followed up with the 1982 World Charter for Nature.[120] The Charter does not make specific reference to a human right to a

---

[117] See *Tingni Community Case*, *supra* note 97, para 173.

[118] See S. Atapattu, "The Right to a Healthy Life or the Right to Die Polluted?: The Emergence of a Human Right to a Healthy Environment under International Law" (2002) 16 Tul. Envtl. L.J. 65, and M. Burger, "Bi-Polar and Polycentric Approaches to Human Rights and the Environment" (2003) 28 Colum. J. Envtl. L. 371. See also R. Picolotti & J. D. Taillant, eds., *Linking Human Rights and the Environment* (Tucson: University of Arizona Press, 2003).

[119] *Supra* note 27.

[120] *Ibid.*

certain level of environmental quality, but its provisions are generally consistent with an existence of such a right. The Charter imposes obligations on States to protect and preserve nature and natural resources much in the same way as the Stockholm Declaration aimed to achieve ten years earlier. In fact, it is arguably more eco-oriented. The Rio Declaration[121] fails to make any specific reference for a human right to a given standard of environmental health, so it can at best be seen as preserving the status quo as established through principle one of the Stockholm Declaration.[122]

This trilogy of soft law declarations on the environment from Stockholm to Rio is, of course, not the only indicator of the evolution of an internationally recognised human right to a clean environment. Other sources to consider would include the interpretation of the Universal Declaration on Human Rights, Regional Human Rights initiatives,[123] developments with respect to indigenous rights to a healthy environment such as the ILO Convention 169 on Indigenous and Tribal Peoples, the Draft Declaration on Indigenous Human Rights approved by the IAHR Commission, and of the Draft Declaration of Principles on Human Rights and the Environment (1994).[124]

A globally recognized, separate right to a clean environment is likely still some time away. The global community is still struggling to define the exact scope of such a human right. Nevertheless, regionally such a right is recognised in some human rights regimes, most obviously in Africa, though it is in the form of a collective right.[125] In the Americas, such a right is recognized by States that have ratified the San Salvador Protocol, and it is considered likely that the IAHR

---

[121] *Supra* note 27.

[122] For an overview of the evolution of human rights to a clean environment under these three soft law declarations, see Watters, *supra* note 112, at 266. At 275 Watters summarises the principles of human rights that are particularly relevant to environmental protection as follows: 1. The right to self determination with sovereignty over natural resources, 2. The right to health including the right to freedom from health threatening environmental degradation, 3. The right to information about the environment, 4. The right to participate in environmental decision-making, 5. The right to free association, 6. The right to preservation and the use of the environment for cultural purposes, and 7. The right to freedom from discrimination and the right of equal protection of the law.

[123] For example, the San Salvador Protocol, *supra* note 28, makes reference to the right to health and the right to a healthy environment in articles 10 and 11.

[124] For an assessment of the Draft Declaration, see N. A. F. Popovic, "Human Rights, Environment and Community: A Workshop: Conference held at University at Buffalo Law School" (1998) 7 Buff. Envtl. L.J. 239, and N.A.F. Popovic, "In Pursuit of Environmental Human Rights: Commentary on the Draft Declaration of Principles on Human Rights and the Environment" (1996) 27 Colum. H.R.L. Rev. 487. See also Karrie A. Wolfe, "Greening the International Human Rights Sphere? An Examination of Environmental Rights and the Draft Declaration of Principles on Human Rights and the Environment" (2003) 13 J.E.L.P. 109.

[125] *African Charter on Human Rights and Peoples Rights*, 27 June 1981, OAU Doc. CAB/LEG/67/3 rev. 5, 21 I.L.M. 58 (entered into force 21 October 1986).

Commission would recognize such a human right in some form, even amongst States that have not yet ratified the Protocol. Claims in the climate change context against Canada or the United States, both of which have yet to ratify the San Salvador Protocol, would directly confront the Commission with the issue of the status of a human right to a clean environment.

(ii) *Substance of a Right to a Clean Environment*

Historically, efforts to consider the state and substance of the right to a clean or healthy environment as a separate human right have focussed on a few well recognized binding and non-binding instruments, such as those previously referred to in other parts of this study. They include the Stockholm and Rio Declarations, various UN human rights instruments, regional efforts to recognize this right, and the 1994 Draft Declaration of Principles on Human Rights and the Environment.[126] While this Declaration may still be the best indication of what the substance of a right to a clean environment might appear to look like, it is no longer the only effort. In fact, in a recent book, Maguelonne Dejeant-Pons and Marc Pallemaerts[127] have put forward the first published comprehensive collection of agreements, instruments and other documents relevant to an assessment of the status or substance of human rights to a healthy environment. The book lists numerous sources of statements that can support the current status and future evolution of a human right to a clean and healthy environment. For purposes of simplicity, the Draft Declaration will be considered here as the primary basis for determining the substance of a human right to a healthy environment.[128]

Based on the 1994 Draft Declaration, the heart of a right to a clean environment is the recognition of the indivisibility of cultural, economic, social, political, and environmental rights. Furthermore the draft declaration embraces the concept of sustainable development and intergenerational equity. It does so in a way that clearly positions the right to development as dependent upon intra-and intergen-

---

[126] *Draft Declaration of Principles on Human Rights and the Environment*, U.N. Doc. E/CN.4/Sub.2/1994/9 Annex. 1 (1994).This Declaration was developed by an international group of experts convened at the request of the United Nations. It is the first global effort to establish the substance of a human right to a healthy environment.

[127] See Dejeant-Pons, *supra* note 28. The authors list and reproduce literally hundreds of provisions that all make some reference to the recognition that there is a link between the rights and needs of individuals and communities on the one hand, and a certain level of environmental protection on the other. Beyond those referred here, the instruments referenced range from specific international environmental agreements to various efforts to build upon the 1994 draft declaration of principles on Human Rights and the Environment.

[128] See N. A. F. Popovic, 'Human Rights, Environment and Community: A Workshop: Conference held at University of Buffalo Law School', *supra* note 123, at 245 The author identifies that the Draft Declaration does in fact have legal status in that it has been recognized by the UN, and cited by scholars and judges.

erational equity. In other words, it clarifies that equity comes first. The right to development is limited by the obligation to ensure sustainable and equitable outcomes for now and into the future. Article 4 reads as follows:

> All persons have the right to an environment adequate to meet equitably the needs of present generations and that does not impair the rights of future generations to meet equitably their needs.[129]

A number of the rights in the Draft Declaration are qualified in such a way as to limit their application to incidents of direct interference with already existing human rights, such as the right to life. Such proposed rights include the right to air, water and food necessary to ensure human health. However there are signs of a shift away from this connection as a precondition. An example would be Article 6, proposing a right to preservation of elements of the environment necessary to maintain biological diversity and ecosystems.

A useful starting point in considering the scope of these rights is the question of what level of environmental health is protected. The answer for most of the substantive rights in the draft declaration appears to be that it is a level of environmental health that prevents interference with the health or well-being of humans. This would appear to be a sensible definition of a human right to a clean environment, though there may be considerable debate about the level of interference that would result in a violation of such a right and the burden of proof associated within each claim. It is also clear however, that the Draft Declaration does not exclude rights that have only an indirect link to human health, such as the right to the preservation of biodiversity or the right to the protection of ecosystems. Perhaps the most interesting question is whether the right to a clean environment protects the rights of future generations to the same standard of environmental health.[130] The answer here, based on Article 4 of the Draft Declaration, is clearly yes, given the reference to equity and the clear limit on the rights of the present generation when it comes to impairing the rights of future generations. While this is still somewhat open to interpretation, the wording of Article 4 appears to clarify that the rights of present generations cannot be given priority over those of future generations. This is an issue left open in most definitions of sustainable development.

Considering comments in the 1997 Ecuador Report, it must be considered likely that the Commission recognizes human rights to a healthy environment in

---

[129] *Stockholm Declearation*, *supra* note 27, art. 4.

[130] For a more detailed discussion on the right to a healthy environment in an intergenerational context, see G. F. Maggio, "Inter/Intra-generational Equity: Current Applications under International Law for Promoting the Sustainable Development of Natural Resources" (1997) 4 Buff. Envtl. L.J. 161.

some form. It is also likely that the Commission will be influenced by the San Salvador Protocol, the Draft Declaration on Indigenous Human Rights, and the Draft Declaration of Principles on Human Rights and the Environment in defining the scope of such a right.

The real question is whether evidence of interference with the climate system, or destabilization of the global climate system will be enough, or whether more specific links will have to be drawn between the impacts felt in Polar Regions and Small Island States and the GHG emissions from a country such as the United States. A separate right to a healthy environment provides new avenues only to the extent that it allows a human rights claim to stand solely on the basis of proving that there is an impact on the environment. As long as a tribunal still requires specific evidence of the human impact resulting from the environmental change as a precondition for finding a human rights violation, the separate right to a healthy environment will remain a hollow right.

The crucial test will therefore be whether the right to a clean environment and its violation in the climate change context will be recognized in the absence of a specific and proven human impact.[131] This will have to be followed closely in the future evolution of the right to a healthy environment, in the ongoing debate over its scope and nature, be it under regional human rights regimes in the Americas or Africa, under national provisions in countries that have enshrined a right to a clean environment in their national constitutions,[132] or in the context of the on-going effort to develop global consensus on the existence, nature and scope of such a right. In the context of the IAHR regime, given past rulings of the tribunal and the substance of the San Salvador Protocol, the existence of some form of right to a certain minimum condition of the environment is likely to be accepted. The debate is therefore likely to turn quickly to the scope of such a right, including whether in specific cases the right to a clean environment exists in the absence of proven any human impact.

(f) Prognosis for Success and Possible Implications

The inevitable conclusion is that a claim brought by an individual or community affected by human-induced climate change can succeed in a human rights claim under the IAHR regime against a Member State that is responsible for that change. Claims can be based on existing human rights recognised in the IAHR

---

[131] For example, is it enough to demonstrate that there is dramatic risk of loss of biodiversity, with a predicted average risk of extinction of 35%? See discussions in Chris D. Thomas *et al*, "Extinction Risk from Climate Change" (2004) 427 *Nature* 145.

[132] See, for example, Jan Glazewski, "Environmental Rights and the New South African Constitution", in Boyle, Alan E., *et al*, eds., *Human Rights Approaches to Environmental Protection* (Oxford: Clarendon Press, 1996), at 177.

regime under a number of instruments binding on some or all of the Member States. Alternatively, a claim could be brought on the basis of a separate right to a clean environment. A middle ground between these two approaches might be a claim that is based on the general recognition that a stable climate, free from human interference, is a pre-condition for many human rights that have been accepted as binding by the international community generally and Member States of the OAS specifically.

Internationally there are indications that we are moving towards recognizing a right to a healthy environment as a separate human right. Regional efforts such as the San Salvador Protocol, and regional efforts in Africa, point in this direction, as do tribunal rulings and academic commentary. More specifically, in the context of the IAHR regime, the Commission, in the 1997 Report on Ecuador, indicated willingness to consider the right to protection from environmental pollution before the San Salvador Protocol came into force.

There are clear indications that the move toward formal recognition of a right to a healthy environment may be influenced by an increasing acknowledgement of indigenous human rights, given their special relationship to the environment and their unique social, cultural and economic dependence on nature. Various draft declarations on indigenous human rights clearly point in this direction, as does the ILO Convention 169 on indigenous rights. Nevertheless, in case of a claim against a State that has not ratified the San Salvador Protocol, such as the United States or Canada, it is questionable whether the IAHR Commission would go as far as formally ruling that these countries are in violation of an independent right to a clean environment as a result of their climate change policies.

Making out a claim under more traditional human rights such as the right to life, the right to health, or the right to property is likely to be successful if and when there is sufficient evidence to substantiate the claim. While the burden of proof remains for the State, the eventual success of a claim will depend upon the credibility of the science presented, and the ability of the claimant to link actions of the State to the impacting changes that are taking place or are predicted to occur in the future. If the findings of the IPCC are accepted, and the reports on changes that have already taken place in the Arctic are accurate, it would seem that a claim brought by northern Inuit communities should be successful even if based on evidence from current impacts alone.

This leaves as the third avenue, the position that a stable climate is a precondition for numerous human rights. This, if accepted, would limit the debate to whether a given State is responsible for the destabilization of the climate as a result of its actions or omissions. On balance, regardless of whether this is formally accepted by the Commission, it is unlikely that the link between climate change and existing human rights such as the right to life, food, property, health, culture, development, security, water, *et cetera* will be a major barrier for any claim.

The crucial issue, from an evidentiary perspective, is likely to be the level of responsibility for climate change that can be attributed to a given State as a result of its actions and policies on the issue. In other words, will accused States be able to avoid liability by claiming that climate change would be happening even if they made more serious efforts to reduce their emissions. In the case of Canada, the argument would be most obvious, given that in spite of its high *per capita* emissions, Canada's total emissions is currently less that 3% of global emissions.

Consistent with the generally more lenient approach to evidence in recognition of the imbalance of power and resources between the claimants and the accused States in the human rights context, it would only be reasonable to conclude that the claimant would not have to prove that the climate change impacts would not occur "but for" the actions of the accused State. As a starting point, there are numerous secondary effects of a State's position on climate change that are not reflected in an assessment of a State's GHG emissions alone. The "but for" test would therefore be an inadequate measure of a State's responsibility for human rights violation, even leaving aside the inequity of placing such an evidentiary burden on an individual or community with often very limited resources.

There are a number of secondary effects to consider. For example, the United States' position on energy efficiency, conservation, public transportation, urban planning and climate change since the 1970s, has not only resulted in *per capita* emissions in the US being twice that in Europe, but it prevented an agreement on binding targets under the UN Framework Convention. Moreover, the United State's role in the climate change negotiations is at least partly responsible for lower targets negotiated under the Kyoto Protocol. Other criticisms include loopholes for State compliance since Kyoto, and significant delays in the implementation of the protocol.[133] In short, the impact the United States has had on global efforts to address climate change goes well beyond GHG emissions in the US.

Even in the context of State responsibility and resulting damage claims before the International Court of Justice, or damage claims before domestic courts using principles such as negligence or strict liability, however, courts have introduced legal principles that would prevent accused States from applying the "but for" test to argue that they did not cause human rights violations associated with GHG emissions. Specifically, exceptions to the "but for" test have been recognized in case of multiple defendants, or multiple Parties who are collectively responsible

---

133 For a more detailed discussion of the Kyoto process and the role of the United States in that process see Section 3 of Chapter 2.

for the commission of a wrong.[134] It is therefore highly unlikely that the IAHR Commission would place the burden of proof on the claimant to show that the impacts would not occur or have occurred but for the policies of the accused State. Rather, a finding that an accused State has, on balance, contributed to the problem rather than to the solution should prove sufficient in this context.

Factually, it is important to keep in mind that the United States and Canada are currently the highest *per capita* contributors to GHG emissions (together with Australia). Both States are historically among the highest contributors for greenhouse gases currently in the atmosphere, and are among the States that have the greatest capacity (in terms of economic strength and domestic opportunities to reduce emission) to address the issue of climate change both domestically and internationally.[135] On balance, it must be considered unlikely that causation would pose a significant problem for a claim against either the United States or Canada.

This leaves one final question for this study. Assuming that a claimant can succeed against a State such as the United States or Canada for their failure to address climate change, what impact is this likely to have on the future evolution of the climate change regime. There is, of course, no straightforward answer to this question. As a starting point, it is important to recall that a claim against Canada or the United States would involve limited remedies under the IAHR regime due to the fact that neither State has accepted the American Convention or the jurisdiction of the IAHR Court. This leaves as the only direct outcome an advisory opinion of the IAHR Commission on the compliance of the accused State with human rights recognized under the IAHR regime. The question therefore becomes whether such an advisory opinion is likely to influence government policy on climate change.

In Canada, if an advisory opinion concludes that Canada is in violation of Inuit human rights, this could have significant long-term impacts. It could make it difficult for those opposed to effective climate change action to continue to publicly oppose action to address this issue. Furthermore, it could permanently shift the debate from whether climate change should be taken seriously to how much and how quickly climate change impacts can be rectified.

Impact in the United States is much more difficult to predict, and a complete response is well beyond the scope of this study. It is clear, however, that there are many factors that have influenced the US position on climate change. In fact, that position has changed over time in all levels of government. From the Clinton to the Bush administration, there has been a shift away from multilateral cooperation

---

[134] See the discussion on causation in P. Barton, "State Responsibility and Climate Change: Could Canada Be Liable to Small Island States?" (2002) 11 Dal. J. Leg. Stud. 65, at 72.

[135] See Chapter 2.

and domestic implementation toward limited bilateral cooperation and domestic action demanding only voluntary measures, research and development for climate change.[136] On the other hand, at other levels of the US government, there seems to be more willingness to take climate change seriously. It is difficult to assess, in this context, what relative influence a finding that the US is violating human rights of the Inuit people would have on domestic policies for climate change. However it may compel the United States to at least appear to be acting on this issue. This may in turn strengthen public support for climate change measures in the United States; at the State level if not nationally.

In this context it is important to consider that the main contributors to GHG emissions are still developed countries. These are countries that tend to pride themselves on their human rights record, and have proven themselves much more susceptible to suggestions that their actions are in violation of human rights. A finding of a human rights violation is therefore likely to have a significantly larger impact in countries such as Canada and the United States than apparently less democratic States, or States that have a less stellar human rights reputation.

4. CONCLUSION

Climate Change is as much about equity as it is about environmental protection and biodiversity. Too much public discourse on this issue has been about economics. Human rights have the potential to provide intra and intergenerational equity aspects of this issue the attention it deserves. While it is debatable whether this is enough, or whether it is more appropriate to consider environmental protection independent of its impact on human rights is a debate for another day. There is some hope that human rights to a clean environment may actually evolve to protect all aspects of nature, not just essential rights associated with meeting immediate human needs. In the meantime, engaging in the debate about the impact of human rights as a result of climate change can only serve to accelerate international progress on this issue.

---

[136] McFarlane, Amy, "Between Empire and Community: The United States and Multilateralism 2001 – 2003: A Mid-Term Assessment: Development: In the Business of Development: Development Policy in the First Two Years of the Bush Administration" (2003) 21 Berkley J. Int'l. L. 521.

8

Linking Climate Change and Other Multilateral Environmental Agreements: From Fragmentation to Integration?

1. INTRODUCTION

The central question posed in this Chapter is whether multilateral environmental agreements (MEAs) such as the Convention on Biological Diversity (CBD)[1] have the potential to make a positive contribution to the future evolution of the climate change regime generally and compliance with the Kyoto Protocol[2] more specifically. To this end, the Chapter explores linkages and overlaps between the climate change regime and other MEAs, with a particular focus on the CBD.[3]

As discussed elsewhere throughout this book, the development of an effective response to the challenge of climate change internationally has so far proven to be difficult at best. Despite early optimism that the climate change issue would follow in the footsteps of Ozone Layer Depletion[4] with rapid international agreement, obvious technological fixes followed by effective domestic implementation, quite the opposite is the case. After a decade of very difficult international

---

[1] Convention on Biological Diversity of the United Nations Conference on the Environment and Development, June 5, 1992, U.N. Doc. DPI/1307, reprinted in 31 I.L.M. 818 [CBD].

[2] United Nations Framework Convention on Climate Change, Intergovernmental Negotiating Committee for a Framework Convention on Climate Change OR, 5th Sess., Annex, UN Doc. A/AC.237/18 (PartII)/Add.1 (1992), 31 I.L.M. 849, online: UNFCCC <http://unfccc.int/resource/docs/a/18p2a01.pdf> [UNFCCC or The Framework Convention], and *Conference of the Parties to the Framework Convention on Climate Change: Kyoto Protocol*, 10 December 1997, U.N. Doc. FCCC/CP/1997/L.7/add. 1, 37 I.L.M. 22 (1998), [hereinafter the Kyoto Protocol].

[3] For a general discussion of linkages between the Kyoto Protocol and MEAs, see: The United Nations University Institute for Advanced Studies, *Global Climate Governance; Inter-linkages between the Kyoto Protocol and other Multilateral Regimes*, Final Report (Tokyo, Japan 1999) at 62.

[4] For an assessment of the international response to Ozone Layer Depletion, see discussion in Chapter 1. For the key international agreements on Ozone Layer Depleting substance, see the *Montreal Protocol on Substances that Deplete the Ozone Layer*, 16 September 1987, amended at London on 29 June 1990, amended at Copenhagen on 25 November 1992, amended at Vienna in 1995, amended at Montreal on 17 September 1997, and amended at Beijing on 3 December 1999, 1522 U.N.T.S. 3, Can. T.S. 1989 No. 42, 26 I.L.M. 1550 (entered into force 1 January 1989), online: United Nations Environment Programme <http://www.unep.org/ozone/pdf/Montreal-Protocol2000.pdf>.

negotiations following the entry into force of the UN Framework Convention on Climate Change, there are few signs of international action, straightforward technological fixes or effective domestic action to address this issue.

Much has been written about the reasons for the relatively slow progress on climate change to date.[5] These challenges may also provide an opportunity to move beyond the fixing of the specific issue of climate change to consider climate change for what it is; one symptom of a broader problem of over-consumption, waste, pollution, alteration of natural systems, loss of biodiversity, and inequity. If we look at unresolved global environmental problems, starting with issues tackled through existing MEAs, as symptoms of a deeper problem, rather than as isolated issues to be addressed one by one, the various obligations and commitments made in MEAs could become mutually supportive rather than competing for the attention of domestic governments strapped for resources. They could be designed to think in terms of years, not decades.

The purpose of this Chapter is to explore overlaps and linkages between climate change and the causes of, and solutions to, biodiversity as an example of some of our other most pressing environmental challenges. The Chapter considers whether an integrated approach to the implementation of international environmental obligations could improve progress on the climate change issue, on other environmental challenges, and put Canada on the path to sustainability. The focus of this study is on the links between climate change and biodiversity. However, to illustrate that there are linkages with other environmental issues inviting similar consideration of integration, linkages to desertification, wetland protection, air pollution, and resource depletion are briefly referenced as well. A detailed assessment of those linkages and the impact of all relevant obligations under MEAs on an integrated approach to climate change in Canada is beyond the scope of this study. Rather, the Chapter has the more modest objective of demonstrating the utility of the general approach.

2. CLIMATE CHANGE AND BIODIVERSITY

A comprehensive assessment of our scientific understanding of climate change,[6] biological diversity[7] and the linkages between the two is not necessary

---

[5] Boyd, D. *Unnatural Law* (UBC Press, 2003) at 66-94. See also discussion in Chapters 1 and 2.

[6] For a general overview of the state of the science on climate change, see Robert T. Watson *et al*, eds., *Climate Change 2001: Synthesis Report: A Contribution of Working Groups I, II and III to the Third Assessment Report of the Intergovernmental Panel on Climate Change* (Cambridge, England: Cambridge University Press, 2002); John T. Houghton *et al*, eds., *Climate Change 2001: The Scientific Basis: Contribution of Working Group I to the Third Assessment Report of the Intergovernmental Panel on Climate Change(IPCC)* (Cambridge, England: Cambridge University Press, 2002); James J. McCarthy *et al*, eds., *Climate Change 2001: Impacts, Adaptation & Vul-*

here, as much has been written on both.[8] A brief overview of the linkages from a scientific perspective is, however, warranted.

As a starting point, it is generally recognized that all significant changes in the climate, whether naturally occurring or human-induced, have an impact on biodiversity. Changes in the climate have contributed to both the loss and increase of biodiversity throughout the evolution of life. The main concern with human-induced climate change is that the rate of change is so much greater that it will make it much harder for existing species and ecosystems to adjust to the changes so as to minimize the net loss of biodiversity. According to a recent article in the journal *Nature*,[9] the average risk of extinction due to the threat of human-induced climate change is about 35% for any given species. This clearly ranks climate change as one of the top threats to biodiversity, together with habitat loss and invasive species.[10]

The linkages between climate change and biodiversity, of course, do not end there. Plant life, soils and the oceans all have a significant role to play in the management of the carbon cycle,[11] the main natural controlling mechanism for the concentration of carbon dioxide in the atmosphere. Given that loss in biodiversity affects plant life, soils, and the health of oceans, there are risks of loss of biodiversity contributing to GHG emissions and to climate change.

Not surprisingly, there are also links in terms of human activities that contribute to each of the problems. Many of the activities that contribute to climate change also have an impact on biodiversity. For example, the design of many of our cities

---

nerability: Contribution of Working Group II to the Third Assessment Report of the Intergovernmental Panel on Climate Change (IPCC) (Cambridge, England: Cambridge University Press, 2002); Bert Metz *et al*, eds., *Climate Change 2001: Mitigation: Contribution of Working Group III to the Third Assessment Report of the Intergovernmental Panel on Climate Change (IPCC)* (Cambridge, England: Cambridge University Press, 2002).

7 For a good general overview of the biodiversity issue from a Canadian perspective, see Boyd, D. *Unnatural Law* (UBC Press, 2003) at 164. See also Elder, P.S., "Biological Diversity and Alberta Law" (1996) 34:2 Alta. L. Rev. 293.

8 For an assessment of links between climate change and biodiversity, see Intergovernmental Panel on Climate Change, *Climate Change and Biodiversity* (Technical Paper, IPCC Working Group II, Technical Support Unit) (IPCC, April 2002).

9 See Chris D. Thomas *et al*, "Extinction Risk from Climate Change" (2004) 427 *Nature* 145.

10 An example of a specific current threat to biodiversity that is considered to be linked to climate change is coral bleaching. See Report on Biological Diversity and Climate Change, Subsidiary Body on Scientific, Technical and Technological Advice under the Convention on Biological Diversity, 30 September 2003, UNEP/CBD/SBSTTA/9/INF/12, Annex III, Page 11 (Ad Hoc Expert Report). See also M. G. Davidson, "Protecting Coral Reefs: The Principal National and International Legal Instruments" (2002) 26 Harv. Envtl. L. Rev. 499. For a discussion of the threat posed by invasive species, see M. Doelle, "The Quiet Invasion, Law and Policy Responses to Invasive Species in North America" (2003) 18 Int'l J. Mar. & Coast. L. 261.

11 *Ibid.* Ad Hoc Expert Report, Annex III, at 44 -46.

results in inefficient transportation, which contributes to climate change, and at the same time directly threatens biodiversity through habitat loss as a result of urban sprawl. Similarly, oil and gas production results in significant GHG emissions from production and end use. At the same time, oil and gas production, depending on the location, can be a direct threat to ecosystems essential for the protection of biodiversity.[12] Similar linkages exist for other human activities in areas such as agriculture, energy production, and others.

A third category of linkages includes mitigation and adaptation measures considered as a response to climate change. These climate change mitigation measures, of course, have the potential to directly affect biodiversity, either in a positive or negative manner.[13] An obvious example would be the use of biomass as a source of energy in place of fossil fuels. Increased use of biomass would reduce GHG emissions from energy production relative to fossil fuels, but would introduce a competing land use with the potential threat to biodiversity through loss of ecosystems to support biodiversity. This category of linkages will be the focus of the analysis below; it is here that the integration of climate change and biodiversity has the potential to influence choices being made for the implementation and future evolution of the climate change regime. First, however, a brief introduction to the Biodiversity Regime is provided.

3. THE BIODIVERSITY REGIME[14]

The Convention on Biological Diversity, coming into force in December 1993, followed the example of the Vienna Convention for the Protection of the Ozone Layer substances in that it is a framework convention that seeks to lay the structural foundation for future substantive negotiations on the environmental challenge identified. In addition to providing a structure for future negotiations, the CBD does establish some general principles to guide future negotiations as well as domestic action on the issue. It is not surprising, therefore, that the Convention has been characterized as an agreement to mutually support national

---

[12] An illustration of the potential link between production and biodiversity loss in Canada is shown in the offshore oil and gas industry. A number of aquatic species are threatened or otherwise at risk. Offshore development has the potential to contribute to the threat to these species. In fact, a number of these species are being considered for protection under the *Species at Risk Act*, 2002 S.C. c. 29. This will raise questions for Canada about the relative priority of oil and gas production and biodiversity.

[13] See Ad Hoc Expert Report, *supra* note 10, Annex III, at 48.

[14] For an overview of the CBD, see C. Tinker, "The Rio Environmental Treaties Colloquium: A New Breed of Treaty: The United Nations Convention on Biological Diversity" (1995) 13 Pace Envtl. L. Rev. 191. For an insightful comparison of the climate change and biodiversity regimes, see J. Boynton, "Issue Salience in Climate Change and Biodiversity Discourses" (Paper prepared for presentation at the 45th Annual Meeting of the International Studies Association, Montreal, Quebec, Canada March 17-20, 2004) [unpublished].

efforts to protect biodiversity rather than an actual international consensus on actions to be taken.[15]

It considers the issue of biodiversity from both an environmental protection and a development perspective, and in the process acknowledges that different Member States may have different interests and responsibilities with respect to biodiversity. The Convention does this by treating biodiversity both as an essential component of ecosystem health and as a potentially valuable resource. It furthermore appears to recognize, at least in a general way, that not all countries and regions of the world can or place the same relative priority on biodiversity protection. Specifically, developing countries that host much of the world's biodiversity are more likely to be focussed on its value as a resource to help meet basic needs and overall development objectives. Conversely, many developed countries who have the luxury of placing higher priority on biodiversity conservation for ecosystem health often do not host natural ecosystems with the same abundance of biodiversity and the same level of significance.

The Preamble of the Convention signals much of this context. It notes, for example, that States are responsible for "conserving their biological diversity and for using their biological resources in a sustainable manner", and at the same time that the conservation of biological diversity is a "common concern of humankind".[16] The Preamble furthermore emphasizes the preference for *in-situ* conservation, the importance of taking a precautionary approach, the importance of regional and global cooperation, and the importance of meeting the needs of developing countries. These various principles are all captured in some form in the objectives of the Convention, as stated in Article 1:

> The objectives of this Convention, to be pursued in accordance with its relevant provisions, are the conservation of biological diversity, the sustainable use of its components, and the fair and equitable sharing of the benefits arising out of the utilization of genetic resources. . .[17]

In terms of specific obligations, the Convention requires Member States to cooperate in their efforts to pursue the objectives of the Convention.[18] Parties commit to the identification and monitoring of "important" components of bio-

---

[15] It has also generally been characterized as even less successful than the UNFCCC in motivating States to action. See, for example, C. Stone, "Environment 2000 – New Issues for a New Century: Land Use, Biodiversity, and Ecosystem Integrity: Land Use and Biodiversity" (2001) 27 Ecology L. Q. 967, and A. Hubbard, "The Convention on Biological Diversity's Fifth Anniversary: A General Overview of the Convention – Where has it been and where is it going?" (1997) 10 Tul. Envtl. L. J. 415.

[16] CBD, *supra* note 1, Preamble.

[17] *Ibid.* CBD, Article 1.

[18] *Ibid.* CBD, Article 5.

diversity, and to establishing a system of protected areas to ensure *in-situ* protection where possible and appropriate.

Integration of biodiversity issues into relevant plans, programs and policies is specifically identified,[19] and of particular interest in the context of the connection between climate change and biodiversity. Furthermore, where a "significant adverse effect" on biodiversity has been identified, Member States are committed to regulate or manage the relevant processes and categories of activities "as far as possible and appropriate".[20] This qualifier is used throughout the Convention with the effect of turning otherwise potentially firm and binding obligations into discretionary commitments.

With respect to the use of biodiversity, such as the proposed use for carbon storage as contemplated under the Kyoto Protocol,[21] Article 10 of the CBD commits Parties to find ways to minimize adverse impacts from the use of biological resources. It further points to the importance of cooperation within States, among governments and in private sectors, to ensure that biological resources are used sustainably. Other commitments of interest include providing incentive measures for conservation and sustainable use,[22] and providing for environmental impact assessment at the project and policy level before decisions are made that may have a significant adverse effect on biodiversity.[23]

With respect to other international conventions, such as the UNFCCC and its Kyoto Protocol, Article 22 holds that rights and obligations of existing conventions are not affected, unless the exercise of those rights and obligations would cause a serious damage or threat to biological diversity. This provision, on its own, would suggest that rights and obligations under the UNFCCC would be affected to the extent that they cause a serious threat to biodiversity. This might place some limits on the ability of States to choose mitigation measures for climate change that cause serious threats to biodiversity. Any such link would have to be based on the UNFCCC, and not on the Kyoto Protocol, as it was clearly not in existence at the time. As such, this protection against climate change mitigation measures harmful to biodiversity is legally minimal and practically non-existent.

A related issue is whether or not a general obligation under the CBD to mitigate climate change because of its threat to biodiversity would be overridden by an implied right of States to emit greenhouse gases up to the levels allowed under

---

[19] *Ibid.* CBD, Article 6(b).
[20] *Ibid.* CBD, Article 8(l).
[21] Kyoto, *supra* note 2, Article 3.3, 3.4.
[22] CBD, *supra* note 1 Article 11.
[23] *Ibid.* CBD, Article 14.

the Kyoto Protocol.[24] Such an interpretation of the relationship between the two agreements is possible, but would be inconsistent with the clear position of many Member States that the Kyoto targets are obligations, not rights, to emit greenhouse gases. This leaves the door open to consider under the CBD whether more climate change mitigation than set out in Kyoto is required to meet the obligations in the CBD, such as Article 5. In a somewhat generous interpretation of the obligations under the CBD, this might suggest that Parties to the CBD may be obligated to reduce GHG emissions beyond the targets set out in the Kyoto Protocol, if such further reductions are necessary to prevent "serious damage or threat to biological diversity".[25]

There appears to be a recognition of the threat climate change poses to biodiversity and the objectives of the Convention.[26] Not surprisingly, therefore, there has been considerable effort under the Convention to seek out cooperation with the UNFCCC on linkages between climate change and biodiversity issues. Some of these efforts have taken the form of letters and submissions from the CBD secretariat submitted at various Conferences of the Parties to the UNFCCC.[27]

Perhaps the most significant effort has been the work of an Ad Hoc Technical Expert Group set up under the Convention to consider linkages between climate change and biological diversity, and more specifically to give advice on how to integrate biodiversity issues into the implementation of the UNFCCC and the Kyoto Protocol. This effort resulted in a report to the Conference of the Parties to the CBD in November 2003.[28] In it, the Ad Hoc Group considers in some detail the various climate change mitigation and adaptation options and their potential implications for biodiversity. Absent from the discussion is a detailed assessment of the overall threat climate change poses to biodiversity, or the potential ways in which loss of biodiversity can contribute to climate change.[29]

---

[24] In case of Canada, this would amount to some 560 MT of CO_2 equivalent GHG emissions per year between 2008 and 2012, plus any additional amounts offset using sinks and the Kyoto flexibility mechanisms. See generally Chapter 2.

[25] While this does not add any new obligations to take action, it makes it clear that the Kyoto Protocol does not relieve Parties of any existing obligation under the CBD. Practically, of course, it would be unusual for these issues to be resolved under the CBD when they have not been resolved under the climate change regime.

[26] See IPCC Third Assessment Report, *supra* note 6, the IPCC Report on Climate Change and Biodiversity, *supra* note 8, and Thomas *et al*, *supra* note 9.

[27] At the 9th Session of the Conference of the Parties to the UNFCCC in Milan, Italy (December 2003) for example, the IUCN, the Deputy Secretary General to the Ramsar Convention, and the Executive Secretary of the CBD each submitted letters encouraging cooperation and exploration of linkages with the climate change regime. Copies of these letters were distributed at the 9th Conference of the Parties and are on file with the author.

[28] Ad Hoc Expert Report, *supra* note10.

[29] The Ad Hoc Expert Report does provide an overview of the impacts of climate change on biodiversity in terrestrial and marine ecosystems, see *supra* note 10, Annex III, at 32 to 38.

4. BIODIVERSITY IMPLICATIONS OF CLIMATE CHANGE MITIGATION[30]

The Expert Group set up under the CBD was given a specific mandate to consider the impact of climate change mitigation and adaptation on biodiversity, the capacity of biodiversity to contribute to mitigation and adaptation, and to consider what measures could help to achieve the objectives of both regimes.[31] A number of other organizations were invited to contribute to the work of the Expert Group. Participants included representatives of the IPCC, the UNFCCC, the Convention on Migratory Species, the Convention on Wetlands of International Importance (RAMSAR), and the United Nations Convention to Combat Desertification.[32]

How climate change is addressed may, in fact, be more important than when and how quickly it is addressed. We can choose to continue the trend of addressing each environmental challenge in isolation, ignoring both opportunities to address other issues at the same time and the real risk of creating new problems in the course of solving climate change. In this context, let us consider the following mitigation options for climate change:

- Energy conservation and improved efficiency;[33]
- Solar energy;[34]
- Wind energy;[35]
- Energy from biomass, including wood and ethanol;[36]

---

[30] While the focus has been on mitigation, there are similar linkages on the adaptation side, how we adapt can similarly affect other environmental issues, and the ability or limits on the ability of ecosystems and species to adapt affects the urgency of mitigation. For a recent assessment of linkages between the two regimes, see: F. Jacquemont & Alejandro Caparrós, "The Convention on Biological Diversity and the Climate Change Convention 10 Years After Rio: Towards a Synergy of the Two Regimes?" (2002) 11 R.E.C.I.E.L. 169.

[31] Ad Hoc Expert Report, *supra* note 10, Annex III, at 12.

[32] Ad Hoc Expert Report, *supra* note 10, Annex III, at 12.

[33] This is also referred to as demand side management. It includes any measures that reduce the consumption of energy. For a discussion of the feasibility of a mitigation strategy focussed on demand side management, see R. Torrie, *et al*, "Kyoto and Beyond: The Low-emission Path to Innovation and Efficiency" (October 2002, David Suzuki Foundation and Climate Action Network Canada), online: David Suzuki Foundation <www.davidsuzuki.org>. The same principle can be applied for non-energy sources of greenhouse gas emissions by focussing on reducing or eliminating human activities that lead to emissions. A good example would be the diversion of organic waste from landfills as a way to reduce or eliminate methane emissions resulting from the anaerobic decomposition of organic material in landfills.

[34] Solar energy options include photovoltaic solar energy, or electricity from solar, thermal solar energy (or the collection of heat from solar energy), and passive solar (or the direct use of the sun's energy for heating).

[35] Ad Hoc Expert Report, *supra* note 10, Annex III, at 68.

[36] Ad Hoc Expert Report, *supra* note 10, Annex III, at 65-66.

- Hydropower;[37]
- Nuclear power;[38]
- Use of forests as sinks;[39]
- Use of soils, such as cropland and grazing land, as sinks;[40]
- Deep sea ocean storage of carbon;[41] and
- Carbon injection in oil wells and other geological formations for long-term storage.[42]

Of the measures listed, conservation and efficiency clearly is an effective way of addressing climate change without any risk of contributing to the loss of biodiversity. This is the case because the focus is on directly reducing the human activity that leads to the emissions rather than finding a replacement process. This means at their worst, these measures are biodiversity neutral. To the extent that the energy sources eliminated or reduced as a result of conservation and efficiency have biodiversity impacts in addition to their climate change impacts, these measures can make a significant contribution to addressing biodiversity as well. This is particularly true if decisions about where to reduce energy supply to reflect reduced demand are made based on what sources of energy are most problematic from a combined climate change and biodiversity perspective.

The fundamental benefit of the use of energy efficiency and conservation as a climate change mitigation strategy is that it reduces the level of human interference with natural systems overall. As a result, it has the potential to contribute to the mitigation of numerous environmental challenges, including climate change, resource depletion, air and water pollution, and loss of biodiversity.[43]

The reduction of GHG emissions with the use of other sources of energy instead of reducing the demand for energy can lead to similar results. The main difference here is that we have to consider the impact of producing the energy from these alternate sources. In other words, the overall benefit can only compete with energy conservation if the alternative source of energy has no negative impact on bio-

[37] Ad Hoc Expert Report, *supra* note 10, Annex III, at 66.

[38] Eligibility of nuclear power is limited for the Clean Development Mechanism, but not for domestic emission reductions. In other words Canada is free to use nuclear power in Canada to meet its Kyoto obligations, but has agreed not to use emission reductions achieved by supporting nuclear power in developing countries.

[39] Ad Hoc Expert Report, *supra* note 10, Annex III, at 50-60.

[40] Ad Hoc Expert Report, *supra* note 10, Annex III, at 61-63.

[41] This offset option is generally recognized not to be eligible under Kyoto rules, see Ad Hoc Expert Report, *supra* note 10, Annex III, at 63.

[42] Ad Hoc Expert Report, *supra* note 10, Annex III, at 64.

[43] See Torrie, *supra* note 33.

diversity. Otherwise, the net impact is the difference between the impact of traditional sources and these alternative sources.[44]

In case of solar energy, the two main areas of concern would be the production of the solar panels and the space required to operate the panels. In case of wind power, there are similar considerations. In both cases, the net impact clearly points to these options being a net contribution to solving both the climate change and biodiversity problems, but because of the impact of production and siting, reducing energy consumption where possible is still a preferable option.

The next group of mitigation measures includes the use of biomass, hydropower, and nuclear energy. In each case, the switch from the use of fossil fuels would reduce GHG emissions, but the reduction would come at a cost. In case of the use of biomass, there are two issues to consider. One is the potential for competing land-use between fuel production from biomass and protection of biodiversity. The other is the pollution from the burning of biomass which may have an adverse effect on biodiversity.[45]

In case of hydropower, there are again significant reductions in GHG emissions compared to the use of fossil fuels for power generation. At the same time, hydropower is clearly a competing land use and a resulting threat to biodiversity. In addition, hydropower leads to mercury and other water contamination which also provides a threat to biodiversity.[46] Finally, nuclear power also has much lower GHG emissions than power production using fossil fuels. It carries with it a significant threat to biodiversity from accidents and from the waste generated by nuclear power plants.

The last category is a group of climate change mitigation measures to offset emissions by taking greenhouse gases back out of the atmosphere. They include the use of forests, soils[47] and oceans[48] to take CO_2 out of the atmosphere and store it in the form of carbon. They also include measures to capture the CO_2 during combustion of fossil fuels and storing it in oil wells and other geological formations. This group of mitigation measures has the potential to contribute to solving

---

[44] For a discussion of the biodiversity impacts of wind and solar energy, see IPCC Report on Biodiversity, *supra* note 8, at 40.

[45] IPCC Report on Biodiversity, *supra* note 8, at 38.

[46] For an overview of impacts from hydro-power, see IPCC Report on Biodiversity, *supra* note 8, at 39.

[47] For an overview of the environmental costs and ancillary benefits of storing carbon in forests and soils as a way to offset greenhouse gas emissions, see Metz, Bert, *et al*, eds., *Climate Change 2001: Mitigation: Contribution of Working Group III to the Third Assessment Report of the Intergovernmental Panel on Climate Change (IPCC)* (Cambridge, England: Cambridge University Press, 2002) at 326. See also IPCC Report on Biodiversity, *supra* note 8, at 35-37.

[48] IPCC Report on Biodiversity, *supra* note 8, at 41.

the climate change problem, but is subject to the issue of permanence. To the extent that the storage is temporary, these measures delay the problem rather than solve it.

With respect to biodiversity, the answer is mixed. On one hand, mitigating climate change through storage of carbon in living things provides an incentive to protect life. Done properly, this can clearly contribute to biodiversity. On the other hand, if the storage of carbon is pursued in isolation from the need to protect biodiversity, climate change mitigation through carbon storage in forests, soils and oceans can be a significant threat to biodiversity. Much again turns on whether countries choose to consider climate change in the form of carbon storage and biodiversity as two separate issues, or whether they become two motivations toward the objective of ensure the health of forests, soils and oceans. In other words, do we look at each service nature provides separately and try to manage that service, or do we protect natural systems overall in the general recognition that overall ecosystem health provides the best hope for nature to continue to perform these services, including biodiversity protection and climate change mitigation?

Let's consider a specific example. There is a general incentive in the Kyoto sinks rules for countries not to deforest currently forested land. This is accomplished by creating a debit under Article 3.3 of the Kyoto Protocol for the loss of capacity of deforested land to take carbon out of the atmosphere. In other words, if a Party to the Kyoto Protocol changes forested land to some other use, it has to make up for the carbon sequestration potential of that forest during the first commitment period. For every ton of carbon that forest would have extracted from the atmosphere between 2008 and 2012, an additional ton of carbon credit has to be generated by that Party.[49]

Whether this incentive will serve to protect biodiversity depends on what actions are taken in response. If this means countries will leave natural forests undisturbed, or manage such forests with a view to protecting the forest for biodiversity and other sustainable purposes, this incentive against deforestation can clearly have significant biodiversity benefits. Conversely, deforestation under Article 3.3 does not include the harvesting cycle of a given forest.[50] This means the incentive is limited to permanent land use changes.[51] The incentive against deforestation does not extend to the harvesting of an old growth forest, as long

[49] See Chapter 2, specifically Section 3 regarding sinks.

[50] See Sierra Club of Canada, *Forests, Climate Change and Carbon Reservoirs, Opportunities for Forest Conservation* (Ottawa: Sierra Club of Canada, September, 2003), at 9-10, on the meaning of deforestation under the Kyoto Protocol. As explained there, deforestation does not include logging, as long as the land is not converted to another use. Logging therefore does not create a debit for emission accounting purposes under the Kyoto rules.

[51] Such as deforestation for suburban sprawl, agricultural purposes, industrial development, etc.

as that forest is allowed to naturally re-grow or is replanted. The use of monoculture tree plantations in place of old growth forests is perhaps the most striking example. In that case, a forest that supports biodiversity is replaced with one that has few sustainability benefits other than to temporarily store carbon by taking it out of the atmosphere and as a source of timber.

Without an integrated approach to its implementation, Kyoto may inadvertently provide an incentive to converting old growth forests with monoculture tree plantations, because the rate of carbon uptake is significantly higher, especially early in the growing cycle.[52] Furthermore, because mitigation measures that involve taking carbon back out of the atmosphere do not prevent the emissions, they also do not reduce the threat to biodiversity resulting from the production or use of energy in the first place, be it through air pollution from the burning of fossil fuel or competing land use and water pollution from hydro-power.

To the extent that the mitigation measures designed take carbon dioxide back out of the atmosphere also involve other alterations of natural ecosystems, they do pose an additional threat to biodiversity. For example, deep sea storage of carbon can be enhanced by the use of enzymes. The resulting impact has the potential to pose a risk to biodiversity in our oceans. In case of forests, to the extent that the increased storage of carbon in forests is achieved by altering natural undisturbed forest ecosystems, this measure can also pose a threat to biodiversity.[53] Increasing carbon content in agricultural soils can be a threat to biodiversity in two ways, through the increased use of pesticides as a result of a move to no-till agriculture, and as a result of increased demand for agricultural land if production from existing land decreases as a result of the effort to increase the carbon content of these soils.[54]

5. LINKAGES TO OTHER GLOBAL ENVIRONMENTAL CHALLENGES

To complete the analysis started here, linkages with other environmental challenges, some of which are already subject to MEAs, will similarly have to be considered. They include desertification, wetlands, air pollution, and resource depletion. For each of these, there is the potential that consideration of linkages

---

[52] For a more detailed discussion of the short-term versus long-term implications of relying on forest plantations, see Opportunities for Forest Conservation, *supra* note 50, at 23. See also Friends of the Earth International, "Tree Trouble: A compilation of testimonies on the negative impact of large scale, monoculture tree plantations", released at the sixth Conference of the Parties of the UNFCCC in The Hague, October 2000, on file with author.

[53] IPCC Report on Biodiversity, *supra* note 8, at 37.

[54] For a more detailed analysis of climate change mitigation tools that are biodiversity friendly, see Keya Choudhury, *et al*, *Integration of Biodiversity Concerns in Climate Change Mitigation Activities* (Berlin: Federal Environmental Agency, 2004). The primary objective of this tool kit is to assist with integration at the national and sub-national level.

will lead to selection of more appropriate short and long-term climate change mitigation strategies and the potential for climate change mitigation to make a more significant contribution to sustainability. A brief introduction to these most obvious environmental challenges with connections to climate change is provided here.

(a) Desertification[55]

The existence of links between climate change, biodiversity and desertification is clear.[56] Climate change will bring with it increases in temperature and reduction in precipitation levels in many dry areas of the world, increasing the risk of desertification. Beyond this, any mitigation measure that introduces a competing land use also brings with it the risk of indirectly contributing to desertification by encouraging the conversion of natural ecosystems for human use, one of the main causes of desertification.

The use of biomass for energy and the use of forests and soils[57] as sinks are therefore all climate change mitigation measures that have the potential to contribute to desertification. This means an integrated approach would suggest either minimizing their use or developing processes and criteria to guard against this risk. Energy conservation and efficiency, by reducing energy production and consumption, alternately, clearly have the potential to reduce the threat of desertification by reducing the demand on land use for energy purposes. Other mitigation measures have the potential to contribute positively or negatively to desertification, depending on where and how they are implemented.[58]

---

[55] Desertification is the subject of other MEA that arose out of the Rio process, the *United Nations Convention to Combat Desertification* (17 June 1994), 33 I.L.M. 1328. The links between climate change and desertification are recognized in the Preamble of the UNFCCC, see *supra* note 2.

[56] See, for example C. Basset, *et al*, "Implementing the UNCCD: Towards a Recipe for Success" (2003) 12 R.E.C.I.E.L. 133, J. Zeidler, *et al*, "The Dry and Sub-humid Lands Programme of Work of the Convention on Biological Diversity: Connecting the CBD and the UN Convention to Combat Desertification" (2003) 12 R.E.C.I.E.L. 164, B. Kjellen, "The Saga of the Convention to Combat Desertification: The Rio/Johannesburg Process and the Global Responsibility for Drylands" (2003) 12 R.E.C.I.E.L. 127, and B. Boer, *et al*, "Legal Aspects of sustainable Soils: International and National" (2003) 12 R.E.C.I.E.L. 149.

[57] Soils store a significant amount of carbon. For land that is already managed for agricultural or forest purposes, management practices can have a significant impact on the amount of carbon stored in the soils. An example would be low or no tillage agriculture. Soil management practices can also have a significant impact on desertification.

[58] There have been efforts under the UNCCD to explore relationships with other MEAs. See, for example, UNCCD Committee for the Review of the Implementation of the Convention, *Review of Activities for the Promotion and Strengthening of Relationships with Other Relevant Conventions and Relevant International Organizations*, (15 October 2002), ICCD/CRIC(1)/9, and the follow-up report to the 6th Conference of the Parties, *Review of Activities for the Promotion and Strengthening of Relationships with Other Relevant Conventions and Relevant International*

(b) Wetlands (RAMSAR)[59]

Similar to desertification, wetlands are both likely to be affected by climate change and certain climate change mitigation measures may pose further threats to wetlands. In terms of direct threats to wetlands, changes in temperature and precipitation are the most obvious climate change threats to wetlands. Competing land use pressures are again the main concern on the mitigation side.

It would therefore seem, again, that energy conservation and efficiency has the greatest potential for addressing climate change while contributing to the protection of wetlands, whereas fuel switching options can be a positive influence depending on where and how they are implemented. Measures designed to offset emissions are likely to pose a net threat to wetland protection, given their tendency to place demands on land use and their failure to prevent collateral impacts of the activities that result in GHG emissions, such as air pollution from the burning of fossil fuels. The overall conclusion would again be that for most mitigation measures, other than energy conservation and efficiency, an integrated approach would require caution in the implementation to ensure co-benefits and avoid creating problems for wetlands in the process of mitigating climate change.

(c) Air Pollution[60]

Air pollution is central to the use of fossil fuels as a source of energy. Any mitigation measure that reduces the reliance on fossil fuels for energy therefore can contribute to addressing air pollution, as long as the measure does not carry with it its own air pollution problem. In this context, reliance on energy conservation and efficiency, solar, wind, and hydro clearly all have the potential to contribute to addressing climate change and air pollution at the same time. Offset measures, on the other hand, because they do not eliminate the burning of fossil fuels, tend to be air pollution neutral at best, with the possible exception of forest sinks that may function as air filters in addition to temporarily taking greenhouse gases out of the atmosphere. Nuclear power carries with it the risk of radiation pollution and the burning of biomass is generally associated with air pollution. The type of biomass and the burning process used may determine whether it is a net benefit compared to the burning of fossil fuels.

Organizations, Institutions and Agencies, in accordance with Article 8 and Article 22, paragraph 2(i) of the Convention (27 June 2003) ICCD/COP(6)/4.

[59] The relevant MEA is the *Convention on Wetlands of International Importance Especially as Waterfowl Habitat* (RAMSAR), 2 February 1971, 996 U.N.T.S. 245, 11 I.L.M. 963.

[60] For an example of an MEA dealing with air pollution, see *Stockholm Convention on Persistent Organic Pollutants* (22 May 2001) 40 ILM 532.

(d) Depletion of Non-Renewable Resources

Given that fossil fuels are a non-renewable resource, any solution that reduces the use of fossil fuels without depleting other non-renewable resources will assist in reducing the rate of depletion of non-renewable resources. Clearly in this category are energy conservation and efficiency, the use of renewable resources such as solar energy, wind, geothermal energy and biomass for energy production. Hydropower and nuclear energy also fall into this category. Offset mitigation measures, such as sinks, clean coal technology, and carbon storage technologies do not address the depletion problem.

6. HOW TO EXPLOIT THE LINKAGES[61]

Applying the above analysis would lead to a conclusion that as much of the mitigation effort as possible should focus on reducing energy consumption through conservation and efficiency. The next obvious choice appears to be switching fuel to solar, wind, and geothermal energy. Beyond these measures we must, at best, acknowledge that other mitigation measures carry with them a significant risk of creating other environmental problems. At worst, we may find that they make a negative net contribution to efforts to become sustainable, in that they may create more problems than they solve.

This would lead to a conclusion that climate change mitigation efforts should focus on reducing energy demand and the promotion of solar, wind and geothermal energy as alternatives to fossil fuels. A consideration of other mitigation measures would have to precede a further analysis of whether and how these measures fit into an overall sustainability strategy. A detailed consideration of criteria that might be applied to such an analysis is not possible here, but considerable thought has been given to the issue of sustainability criteria elsewhere.[62]

There are two basic options for the exploitation of these overlaps and implementing solutions that maximize co-benefits. This can take place at the international level, or it can be left to individual countries to determine how to integrate their various international obligations and commitments into an overall strategy for sustainability. We can see from the work of the Ad Hoc Expert Group under

---

[61] See Ad Hoc Expert Report, *supra* note 10, Annex III, at 86.

[62] See C. George, "Testing for Sustainable Development through Environmental Assessment" (1999) 19:2 *Environmental Impact Assessment Review* 175. The article lists 18 criteria to test the sustainability of proposed development, including equity in various contexts, social impacts, public participation, precaution, biodiversity, climate change, and overall local and global impacts.

the CBD that there are some informal efforts to identify overlaps.[63] What is equally clear from the formal negotiations, particularly under the UNFCCC and the Kyoto Protocol, is that there is resistance to making formal links within individual international regimes. In particular, in the context of the climate change negotiations, Parties considered and rejected opportunities to restrict the use of specific mitigation measures so as to prevent negative impacts on biodiversity, air pollution, and wetlands.[64]

The main obstacle to such an international approach to these linkages is that not all countries are Party to the various MEAs, and it may only take one country who is not a Party to prevent agreement on a formal link between two regimes. For example, when the issue of linkages to the CBD arose in the context of the climate change negotiations leading up to the Marrakech Accords under the Kyoto Protocol, a number of countries who were not Party to the CBD objected to any proposal to provide a formal link between the two agreements and obligate Parties to the Kyoto Protocol to implement climate change mitigation measures[65] in a manner consistent with the CBD.

Furthermore, it is difficult to develop global priorities in light of varying national circumstances. So while there are clearly opportunities for better coordination between international environmental regimes, the basic approach internationally has been to generate commitment on an issue by issue basis, and to leave it up to individual countries to find ways to implement issue-specific commitment in an integrated manner consistent with its vision for a sustainable future. In the absence of global governance on the environment, this is unlikely to change significantly.[66] It remains to be seen whether there will be much progress in establishing international rules to mandate countries to consider linkages.[67]

The alternative, of individual countries taking responsibility for integration, is considered here. This means State Parties would be left to find ways to explore

---

[63] What is interesting is that while the Ad Hoc Expert Group talks about the biodiversity risks of the various options, and provides considerable guidance on how to reduce that risk where possible (particularly with respect to the use of sinks), it does not rank the various measures in terms of their overall performance toward the combined objectives of the conventions being considered. That is left to individual countries.

[64] The Preamble of the UNFCCC, and Article 2 of the Kyoto Protocol also clearly puts the obligation to ensure climate change mitigation and adaptation measures contribute more broadly to sustainability rests with individual Member States, not the international community. See Kyoto Protocol, and UNFCCC, *supra* note 2.

[65] Specifically with respect to the use of sinks and the Clean Development Mechanism.

[66] See also discussion of integration in the Epilogue.

[67] Note, however, that there is considerable potential for integration at the international level at two other stages, one is at the stage of defining the problem and the range of responses to be negotiated. The other stage is the implementation stage, including the process of interpreting obligations under individual regimes. These issues are considered further in the Epilogue.

these linkages, and to develop implementation strategies that take advantage of co-benefits and avoid measures that address one, while contributing to other environmental challenges. Seen in this light, each international instrument that motivates, commits or obligates countries to take action on a particular environmental issue is a potential tool to be used by countries in their efforts to sustainability. Whether these individual instruments contribute to an integrated approach that explores co-benefits would be up to each State.

Using Canada as an example, it is therefore up to various levels of government responsible to choose to implement Kyoto in isolation and risk spending effort on measures that may cause more problems than they solve, or to see Kyoto as just one more motivation to make the shift to sustainability by taking a broad look at the relationship between human activity and nature, leading to an overall consideration of how the needs and aspirations of Canadians can be met in a more sustainable manner. The extent to which Canada has embarked on this in the implementation of our Kyoto commitments is considered in the following Section.

7. SIGNS OF INTEGRATION IN CANADA?

Having considered what measures would appear to be favoured under a more integrated approach to the major environmental challenges facing us, let us turn to the Climate Change Plan for Canada, released by the federal government in the lead up to Canadian ratification of the Kyoto Protocol in December, 2002.[68] Three separate questions are now explored to determine the extent to which the federal plan shows signs of integration:

- Does the Climate Change Plan for Canada focus on mitigations measures identified in this paper as most likely to lead to co-benefits?
- Does the Climate Change Plan provide for a process, substantive criteria, or both to ensure that climate change mitigation measures allowed under the plan are implemented in a manner that maximizes opportunities for co-benefits and minimizes the risk of working at cross purposes with efforts to address other pressing environmental challenges?
- Does the Climate Change Plan give any other indication that linkages to other environmental issues were considered in designing Canada's climate change plan?

---

[68] See Government of Canada, *Climate Change Plan for Canada*, (Ottawa: Government of Canada, 2002) The plan was further supplemented with Government of Canada *Project Green, Moving Forward on Climate Change, A Plan for Honouring our Kyoto Commitment*, (Ottawa, Government of Canada, 2005), both available at <http://www.climatechange.gc.ca>. However, the updated plan does not propose any significant changes in direction, it mainly provides additional funding to achieve the objectives of the 2002 plan. The analysis here is therefore based on the 2002 plan.

(a) Does The Climate Change Plan Focus on Conservation, Efficiency and Renewable Energy?

The gap between business-as-usual and the Kyoto target is estimated to be about 240 megatonnes (MT) of GHG emissions.[69] This means Canada has to take measures to reduce GHG emissions by 240 MT by the year 2012. The federal plan released in the fall of 2002 includes measures for 175 MT, with further measures to achieve the remaining 60 MT to be announced in the future. Of the 175 MT of measures announced, about 19 MT assigned to reductions from the transportation sector appear to have a good chance of coming from energy conservation, efficiency and fuel switching to renewable sources.[70] Eight MT of reductions in the building sector are linked to energy conservation and efficiency.[71] In the category of renewable energy and cleaner fossil fuels, about 7 MT appear to be allocated to fuel switching to renewable sources.[72] Finally, about 2 MT are allocated to the capture of methane gas from landfills, for a total of 36 MT.

On the other side, there are 89 MT allocated to industrial emitters that are not specifically linked to conservation, efficiency or renewables, but that have the potential to come in part from these areas. There are a minimum of 12 MT allocated to international credits. Finally, 38 MT are allocated to the removal of GHGs through the use of sinks, for a total of 139 MT. All these measures either directly promote mitigation that poses a risk to biodiversity, or the measure as proposed in the plan would not appear to preclude the risk to biodiversity.[73]

Assuming these mitigation measures all fall into the category of providing a clear net benefit as discussed above, there are 36 MT of mitigation measures out of 175 MT in this category. The remaining 139 MT are either measures where there appears to be little or no control over the specific mitigation measure chosen to achieve the reduction,[74] or the measures chosen were identified above as a potential threat to biodiversity, wetlands, etc.

---

[69] *Ibid.* at 12. GHG emissions are represented in CO_2 equivalent.

[70] Climate Change Plan for Canada, *supra* note 68, at 20. A further 2MT are allocated to biodiesel and ethanol. Given the risk of competing land use, these are not included in measures with clear co-benefits. Whether and how these measures contribute will depend on the existence and effectiveness of criteria or processes to ensure co-benefits are realized, and harm is avoided.

[71] Climate Change Plan for Canada, *supra* note 68, at 25.

[72] *Ibid.* at 33.

[73] *Ibid.* at 13.

[74] Such as emissions trading, purchase of international credits, and domestic offsets.

(b) Does the Climate Change Plan Propose a Process or Criteria to Ensure Co-Benefits?

The next question explored is whether there are criteria or processes proposed in the Climate Change Plan for Canada to safeguard against side effects of these measures. Three areas are particularly relevant here, access to international credits, large industrial emitters, and the use of sinks for compliance. As a starting point, there is no indication that there are restrictions on the access to international credit. At the same time, some of the credits will be purchased by the federal government itself,[75] which preserves the opportunity to limit the purchase to sources of credits that ensure co-benefits.

With respect to large industrial emitters, the main tool identified in the Climate Change Plan for Canada is a covenant to be entered into with each of the sectors involved in combination with a financial or regulatory enforcement mechanism to ensure compliance with the agreed upon commitments. The sectors in question include power generation, the oil and gas sector, mining, pulp and paper production, the chemical industry, iron and steel production, smelting and refining, cement and lime production, and glass and glass container production.[76] To answer the question posed here; whether the reductions will be achieved in a manner that maximizes positive and minimizes negative impacts on other environmental challenges, we have to look at the terms of the covenants being negotiated with these industry sectors.

The first memorandum of understanding (MOU) was negotiated between the federal government and the Pulp and Paper Industry in November 2003.[77] The MOU is consistent with general commitments made by the federal government to industry during the period leading up to the ratification of the Kyoto Protocol. It should therefore serve as an accurate signal of the rules under which large industrial emitters will be permitted to achieve their emission reduction targets. The highlights of the MOU are as follows:

- The emission reduction target for the Pulp and Paper industry will be based on a 15% reduction in emission intensity compared to business-as-usual rather than an absolute target. In other words, any target will be linked to economic output of the industry, and there will be no incentive to produce less.
- There is a commitment to ensure that companies who took early action will not be penalized, in other words business-as-usual will not include voluntary action to reduce emissions before the signing of the agreement. This is also referred to as baseline protection.

---

[75] Climate Change Plan for Canada, *supra* note 68, at 13.

[76] *Ibid*. at 30.

[77] Online at: <http://www.climatechange.gc.ca>.

- The agreement identifies biomass and combined heat and power technology as two areas with particular opportunities for emission reductions.
- The agreement guarantees industry access to domestic offsets such as sinks credits generated in Canada. This means the purchase of sinks credits can replace emission reductions in the Pulp and Paper sector.
- The agreement provides for access to international credits. This means the purchase of international credits can replace emission reductions in this sector.
- The agreement provides a price guarantee for international credits of $15 per ton carbon dioxide equivalent, which means any emission reduction measures beyond this price may not be considered regardless of other benefits.
- The agreement recognizes the importance of healthy forests for the Pulp and Paper industry and in a general way commits the Parties to protecting Canadian forests from the impacts of climate change.

The overall conclusion here is that there is little in terms of substantive criteria to ensure that mitigation measures are selected and implemented to maximize overall benefits. Furthermore there is no indication of any further process to select the most appropriate measures from the menu of mitigation measures permitted under this agreement. This means that there is little control over the choices, which are essentially left to each industry sector.

This leaves the issue of sinks.[78] There is some reference to co-benefits in the Climate Change Plan for Canada,[79] such as the reduction of soil erosion by planting trees around farms. There is little, however, to suggest either substantive criteria or a process that would ensure that sinks projects are only approved if they make an overall net contribution to climate change and biodiversity, let alone the range of environmental challenges we are facing. On the contrary, the suggestion is that "forest plantations" may be a big part of the solution, suggesting that the carbon sequestration potential may override issues of competing land use and biodiversity.[80]

(c) Does the Climate Change Plan Show Other Signs of Integration?

Finally, are there other signs that the plan is a product of an integrated approach to climate change mitigation? Perhaps a good starting point would be the principles proposed by the provinces and territories, which were accepted by the federal government as the basis for the plan.[81] The principles proposed include collabo-

---

[78] For a general discussion of sinks, see Chris Rolfe, *Sink Solution* (2001) online: West Coast Environmental Law Association <http://www.wcel.org/wcelpub/2001/13458.pdf>.

[79] Climate Change Plan for Canada, *supra* note 68, at 39-41

[80] *Ibid.* at 40

[81] *Ibid.* at 1.

ration among affected jurisdictions, regional fairness, minimizing mitigation costs while maximizing benefits, promoting innovation, and limiting uncertainties and risks.

While at least a couple of these principles are sufficiently broad to allow for a consideration of linkages and co-benefits with respect to biodiversity and other environmental challenges, there is little indication throughout the document to support such an interpretation of these principles. In fact, the results of the economic analysis would suggest that much of the decision making was based on traditional economic analysis,[82] which is generally recognized not to internalize costs such as the loss of biodiversity, wetlands, desertification, depletion of resources and air pollution.[83] In other words, it would appear that cost, in the context of this plan, means short term financial cost, and benefit means short term reduction in GHG emissions and financial benefit.

On balance, it would appear that there was little consideration given in the development of the Climate Change Plan for Canada to maximizing co-benefits with other environmental challenges; or at least to minimize the potential for conflict. This leaves short term economic costs and benefits as the most likely driving force on the selection of mitigation measures, rather than a selection based on maximizing co-benefits or an integrated approach to the implementation of environmental commitments made under various MEAs. This would suggest that the incentives created through international commitments and obligations are currently not being used in Canada in any coordinated or integrated manner to assist in our efforts to become a sustainable society. Instead, the individual silos created within Canada on these issues are left to battle for attention and priority,[84] with little hope of integration.[85]

---

[82] Climate Change Plan for Canada, 57-67.

[83] For a more detailed discussion of the limitations of the market to promote good public policy on social and environmental issues, see. Lily N. Chinn, "Can the Market be Fair and Efficient? An Environmental Justice Critique of Emissions Trading" (1999) 26 Ecology L.Q. 80.

[84] On the issue of biodiversity, for example, it will be left to those in charge of the implementation of species at risk legislation to fight for selection of climate change mitigation measures that are consistent with the protection of biodiversity. It seems unlikely that the new *Species At Risk Act*, S.C. 2002 c. 29 is equipped for this battle. For a critical review of this Act, see Sierra Legal Defence Fund, *A Guide to Canada's Species at Risk Act* (Vancouver: Sierra Legal Defence Fund, May 2003).

[85] The most obvious domestic tool for integration would be the environmental assessment processes. To date, the various processes implemented in Canada at federal and provincial levels have proven unsuccessful in contributing to this in any meaningful way. For a general discussion of environmental assessment processes in Canada, see Boyd, *supra* note 5, at 148.

8. CONCLUSION

To conclude that taking biodiversity seriously makes a difference to how States might mitigate climate change is perhaps stating the obvious. The threat to biodiversity is an important reason to take climate change seriously, and it affects the choice of mitigation measures. The biodiversity issue also reminds us that climate change mitigation is not an all or nothing proposition, but that it is at least in part about buying time for all living organisms, including humans, to adjust. It is about slowing down the rate of change, at least for now.

The real question is when and how the integration may take place. As was demonstrated in the context of climate change and biodiversity, it is possible to go through the integration process even before international obligations or commitments are negotiated. Going through such a process before the negotiation of the Kyoto Protocol might have resulted in a Protocol that instead focusses on energy conservation, efficiency and the promotion of renewable energy. The secretariats of the various MEA regimes might be in the best position to ensure that as a particular regime moves toward establishing its substantive direction, the options pursued are already integrated with other MEAs.

Of course to suggest that integration may take place at the international level raises as many questions as it answers. Conflicting memberships, and sovereignty issues will certainly be significant obstacles toward integration at that level. Can and should integration take place before international obligations are negotiated? Does integration have a useful role to play during the implementation phase? What are the mechanisms through which integration may happen internationally, without further slowing the process of international negotiations? These issues will be considered further in the Epilogue.

Integration at a national or sub-national level is also possible. It has the advantage of allowing States to pursue integration in a manner most suitable to local circumstances. The challenge is to ensure that individual States are actually sufficiently motivated to fully integrate their responses to various international obligations to ensure an overall move toward sustainability. As the example of Canada's Climate Change Plan demonstrates, this is likely to be a considerable challenge.

MEAs such as the CBD have the potential to influence individual State behaviour with respect to climate change, in terms of acceptance of Kyoto commitments, in terms of the mitigation path chosen to meet those commitments, and in the negotiating position of individual States for future commitments to mitigate climate change. An important precondition for this influence is a greater level of appreciation of the linkages and overlaps, and an effort in individual countries such as Canada to explore and take advantage of co-benefits and consider integration at three stages in the ongoing evolution of the climate change regime; at the negotiation stage, at the ratification stage, and during implementation.

Kyoto has introduced an additional motivation to address the root causes of many environmental problems. In order to become a meaningful driver for change, two things need to happen with the climate change regime. First, countries need to negotiate more meaningful targets. Second they need to "exploit the overlaps" to direct or channel the motivation created through Kyoto in the direction of addressing root causes rather than treatment of the symptoms. The lesson from Kyoto is that we have to recognize the limitation of issue-specific international regimes. They provide motivation at the national level to act, but they do not on their own put the individual motivations together into an overall strategy for sustainability. The integration of environmental issues such as biodiversity, if taken seriously, will provide the motivation to do both.

For now, it is clear that more integration is needed at the national and international level through the implementation phase of Kyoto. Internationally, respective secretariats are starting to make progress on this issue.[86] Regardless of whether significant progress on this issue of integration can be or should be made internationally, it is time to act domestically in a coordinated and integrated manner. Doing this will make Kyoto more meaningful. It will ensure that Parties spend resources to come into compliance with Kyoto in a manner that puts them on a path to sustainability. Kyoto does not have to be inadequate; it only becomes meaningless if States choose to waste the motivation and incentive it provides, and fail to use it to get onto a path to sustainability. It only becomes meaningless if States choose to go the road of least short-term cost without regard to long-term objectives.

To conclude, there are significant opportunities for MEAs generally, and the CBD specifically, to influence compliance with the Kyoto obligations and the future evolution of the climate change regime. Much depends on the normative influence other MEAs have on Parties to the climate change regime, especially in light of the fact that most of the commitments contained in these MEAs are of a moral and political, not of a legal, nature. To be clear, their influence is normative. Most MEAs do not contain legal obligations sufficient to serve as an enforcement tool to motivate action on climate change. MEAs still have significant potential to provide additional motivation to comply with Kyoto and to push for an effective climate change regime in the years to come. MEAs can also influence the choice of mitigation measures to meet Kyoto obligations.

---

[86] See, for example, *Options for Enhanced Cooperation Among the Three Rio Conventions*, FCCC/SBSTA?2004/INF.19 (2 November 2004).

PART III

THE FUTURE

Parts I and II assessed the climate change regime, its compliance system and some key external influences on the regime and on compliance with obligations under the Kyoto Protocol. In the final Part, the implications of this assessment for compliance with Kyoto, for the future of the climate change regime, and for international environmental law generally are considered. Chapter 9 focuses on the climate change regime itself.

9

The Future of the Climate Change Regime

1. INTRODUCTION

Up to this point, the focus has been on the current state of the climate change regime and its interaction with other key international law regimes and norms. In this Chapter, the focus shifts to the end of the first commitment period in 2012. The overall question posed here is whether the understanding gained on the climate change regime and some of its key external influences provides a basis for predicting its future evolution. To do this, the key conclusions on the state of the regime and its interaction with the WTO, UNCLOS, International Human Rights Norms, and MEAs are first summarized. This is followed by an assessment of various proposals for its future evolution currently being considered by Parties, other stakeholders, and academics. Based on this, some thoughts are offered on how the climate change regime may evolve in the years to come.

2. THE ROLE OF THE "INTERNAL COMPLIANCE SYSTEM"

The conclusion reached at the end of the assessment of the internal compliance system in Chapter 4 was that the Kyoto compliance system was strong relative to those in other MEAs. It will likely stand up to anticipated challenges, such as the failure of several countries to meet their targets through measures implemented before and during the first commitment period. Measures implemented to meet these obligations will, for most Parties, include a combination of domestic action and the purchase of international credits.

What is less clear is what happens if a sequence of unexpected events takes place by the end of the first commitment period in 2012. What, for example, if new scientific discoveries reverse the trend of increased confidence that humans are responsible for more and more rapid changes in the climate system? What if key States initiate the process of revoking their ratification? What if there are no second commitment period targets negotiated by 2012? Under some of these circumstances, one may conclude that compliance with the first commitment period is meaningless, given that the climate change regime has no future as is, and the first commitment period alone is not an effective response.

There are two lines to consider in the context of future changes that may affect compliance with the climate change regime. One is the line beyond which the compliance system will not be able to motivate compliance. The other is the line beyond which compliance is no longer relevant for the future of the climate change regime. A key question is to what extent these two lines coincide. In other words, will the compliance system start to fail before or after circumstances have made compliance meaningless for the effectiveness of the regime? Given that both lines are somewhat subjective, and subject to speculation on unpredictable future events, there is likely to be no agreement on the answer to this question.

There is certainly a risk that absence of a financial incentive to comply, the lack of guaranteed transparency, and the ability to separate binding consequences from substantive obligations in the Kyoto Protocol may combine to expose the limitations of the internal compliance system. As discussed in Chapter 4, compliance with the first commitment period obligations under Kyoto, while not enough for an effective climate change regime, may turn out to be a crucial first step. In this sense, the ability of the Kyoto Compliance System to stand up to the challenges ahead, expected or unexpected, will be crucial.

3. THE ROLE OF "EXTERNAL INFLUENCES"

This Section provides an opportunity to reflect on the potential influence of the external international law factors considered in Chapters 5 to 8 on compliance with Kyoto and on the future of the climate change regime. The WTO and UNCLOS Chapters illustrate the opportunity to move to a legislative approach to international law, to encourage international tribunals to consider their mandate in a larger context of international treaties, customary law, soft law principles and other indications of evolving international norms. In short, illustrating that non-compliance with obligations under the climate change regime may also result in violations of other international rules such as those under the WTO or UNCLOS may provide additional motivation for States to meet their Kyoto obligations and to cooperate on more effective climate change mitigation measures in the future.

The WTO Chapter also illustrates the opportunity for States who have decided to take a leadership role on climate change to utilize existing international rules to protect themselves from some of the economic implications of leading on this issue, and to shift at least some of that burden onto States who have not taken the issue as seriously. There are examples of this happening in the past. Ron Mitchell's article on vessel source pollution and the US role in forcing changes to ship designs provides a good illustration.[1]

[1] See Ronald B. Mitchell, "Regime Design Matters: International Oil Pollution and Treaty Compliance" (1994) 48 International Organization 425.

The Human Rights Chapter is an example of how international law can be used to demonstrate when State action on an issue such as climate change deviates from accepted norms, such as the basic principles of human rights. Such action can help change the context for decision making, and thereby influence States' decisions about compliance efforts and future commitments alike.

The MEA Chapter illustrates the opportunities integration can offer. When objectives and obligations under various MEAs are considered together, rather than in isolation, the path forward becomes much more clear, and many efforts that appear to be a rational response to climate change, turn out to be of questionable value when considered together with other global or national objectives. The compliance connection is that an integrated approach will identify collateral benefits and may therefore increase motivation to comply. The MEA Chapter raises a fundamental question with respect to integration. To what extent should the integration take place at the international level, and to what extent should it happen domestically? This question is dealt with in the Epilogue.

4. FUTURE OF THE CLIMATE CHANGE REGIME

This Section considers the future of the climate change regime, reflecting in particular on the current state of the regime described in Chapter 2, the role of compliance with obligations under the Kyoto Protocol considered in Chapters 3 and 4, and the external influences on the climate change regime identified in Chapters 5 to 8. It is important to recall that the consideration of external influences, which has included the WTO, UNCLOS, International Human Rights, and MEAs, is limited to international law influences, and even within this context is far from complete.

The objective is to illustrate key international law influences, rather than to provide a comprehensive assessment of external influences on the climate change regime. The purpose is to demonstrate that not all hurdles on the way to an effective climate change regime will necessarily be resolved through the process of future negotiations under the UNFCCC. Rather, the future evolution of the UNFCCC can be significantly influenced by other international institutions and regimes.

In considering the future of the climate change regime, it is also important to consider that the current level of commitment to this issue is a reflection of the level of uncertainty about the future impact of climate change, and likely also the fact that the major impacts are not expected for some time to come. If this is the case, and the trend in the science continues to be toward greater certainty that human activities are causing climate change, one would expect that by 2012 to 2015, the perception about the importance of climate change, the urgency of addressing climate change, and the consequences of not addressing climate change in an internationally coordinated manner will be much higher than they are today.

This means that the pressure on countries to meet their obligations will likely be much higher in 2012 to 2015 than it is now. As a result, decisions about whether to purchase credits during the true up period to come into compliance with first commitment period obligations, for example, will likely be made in a very different context than the context within which the Kyoto Protocol and the compliance system were negotiated.

As we move from compliance with Kyoto to the future of the climate change regime, any analysis will have to consider the principles on the basis of which future rights and obligations might be distributed. This was an issue that was heavily debated during the negotiations on the Kyoto Protocol, but never resolved, resulting in pledge-based rather than principle-based targets for Kyoto.[2] The principles that have dominated this debate over time are introduced in the next Section.

(a) The Principles of Responsibility, Potential, and Capacity

Any discussion of principles to guide the future direction of the climate change regime, to have legitimacy in the context of the UNFCCC, ideally would have its origins in the Framework Convention itself. Principles in the UNFCCC that are particularly relevant in this regard include the concept of "common but differentiated responsibilities",[3] recognition that developed countries need to go first,[4] the goal of the Convention as articulated in Article 2, equity for current and future generations,[5] and agreement to take a precautionary approach.[6]

Not all commentators have founded their discussion explicitly within this context. There is, however, a general recognition that the preamble and the first three articles of the UNFCCC provide the starting point for any discussion about the principles that the Parties intended to guide future negotiations. Three key

---

[2] For a discussion of the implication of pledge based targets, see M. Doelle, "The Kyoto Protocol; Reflections on its Significance on the Occasion of its Entry into Force" (2005) 27:2 Dal. L.J. 555.

[3] United Nations Framework Convention on Climate Change, Intergovernmental Negotiating Committee for a Framework Convention on Climate Change OR, 5th Sess., Annex, UN Doc. A/AC.237/18 (PartII)/Add.1 (1992), 31 I.L.M. 849, online: UNFCCC <http://unfccc.int/resource/docs/a/18p2a01.pdf> [UNFCCC or The Framework Convention], Article 3. See also *Conference of the Parties to the Framework Convention on Climate Change: Kyoto Protocol*, 10 December 1997, U.N. Doc. FCCC/CP/1997/L.7/add. 1, 37 I.L.M. 22 (1998), [hereinafter the Kyoto Protocol].

[4] *Ibid.* UNFCCC, Preamble.

[5] *Ibid.* Article 3.

[6] *Ibid.* Article 3.

principles have emerged from this, the principle that obligations should reflect responsibility for the problem, the principle that obligations should reflect potential to contribute to the solution, and the principle that obligations should reflect the capacity to contribute to the solution.[7]

The principle of responsibility is based on historical responsibility. Accordingly, each country would be assigned responsibility for a percentage of the GHG concentrations in the atmosphere above the pre-industrialization baseline of 280 parts per million by volume (ppm). Each country is then responsible for mitigation and adaptation costs in proportion to that percentage. This principle is based on rights and responsibilities rather than on more practical considerations of who can afford to do something about climate change, and where the efforts to mitigate will yield the best results. In basic terms, historical responsibility is reflected in the accumulated GHG emissions attributed to a given State based on past emissions, but without distinguishing between emissions before and after climate change was identified as a potential global threat.[8]

The principle of potential is based on a State's mitigating potential, or its technical potential to reduce GHG emissions. Proposals for GHG emission reductions founded on this principle generally use *per capita* emission, emission intensity or emission growth rates to identify where emission reductions should take place. The basic idea is that reductions should take place where there is the most room to achieve them, in countries that are the least efficient, that have the highest emissions per capita, etc. Current *per capita* emissions are often used as a simplified approximation of mitigating potential.[9]

One would expect any proposal based on the principle of potential to focus on how to achieve emission reductions in the countries that have the most room to reduce emissions. This would suggest a focus on domestic emission reductions, unless the principle is combined with the capacity principle, and those with greater ability may pay for reductions in countries with greater potential for reductions. Another interesting issue here is whether to define mitigating potential as the potential to reduce, or also as the potential to avoid increases. This is particularly relevant with respect to States such as China, India, and Brazil, where potential to actually reduce emissions may be low, but the potential to avoid future increases in emissions may be greatest.

---

[7] Hermann E. Ott *et al*, "South-North Dialogue on Equity in the Greenhouse" May 2004, Wuppertal Institute, Germany, online: Wuppertal Institute <www.wupperinst.org/download/1085_proposal.pdf>.

[8] *Ibid.* at 3.

[9] *Ibid.* at 3.

The principle of capacity is based on a State's ability to pay for mitigation, its access to technology to reduce GHG emissions, and its ability to adapt. Proposals that apply this principle generally rely on Gross Domestic Product (GDP) or the Human Development Index (HDI) to determine the relative capacity of States to contribute to mitigation and adaptation efforts. The basic idea is that the richer, better off countries should do more. Under this principle, obligations are assigned according to some measure of a State's ability to assist in the transition.[10]

The principle of capacity, if implemented on its own, would sidestep the question of responsibility for creating the problem to date, and instead focus on States' ability to devote resources, technology, knowledge and other forms of capacity to climate change mitigation and adaptation. Similarly, on its own, it would ignore where the greatest potential for emission reductions is. To be implemented effectively as the overriding principle, Parties would presumably have to rely on trading, joint implementation, and other forms of cooperation among nations. Otherwise, the mitigation effort might be spent overwhelmingly in rich countries. Emissions in poor countries with limited capacity but great potential might otherwise be ignored.

What implication does the relationship between these three principles have for the allocation of State responsibility for climate change mitigation and adaptation? Assuming a commitment to address climate change,[11] the starting point would be a global level of emission that would ensure stabilization of GHG concentrations at levels with a low risk of causing irreversible harm to the climate system. Given the constant evolution of our understanding of the science, that same global level of emissions would have to be updated regularly. The principles would then either individually or collectively guide the allocation of the obligation to reduce emissions to meet the global target. At the same time, responsibility to assist with adaptation would presumably be resolved applying the same principle or mix of principles.

Where do key individual States stand with respect to each of the three principles? States with high responsibility, high capacity and high potential are generally recognized to include the United States, Canada, and Australia. Europe and Japan are in a similar situation, except that the potential for emission reduction is lower given that these States are already much more efficient in their use of energy. In terms of responsibility, Europe, because it started industrializing first, would rate slightly higher than North America, Australia and Japan. It is not surprising that developed countries generally have favoured a mix of capacity and potential as the dominant principles for long-term emission reduction and adaptation obligations.

---

[10] *Ibid.* at 3.

[11] Perhaps with a recognition that a precautionary approach is warranted.

Key developing countries such as China, India, and Brazil have high potential to avoid future emissions, but much lower capacity and responsibility than developed States. Other developing States, such as OPEC members, Korea, and Signapore, are generally recognized to have the highest capacity of the developing world. Most other developing States rate low under all three principles. Not surprisingly, developing States generally have advocated strongly for historical responsibility as the basis for allocating future obligations.

With this in mind, how would each of the three principles influence mitigation and adaptation obligations of key States? With respect to a number of areas of global cooperation under discussion, the response seems uncontroversial. As a starting point, any of the three principles would support the need to make domestic emission reductions in developed States. There will be some variations in the relative amount of reductions, but no fundamental disagreement on the need for action on this front. Similarly, one would expect developed States to generally look after their own adaptation needs. It is not surprising, therefore, that emission reduction in developed States is at the heart of the Kyoto Protocol. Support for emission reduction and adaptation efforts in developing States can also be justified on the basis of differences in historical responsibility, differences in capacity and differences in potential.

There are two areas, however, that are much more controversial. One is the question of liability for climate change impacts that the global community is unable to avoid through mitigation. The other is the expectation that developing countries will commit to a low emissions development path in return for getting help from developed countries on mitigation and/or adaptation. Both issues are controversial for the same reason; their resolution depends on whether historical responsibility or current capacity is the driving force behind the resources transferred from developed to developing nations to deal with climate change.

If the dominant principle is historical responsibility, developed States clearly are liable for future impacts in developing States. Under this scenario, it is difficult to see how developed States can impose conditions in return for assistance offered either on mitigation or adaptation. If, however, the basis for assistance is capacity, and there is no basis for liability for future impacts, it is then reasonable for developed States to expect some commitment to a low emissions development path in return for offering mitigation and adaptation help. One would expect this to be a key difference in perspective between developed and developing countries. How these issues have affected the negotiations for the period after 2012 is considered in the following Section.

(b) The State of Negotiations Under the UNFCCC

Discussions among the Parties to the UNFCCC on what should happen after 2012 have been underway informally since COP 7. By the time the EU tried to move to a more formal phase of negotiating future commitments at COP 8 in New Delhi in 2002, it became clear that these negotiations would be difficult at best.[12] No significant progress was made in either Milan or Buenos Aires,[13] leaving discussions on future commitments stalled. The current state of thinking on the future can be summarized by highlighting the different perspectives on these issues in developed and developing countries.

(i) *Developed Country Perspectives and Issues*

- Developed countries are generally prepared to accept some responsibility to assume leadership.[14]
- The mitigation approach proposed is generally based on the concept that economic growth and GHG emissions must be decoupled, so that economic growth can continue while GHG emissions go down.[15]
- There is limited willingness to sacrifice economic growth in developed countries for GHG emission reductions.[16]
- There is a recognition that some way has to be found to ensure strong action to reduce GHG emissions is taken in developing countries (some more from a fairness or competitiveness perspective, others more based on environmental effectiveness).[17]
- Strong and effective leadership from the EU will be crucial. So far the EU

---

[12] Benito Mueller *et al*, "Framing Future Commitments; A Pilot Study on the Evolution of the UNFCCC Greenhouse Gas Mitigation Regime", Oxford Institute for Energy Studies, EV32, June 2003, at 4-3.

[13] *Report of the Conference of the Parties on its Tenth Session*, Conference of the Parties, United Nations Framework Convention on Climate Change (UN FCCC), 6-18 December 2004, FCCC/CP/2004/10/Add.1(Decisions 1/CP.10 - 11/CP.10), FCCC/CP/2004/10/Add.2(Decisions 12/CP.10 - 18/CP.10), online: UN FCCC <http://unfccc.int/2860.php> [hereinafter COP 10], *Report of the Conference of the Parties on its Ninth Session*, Conference of the Parties, United Nations Framework Convention on Climate Change (UN FCCC), 1-12 December 2003, FCCC/CP/2003/6/Add.1(Decisions 1/CP.9 - 16/CP.9), FCCC/CP/2003/6/Add.2(Decisions 17/CP.9 - 22/CP.9 & Resolution 1/CP.9), online: UN FCCC <http://unfccc.int/2860.php> [hereinafter COP 9], *Report of the Conference of the Parties on its Eighth Session*, Conference of the Parties, United Nations Framework Convention on Climate Change (UN FCCC), 23 October – 1 November 2002, FCCC/CP/2002/7/Add.1(Decisions 1/CP.8 - 20/CP.8), FCCC/CP/2002/7/Add.2(Decisions 21/CP.8 - 25/CP.8 & Resolution 1/CP.8), online: UN FCCC <http://unfccc.int/2860.php> [hereinafter COP 8].

[14] Mueller, *supra* note 12, at 3-1.

[15] *Ibid.* at 3-1.

[16] *Ibid.* at 3-1.

[17] *Ibid.* at 3-1.

has been largely outwitted by OPEC and the US on the long-term issues, in spite of its success in reviving the Kyoto process after the US pull-out. A key challenge for the EU has been its inability to adjust its position during negotiations, in part due to its process for developing EU positions. This may become even more difficult with the recent expansion of the EU.[18]
- There appears to be no readiness to take on the issue of liability.[19]
- The specific needs and perspectives of EITs, especially Russia and Ukraine, and their place among developed States, need to be taken into account.[20]

(ii) *Developing Country Perspectives and Issues*

- The G-77 is a coalition formed because of the power imbalance between developed and developing countries, capacity issues, and the common demand of developing countries for development assistance. The G-77 coalition is not based on a common interest to prevent climate change.[21]
- Subgroups, such as AOSIS, OPEC, LDC's, Emerging Economic Powers (India, China, Brazil), Africa, South America, and Asia all have very different interests on this issue.[22]
- The lack of adequate negotiating capacity of subgroups and individual States is a key factor in the survival of the G-77 in spite of its diverging and often conflicting interests with respect to climate change.[23]
- Developing countries essentially argue that the three principles of responsibility, potential and capacity all point to Annex I country action, not developing country action. The only one that is likely to change in the near future is potential, as GHG emissions in countries like China and India begin to grow significantly. There is no real opposition in the G-77 to reduction efforts in developing countries, but those efforts are seen as the responsibility of Annex I countries.[24]
- There is strong opposition to any discussion about Non-Annex I commitments to reduce or curtail GHG emissions. The G-77 view is that the way to achieve GHG emission reductions in Non-Annex countries is through Annex I countries leading by example, and through sustainable development assistance in the form of technology transfer, capacity-building, and resources for adaptation.[25]

---

[18] *Ibid.* at 3-3.
[19] *Ibid.* at 3-4.
[20] *Ibid.* at 3-4.
[21] *Ibid.* at 4-1 to 4-5.
[22] *Ibid.* at 4-1 to 4-5.
[23] *Ibid.* at 4-6 to 4-21.
[24] *Ibid.* at 4-3.
[25] *Ibid.* at 4-3.

It is clear that expectations between developed and developing countries are conflicting on a number of crucial points. In addition, the United States has opted out of the Kyoto process citing economic concerns and that the US will only take on binding targets if developing countries take on mitigation commitments as well. In the following Section, various proposals to overcome these challenges are assessed.

(c) Selected Proposals For Future Negotiations

Over the past few years, as the focus has started to shift from the Kyoto rules to the period after 2012, a number of proposals have been put forward for the long-term allocation of rights and responsibilities with respect to both mitigation and adaptation. The following review of some key proposals is carried out to identify opportunities to overcome the current impasse and move toward agreement on principles to guide the long-term allocation of rights and obligations.

The various proposals are considered in the context of the three principles described above, responsibility, capacity, and potential. A number of principle-based approaches have been proposed. They differ mainly in their perspectives on the relevance or relative weight to be given to each of the three principles. Some clearly choose one principle as dominant in allocating rights and responsibilities. Others seek to blend principles. Others yet seek to explore the impact of the various principles on the actual allocation of emission reduction responsibilities and to focus debate on bridging gaps in the results as opposed to differences in terms of the relative legitimacy of the three principles. A few key proposals are briefly summarized below.

(i) *Continuing with the Kyoto Approach of Pledge-based Obligations*

This approach would see the continuation of the basic approach taken in the negotiations leading to the Kyoto Protocol in 1997.[26] The advantage of this approach is that it does not require agreement on principles, and it is easily accommodated under the current regime. It allows for progress to be made based on individual States' willingness to commit to reductions while the discussion on guiding principles for long-term targets continues.

Perhaps the most apparent problem with the approach is that it appears to have led the regime to a significant impasse. The absence of agreement on principles to guide long-term targets in combination with the recognition of the challenge

---

[26] See Kevin A. Baumert *et al*, eds., *Building on the Kyoto Protocol: Options for Protecting the Climate* (Washington, D.C.: World Resources Institute, 2002), online: WRI <http://pdf.wri.org/opc_full.pdf>, at 31-60.

of achieving reductions are likely to make future negotiations difficult. Specifically, this approach provides little incentive for a State to agree to do anything more than what that State was willing to do without international cooperation. The experiences with the UNFCCC in 1992 and with the Kyoto Protocol in 1997 have already demonstrated that it is difficult under such circumstances to achieve meaningful reduction targets.[27]

The Kyoto process as the basis for distributing long-term responsibility for mitigating and adapting to climate change is problematic more specifically because it uses the grandfathering principle by requiring only modest reductions relative to emissions in 1990. In other words, it suggests some form of entitlement by developed States in high levels of emissions while asking developing States to reduce their already relatively low emissions. This approach can therefore be expected to perpetuate concerns in developing countries that climate change mitigation will limit their ability to develop[28], and make them even more reluctant to take on emission reduction commitments themselves.

Some have suggested that tinkering with Kyoto may be enough to address these concerns. One solution offered to this particular challenge is the use of voluntary commitments as a first step to bring key developing countries into the fold on mitigation[29]. Another way to complement the current pledge-based Kyoto approach to GHG emission reductions is to extend the application of the CDM as the mechanism to engage developing countries. The basis idea is to allow for CDM credits to be generated in developing countries in a more efficient and systematic way than possible under the project-based CDM. The sector based CDM would encourage whole sectors to be upgraded to reduce emissions.[30]

This proposal holds considerable promise to expand the use of the CDM. It is not likely, however, to play a significant role in overcoming the current impasse. It is not likely that developed countries will see this as a form of commitment by developing countries. Similarly, it will not likely be seen by developing countries as sufficient to demonstrate that emission reductions can be made without limiting development.

In the end, the current pledge-based approach is unlikely to be able to either motivate developed States to take on meaningful emission reduction targets or to take on mitigation commitments in any form. It seems clear, therefore, that a new approach is needed to move toward an effective global response to climate change.

---

[27] The absence of meaningful reductions in developed States in turn makes meaningful developing country participation more difficult.

[28] Baumert, *supra* note 26, at 31, 175.

[29] *Ibid.* at 135-156.

[30] *Ibid.* at 89-108.

(ii) *Responsibility as the Guiding Principle: The Brazilian Proposal*[31]

The original Brazilian Proposal was formally introduced during the negotiations leading up to the Kyoto Protocol as a basis for implementing the concept of common but differentiated responsibilities. Specifically, in 1997, Brazil proposed a series of 5-year commitment periods for developed countries with the targets for each determined based on that country's cumulative contribution to GHG concentrations in the atmosphere. By limiting the application of the principle to developed countries, the 1997 proposal reflected a combination of the responsibility and capacity principles.

In terms of burden sharing among developed countries, the impact of the Brazilian proposal was that targets were not necessarily reflective of current emissions or potential for reduction, but rather placed higher burdens on nations that had industrialized earlier and had therefore contributed more to GHG concentrations over time. The fairness of the proposal very much depends on whether it is appropriate to consider emissions before anyone understood the climate change consequences of those emissions.

From a punitive perspective, it is difficult to see the fairness of holding someone responsible for something no one knew to be wrong. From an unjust enrichment and compensation perspective, the answer is less clear. A case can certainly made that nations generally benefited economically from the activities that led to the increase in GHG concentrations over time. More questionable is whether current generations in those countries benefit. In other words, do citizens of the UK benefit from the fact that industrialization occurred earlier there than in the United States?

Not surprisingly, the proposal was not accepted in 1997, and Parties instead opted for a pledge-based approach without reaching any consensus on principles.[32] It is difficult to see how this principle on its own will gain acceptance as a basis for allocating future obligations. The proposal was revised in response to these and other questions raised about the methodology.[33] The updated version was submitted in 1999. A question that has been raised repeatedly with respect to the Brazilian proposal is whether historical emissions are a good basis for allocating future responsibility. Nevertheless, the proposal and the principle of responsibility will likely continue to play an important role in future discussions about responsibility for climate change mitigation and adaptation. In the end, the principle may turn out to be most relevant to the issue of liability. As such, it could provide considerable motivation for developed States to cooperate on mitigation.

---

[31] *Ibid.* at 157.

[32] *Ibid.* at 160.

[33] *Ibid.* at 160.

(iii) *Mitigation Obligations Based on Per Capita Entitlement*

Per capita emissions can be used in two different ways to determine long-term mitigation obligations. They can be used as an indicator of potential, or they can be used more directly as a basis for allocating entitlement. So far, in the UNFCCC context, *per capita* emissions have been used as an indicator, but not as a basis for allocating entitlement. What is explored here is the use of *per capita* emissions as a basis for allocating entitlements to emit or, conversely, obligations to reduce emissions.[34]

A number of proposals have been put forward that build on the basic idea that emission reduction obligations should be allocated on a *per capita* basis. The basic concept is that there is a sustainable or otherwise acceptable level of global GHG emissions that can be determined based on our scientific understanding of climate change. Based on this, the right to emit up to that level is then distributed equally or equitably based on some variation of *per capita* entitlement.

Proposals to implement this basic concept differ mainly in three respects. One is the level of emissions considered to be sustainable or otherwise acceptable. Another is how to get from the current distribution of emissions to the equal or equitable distribution of those emissions considered to be acceptable on a long-term ongoing basis. This is a particularly important issue considering that some States have *per capita* emissions that are 30 times those of other States. The third variant is whether the allocation is based strictly on equal *per capita* entitlement, or whether there is some consideration of factors that justify deviating from equal. This generally includes discussion of equitable rather than equal distribution of entitlement, and raises the further question what differences in circumstances might justify different *per capita* emissions.

The first issue is one of the adequacy of the overall mitigation effort. It is primarily about the science of climate change, how to deal with uncertainty, and about practical limitations, such as technical, political, and economic constraints on the ability to achieve the reductions necessary. On one hand, it requires consideration of what level of GHG emissions can be permitted without compromising the ability to meet the fundamental objective of the UNFCCC as set out in Article 2. On the other hand, it raises questions about the lowest level of GHG concentration that is still achievable.[35]

---

[34] *Ibid.* at 175.

[35] For an analysis of this balance between what is practical and what is necessary to meet the objective in Article 2, see Elzen (den), M. G. J., *et al*, "Meeting the EU 2 Degree Climate Target: Global and Regional Emission Implications" (2005) Report 728001031/2005, Netherlands Environmental Assessment Agency, online: <www.mnp.nl>.

The second question is primarily about process and transition to the desired end point of equitable distribution of entitlement to emit greenhouse gases. How can States with high *per capita* emissions be motivated to implement effective mitigation measures, and what responsibility do such States have to developing States whose emissions are much lower, perhaps even below the ultimate *per capita* emissions target? Different proposals have used different methods to motivate States to make the transition. Most proposals have dealt in some form with equity issues for the transition period,[36] most commonly through the allocation of emissions credits that can be traded to States with high *per capita* emissions.

The third question is about equal versus equitable treatment. It raises issues about national circumstances that may justify higher or require lower *per capita* emissions. How should factors such as population density and the related issue of urban versus rural populations affect *per capita* emission targets? What about energy exporting versus importing States? Should factors such as States' climates and resulting demand for space heating or cooling be considered? Does the approach cover all emissions or only some sources to avoid having to resolve some of these issues?

The implications of a *per capita* based entitlement approach to long-term mitigation obligations are discussed in some detail by Baumert in "*Building on the Kyoto Protocol*".[37] He summarizes the implications of this approach as follows:

(A) Merits

- Simplicity of concept
- Strong ethical basis
- Flexibility to accommodate changing scientific evidence
- Enhancement of efficiency of global trading
- Offer of incentives for developing-country participation
- Consistency with major guiding principles of the UNFCCC
- Amalgamates well with the Kyoto architecture

(B) Demerits

- Limited global acceptability
- Limited flexibility to accommodating varying country circumstances
- Linkage with trading essential for success

---

[36] That is, compensating low per capita emitters and/or penalizing high per capita emitters during the transition period.

[37] Baumert, *supra* note 26, at 175-198.

- Associated issues of hot air and obligation costs[38]

One particular proposal that has attempted to tackle the issues raised here is the contract and converge concept originally put forward by the Global Commons Institute in 2000.[39] The first step in this proposal is to agree on the long-term GHG concentration stabilization level that the global community is committed to achieving. The second step involves an allocation of emissions entitlement annually that is based on the idea of taking everyone from their baseline to a *per capita* share of the total entitlement in time to avoid exceeding the GHG concentration limit set under step one.

Contract and converge thereby forces a transition from emission allocations based on historical emissions to an allocation based on *per capita* entitlement over a period of time that is dictated mainly by the global GHG concentration limit accepted as necessary to meet the objective set out in Article 2 of the UNFCCC. To deal with fairness of the transition for low *per capita* emitting States, and potential technical, political and economic limitations in high *per capita* emitting States to meet steep reduction obligations, the contract and converge approach is usually complemented with a trading system that allows some flexibility in meeting the targets for high emitting States and provides funding mechanisms for low emitting States.[40]

(iv) *Sustainable Development Policies and Measures (SD-PAMs)*[41]

The focus of this approach is on how to engage developing States in climate change mitigation. In this sense the SD-PAMs approach is not a comprehensive alternative proposal to the Kyoto framework, but rather a proposal on how to address one particular challenge, that of developing country engagement in mitigation. The basic concept can be implemented in a variety of ways.

Harald Winkler and others propose a multi step approach in their Chapter on SD-PAMs in Baumert's book "*Building on the Kyoto Protocol*".[42] First, a developing State interested in SD-PAMs has to identify development objectives, such as developing transportation infrastructure, addressing housing needs, providing

---

[38] *Ibid.* at 196.

[39] Global Commons Institute, online: GCI <http://www.gci.org.uk/>, for technical support and information concerning "Contraction and Convergence" a planning model, "Contraction and Convergence Options," is also available for download.

[40] See Niklas Hoehne *et al*, "Evolution of Commitments under the UNFCCC: Involving Newly Industrialized Economies and Developing Countries" Federal Environmental Agency, Germany, February 2003, Research Report 201-41-255 UBA-FB 000412, at 41.

[41] See Baumert, *supra* note 26, at 61.

[42] *Ibid.*

better access to clean water and food, improving access to energy, or creating employment. Alternatively, these objectives could flow from an overall national development strategy.

The second step involves identifying ways to make the development path more sustainable. In the context of climate change, the focus would be on identifying ways to meet the development path in a manner that both reduces GHG emissions and results in an overall change to the development path that is more sustainable than the starting point. At this stage, it is therefore important to quantify the benefits of the alternate development paths identified, both generally within the umbrella of sustainability, and more specifically with respect to GHG emission reductions as compared to the pre-existing development path.

At this stage, the net impacts of the alternate development path for general sustainability and for climate change mitigation have to be quantified in some form. Up to this point, the host country can be expected to be in control of the process. When it comes to quantifying the sustainability and climate change mitigation benefits, however, there would likely have to be a process in place to ensure the net benefits are determined based on consistent and accepted methodologies, much like the process for certifying CDM credits under the Kyoto Protocol.

It is interesting that the SD-PAMs proposal put forward by Winkler essentially stops here. There is only some vague reference to funding or credits being available for these benefits.[43] Open questions in Winkler's proposal include whether the commitments made are voluntary or mandatory, how the change in development policies would be funded, and how the net benefits to sustainability and climate change mitigation would be quantified, reported and verified.

The real opportunitiy of the SD-PAMs approach likely lies in the potential to combine the development agenda of developing States with the climate change mitigation agenda of the developed world. To do this, there would have to be a clear link between development assistance in the form of resources, expertise, capacity-building and access to technology and key choices made by developing States about their development paths. The SD-PAM's approach provides that opportunity. There would also have to be an appreciation in developing countries that the low emission sustainable development path is a better path to prosperity.

For example, the development path China chooses on energy and transportation in the years to come has tremendous implications for the rest of the world. A well

---

[43] One way to facilitate this process might be through an expanded, sector based CDM, but this could also take place outside the current Kyoto process, by setting up separate funding mechanisms for SD-PAMs.

designed SD-PAMs mechanism, either as part of the Kyoto process or outside, could facilitate the process of identifying what the developed world is prepared to contribute to motivate China to leapfrog to renewable energy and public transportation. At the same time, that same process would have to identify what is required to be able to expect China to reject a development path based on private transportation and fossil fuels that was freely available to the developed world, and that is largely responsible for their current state of development. The key question may be whether any level of assistance will entice developing States such as China to make binding commitments to give up a development path that has been used in Europe and North America to produce incredible short term benefits.

(v) *Flexible Targets, Such as Intensity Targets*

The basic concept behind the flexible target approach is that fixed targets, such as the first commitment period targets in the Kyoto Protocol, may be too rigid, especially considering fluctuations in economic activity.[44] Flexible targets are therefore based on the premise that we have to reduce the GHG emission without limiting economic activity. In other words we have to decouple economic activity and GHG emission. This approach essentially chooses economic certainty over environmental certainty, by making the environmental goal of reducing GHG emissions the dependent variable.

This approach furthermore implicitly assumes that the level of economic activity itself is not the problem, or that the climate change challenge does not require a stabilization or reduction in global economic activity. Based on this premise, the dual intensity target approach therefore seeks to provide flexibility that is linked to these factors and thereby strives to encourage States to accept targets they might otherwise object to, due to uncertainties about future circumstances. It requires States to improve the ratio of economic output per unit of GHG emissions rather than to impose firm emission limits.

Intensity targets thereby tend to err on the side of uncertainty with respect to the environmental benefits in return for more certainty on the economic cost of climate change mitigation. It is these aspects of the intensity target approach that have made them controversial for use in developed States. This approach has been expressly rejected in the Kyoto Protocol. At the same time, it is this approach that has been implemented domestically in the United States. It has also formed a part of the implementation strategy for some States who are Party to the Kyoto Protocol, such as Canada's Large Industrial Emitters Programme.[45]

---

44 For an overview of flexible versus fixed targets, and a specific proposal in the form of dual-intensity targets, see Baumert, *supra* note 26, at 109.

45 See Climate Change Plan for Canada (2002), available at <http://www.climatechange.gc.ca>.

The point made by Baumert is that the intensity target approach may nevertheless offer some opportunities in the context of developing States. Economic uncertainty is generally high in developing countries; much higher than in more developed States. This level of uncertainty, in combination with the particular concern in developing countries that climate change mitigation amounts to a limit on development may make intensity based mitigation targets more palatable in developing States, and may therefore provide a way forward on future mitigation commitments.[46]

(vi) *Equity as the Guiding Principle*[47]

The central theme of the proposal put forward by Hermann Ott and others in a report entitled "*Equity in the Greenhouse*" is that equity needs to play a central role in developing the long-term direction for the climate change regime. Ott identifies four imperatives for the reliance on equity in developing a global response to climate change.[48] The legal imperative arises out of Article 3 of the UNFCCC by establishing equity as a guiding principle.[49] Ott points out that there are moral and political imperatives as well.

It is the practical imperative that Ott then focuses on. He argues that the legal, moral and political imperatives for equity have so far not produced an equitable response to climate change. Rather, it has been a response that disproportionally places the burdens on those with little or no responsibility, capacity, or potential to act. The point made is that in the absence of equity, a global consensus on how to respond to this challenge is simply not possible. Assuming this is the case, and that climate change requires a global response, equity is the only option.

The authors conclude that the equity imperative requires Parties to move forward with both mitigation and adaptation. The central question tackled is how to distribute the responsibilities and benefits of climate change mitigation and adaptation efforts necessary. The basic premise here is that the negotiations need to move beyond the developed/developing State divide to identify where individual States fit relative to each other with respect to the three basic principles of responsibility, capacity and potential. Once this is determined, obligations and

---

[46] The concept of dual intensity targets is considered in some detail by Baumert in the context of creating commitments for climate change mitigation in developing countries that are not seen as limiting development, and that at the same time send a sufficient signal to developed States that the effort on mitigation is a global effort.

[47] See Ott, *supra* note 7.

[48] *Ibid.* at 15.

[49] UNCCC, *supra* note 3. Article 3(1) requires Parties to protect the climate system "on the basis of equity and in accordance with their common but differentiated responsibilities and respective capabilities".

benefits on climate change mitigation and adaptation can be distributed accordingly.

There is some discussion of the relative influence of the three principles, but no clear resolution. The authors accept the concept that potential to mitigate is the dominant principle for determining which States have to carry out mitigation measures domestically, or where the reductions should be achieved. The more difficult question is on the relationship between responsibility and capacity. The report focuses more on responsibility, but what is missing is a clear connection between the general principle of equity and the question of whether mitigation or adaptation is more appropriately funded based on current capacity to pay, or based on historical responsibility for the increase in GHG concentrations in the atmosphere.[50]

The authors may be excused for sidestepping this issue, given that most countries with high capacity also share significantly in responsibility.[51] The report does suggest that high responsibility should be more directly linked to domestic emission reductions, and capacity more directly connected to funding mitigation and adaptation in States with less responsibility and capacity. However, there is no indication that this distinction is derived from anything more than a practical perspective that equates capacity with ability to provide resources; making allocation of an obligation to provide funding and other resources on the basis of capacity as opposed to responsibility perhaps more palatable.[52]

Although equity as a guiding principle for allocation of obligations for climate change mitigation and adaptation may have considerable appeal, a key practical question remains. How do international negotiations progress to a point where obligations are distributed based on equity? Is this as simple as pushing equity as the accepted norm and demonstrating that failure to follow this approach is in violation of this norm? Is equity an internationally accepted norm? Is it as simple as pointing to Article 3 of the UNFCCC?

The authors of the report appear to conclude that this comes down to political leadership in countries who are committed to responding effectively to climate change. The report points out that historically this has meant the EU, but that this is also a key question for other States, in particular developing countries, who could be (and perhaps need to be) playing a leadership role. This brings up

[50] *Ibid.* at 31, where the authors suggest that both contribute, but without addressing the relative contribution responsibility and capacity each should make.

[51] See Baumert, *supra* note 26, at 214, table 9.5, where *per capita* emissions targets for different states are compared based on different allocation principles. The last two are generally based on responsibility and capacity respectively, and the variations are relatively minor.

[52] See Ott, *supra* note 7, at 32, Box 2.

questions of why developing States have not taken more of a leadership role. These issues, which are not explored in this report, require a consideration of negotiating capacity of developing States and the G-77 alliance. These issues are explored in the following proposal.

(vii) *Overcoming the Taboos of Liability and Mitigation*[53]

In a report released by the Oxford Institute for Energy Studies at the climate change negotiations in Bonn, Germany in 2003, Benito Mueller and others seek to identify the key challenges facing the climate change negotiations and then develop strategies for overcoming them. The focus of the report is on what Meuller calls the twin taboos of the climate change negotiations; the refusal by developed States to discuss liability, and the refusal by developing States to discuss obligations to mitigate climate change. He relates these taboos with key concerns of the two sides. For developing States, the key concern is with respect to impacts and adaptation. For developed States it is that mitigation needs to take place globally for climate change to be slowed effectively. His proposal lays out a very detailed plan of building trust and confidence that both Parties are prepared to move away from their taboos as long as their key issues are resolved in the process. Essentially, Mueller's point is that at the moment, both sides are holding out for their priority, and in the process, little is being done.

His proposal involves each side taking small steps toward addressing the key issues of the other as a way to build the trust needed to enable both sides to agree to move forward with mitigation and adaptation in parallel. Mueller therefore advocates a focus on developing first, on funding mechanisms for impacts and adaptation, and on developing a mitigation approach that recognizes the wide range of circumstances that exist in developing States.[54] Mueller's proposal then considers mechanisms for making technology available in developing States, including an expanded Clean Development Mechanism as a way to encourage mitigation in developing States. For more advanced developing States, Mueller proposes that they take on mitigation commitments of some form. He then discusses how such commitments might be structured given the current position against formal commitments in key developing States such as China and India.[55]

Mueller also addresses the need to re-engage the United States.[56] The preconditions for re-engagement identified are domestic mitigation efforts in the US, a willingness by the US administration and the federal government to engage internationally, and a willingness by the international community to adjust the

[53] See Mueller, *supra* note 12.
[54] *Ibid*. at 6-3 to 6-7.
[55] *Ibid*. at 6-15.
[56] *Ibid*. at 6-16.

Kyoto framework to address key concerns of the US. Based on Mueller's analysis, the combination of the international community moving ahead and the considerable domestic action at the state level in the US will force the US administration to play a coordinating role. If the pressure for action at the state level continues, so will the pressure to develop consistent rules within the US.

Once there is a clear need for national coordination of climate change law and policy in the US, Mueller argues that it would be in the best interest of the US industries for the US to develop rules consistent with international rules under Kyoto. This would eventually result in industry pressure to re-engage internationally. It is important, however, to separate motivation to re-engage internationally from motivation to move the climate change regime forward. The factors Mueller explores do point to a possible re-engagement, but they do not illustrate whether and why the US would re-engage in a manner that encourages an effective global response to climate change rather than further delay.

Mueller concludes that the re-engagement of the US, engagement of key developing countries such as China and India in mitigation efforts in combination with the building of mutual trust will be key. Related to this, the study identifies the need to build negotiating capacity in developing Parties, and the need to make progress on the issue of responsibility for impacts and adaptation in developing States.

What is missing is a convincing argument that the step-by-step movement past the twin taboos of liability and mitigation will actually motivate any of the key States, such as China, India, and the US. This is particularly problematic with respect to the US, where it is difficult to see any significant short-term domestic benefit other than the contribution to solving a global problem; admittedly one that will have local consequences. Assuming the motivation there is based on an acceptance that global cooperation on the issue of climate change is in the interest of all, Mueller's approach does provide a possible roadmap past the current impasse.

(viii) *The Multi Track Approach*

There is a general recognition that no one approach, in isolation, is likely to be at the same time the most environmentally effective, equitable, economically and politically acceptable, and practically achievable. In particular, there is an emerging recognition by academics, researchers and negotiators that approaches may have to differ for adaptation and mitigation. Furthermore, even on the mitigation side, it is becoming clear that different approaches may be needed depending on a combination of the relative level of responsibility and capacity of a given Party State. This recognition was implicit in a number of approaches described above, and expressly recognized in others.[57] The need for multiple solutions is illustrated in a proposal put forward by the Climate Action Network (CAN) at a side event at the 9th Conference of the Parties in Milan, Italy in December, 2003.[58]

The CAN Proposal identifies three tracks for post-2012, the Kyoto track, a decarbonization track and an adaptation track. The Kyoto track essentially continues with fixed emission reduction targets for Annex I countries. It is expected that with time, some but not all Non-Annex I countries will join the Kyoto track. The proposal suggests that future commitments should be determined on the basis of historical responsibility, capacity, and potential for reductions.[59] This track is also seen as the mechanism for implementing the principle that developed countries need to go first, a crucial part of the concept of common but differentiated responsibility.

The decarbonization track is a second mitigation track that is designed to ensure that some form of mitigation action takes place in most of the States that are not part of the Kyoto track. The CAN proposal advocates for decisions about the choice of track based mainly on a Party's state of development. The most developed States would have mitigation obligations under the Kyoto track. The majority would fall under the decarbonization track.[60] The least developed States would not have any obligations to mitigate, and would therefore be exempt from both the Kyoto and the decarbonization track.

The proposal does not identify in detail how individual Party's mitigation obligations or commitments would be determined. It does exempt those Parties from fixed emission reduction obligations. The SD-PAMs approach described above is referenced as a possible way to implement the decarbonization track. The key component of the proposal is that the nature and extent of the mitigation commitment and obligation for Parties in this track will vary. It will depend on capacity, such as available resources and technologies. For more developed Parties, firmer obligations would be expected. For the least developed States, any effort on mitigation would be purely voluntary.

---

[57] See, for example, Ott, *supra* note 7, which builds on the premise that the obligations and benefits for climate change action be distributed based on Parties' relative capacity and responsibility.

[58] Climate Action Network (CAN) "A Viable Global Framework for Preventing Dangerous Climate Change" (2003) CAN Discussion Paper, online: Climate Action Network <http://www.climatenetwork.org/pages/publications.html>.

[59] See Section 4 on the three principles above.

[60] Initially, this might include all developing States other than the least developed States. Over time, some of the more developed States, such as Signapore, South Korea, and China would be expected to take on binding targets.

The third track is referred to as the adaptation track. It serves to ensure that those with higher responsibility and capacity provide adaptation assistance to those with lower capacity and responsibility. The focus of the proposal is on providing adaptation assistance to the least developed and the most threatened Parties. Obligations to contribute appear to be based on responsibility. The proposal contemplates that Parties eligible for adaptation assistance under this track may also take on some mitigation obligations. One would anticipate that such obligations would generally fall within the decarbonization track, though the proposal leaves open the possibility that a Party that takes on Kyoto type targets might still be eligible for adaptation assistance. A well-developed small island State with low historical emissions may provide an example.

The basic concept appears to be that the respective roles a given Party will be asked to take on will depend on its relative responsibility, capacity and potential at a given time. Equity, in particular intergenerational equity, is identified as a guiding principle, as are the principles of responsibility and capacity. The proposal places less emphasis on the principle of mitigating potential. Clearly, these efforts to combine various approaches to develop an overall long-term strategy for moving from the current state to an equitable distribution of GHG emissions globally at overall levels that are sustainable is still in its infancy. They do show considerable promise in addressing at least some of the key concerns that have been expressed by Parties during the course of negotiations over the past few years.[61]

(ix) *Other Efforts*

In addition to these more formal proposals, there are of course numerous efforts under way at any given point within and outside the Kyoto process that have the potential to influence the future evolution of the negotiations and the climate change regime. Examples include the EU position on future targets,[62] proposals made by Italy at COP 10 to move toward intensity targets to bring the US back on board,[63] various efforts by British Prime Minister Tony Blair to bring the US back to the negotiating table on climate change,[64] and regional efforts within the

---

[61] For another proposal built upon the same basic principle, see Niklas Hoehne *et al*, "Options for the Second Commitment Period of the Kyoto Protocol" Federal Environmental Agency, Berlin, Germany, February 2005, ISSN 1611-8855, online at: <http://www.umweltbundesamt.de>.

[62] "Winning the Battle Against Global Climate Change", Commission of the European Communities, Brussels, 9 February 2005 {SEC(2005) 180} online: <http://72.14.207.104/search?q=cache:crDpXngSfK8J:europa.eu.int/comm/environment/climat/pdf/comm_en_050209.pdf/>.

[63] *Report of the Conference of the Parties on its Tenth Session*, Conference of the Parties, United Nations Framework Convention on Climate Change (UN FCCC), 6-18 December 2004, FCCC/CP/2004/10/Add.1(Decisions 1/CP.10 - 11/CP.10), FCCC/CP/2004/10/Add.2(Decisions 12/CP.10 - 18/CP.10), online: UN FCCC <http://unfccc.int/2860.php> [hereinafter COP 10].

[64] Such as efforts to make climate change a priority at the 2005 G-8 Summit chaired by Tony Blair.

context of the CEC to motivate the US to take domestic mitigation action. In a recent communication from the Commission of the European Communities, the possibility of a separate and complementary process involving the G-8 and key developing countries such as China and India is raised.[65]

Another effort worth noting is the push for international cooperation on renewable energy that arose out of the World Summit on Sustainable Development in Johannesburg, South Africa, in 2002 (WSSD).[66] The United States opposed the creation of an international renewable energy agency at the WSSD, as well as a renewable energy target. States nevertheless agreed to cooperate to promote the use of renewable sources of energy. The first step following the WSSD toward global cooperation on renewables was the 2004 Conference in Bonn, Germany.[67] In the absence of concrete outcomes, it is too early to predict what form this effort will take, let alone what results it will achieve. It does, however, serve as an example of an alternative and possibly complementary approach to Kyoto, one that has the potential to provide a positive external influence on the climate change regime.

If this effort is able to make renewable energy competitive on a broad scale globally, and if it can achieve this without universal support, but rather with a coalition of the willing, this effort on renewables may do as much for climate change mitigation and sustainable development as any binding obligations under the climate change regime. It remains to be seen whether a sufficient level of cooperation on renewables is achievable, or whether the same interests that are preventing more progress on climate change will also prevent meaningful global cooperation to assist with the breakthrough of renewables and other technologies that are part of the solution.

It is important to consider the limitations of such efforts. Renewables initiatives, for example, are not likely to address broader development concerns of developing countries, nor will they address adaptation. Even with respect to climate change mitigation, renewables are clearly only a modest part of the

---

[65] See "Winning the Battle Against Global Climate Change", Commission of the European Communities, Brussels, 9 February 2005 {SEC(2005) 180} online: http://72.14.207.104/search?q=cache:crDpXngSfK8J:europa.eu.int/comm/environment/climat/pdf/comm_en_050209.pdf/, at 5.

[66] For a general discussion of the role of the WSSD in the development of international environmental law, see Nicholas A. Robinson, "Befogged Vision: International Environmental Governance a Decade After Rio" (2002) 27 Wm. & Mary Envtl. L. & Pol'y Rev. 2, and George Pring, "The 2002 Johannesburg World Summit On Sustainable Development: International Environmental Law Collides With Reality, Turning Jo'Burg Into 'Joke'Burg'" (2002) 30 Denv. J. Int'l L. & Pol'y 410.

[67] For more information on the 2004 Renewable Energy Conference in Bonn, see <http://www.renewables2004.de/>.

solution. As discussed in Chapter 8, finding ways to reduce energy consumption is likely to be the single most important climate change mitigation measure.[68]

(d) Key Challenges for the Negotiations

The following key actions need to be taken by some or all States to ensure an effective response to the climate change challenge:

- Domestic Action: Many States will have to make significant emission reductions domestically, either relative to current emissions or compared to business-as-usual projections.
- International Assistance: States with greater capacity will have to support reduction and adaptation efforts in States with lower capacity.
- International Liability: Some States will have to compensate other States or some of their inhabitants for the impacts of climate change.
- Commitment to Low Emissions Path: Some States will have to commit to a low emissions development path in return for receiving assistance on mitigation and adaptation specifically, and sustainable development more generally.

The question posed here is what stands in the way of these key actions, and what can be done to overcome those barriers?

The need to make domestic emission reductions in developed countries is in principle not controversial. The differences between allocation of responsibility to make domestic emission reductions based on historical responsibility and current capacity are not major, suggesting that this should not stand in the way of agreeing on long-term emission reduction targets in developed countries. Compliance with the Kyoto first commitment period obligations should provide the basis for more meaningful reduction targets in the future for two reasons. It will provide better information about the relative economic consequences of making reductions. It will also ensure a level of trust that all developed countries are prepared to carry their share of the burden.

One challenge will be the engagement of the United States in some form of international commitment to make domestic emission reductions. Clarification of liability rules may provide some motivation for re-engagement. Acceptance of historical responsibility as the dominant principle might also provide an avenue for engagement, given that this principle would place a relatively higher burden

---

[68] See, for example, R. Torrie, *et al*, "Kyoto and Beyond: The Low-emission Path to Innovation and Efficiency" (October, 2002, David Suzuki Foundation and Climate Action Network Canada), online: David Suzuki Foundation <www.davidsuzuki.org>.

on developed countries that are already doing more, such as a number of EU States. Furthermore, it is important to recognize that there has been considerable effort in the US to begin to reduce domestic emissions, what is missing is the international commitment to do so, and the cooperation with other developed States toward more meaningful, effective targets in the future.[69]

With respect to domestic emission reductions in developing countries, however, the issue is more complicated. Domestic emission reductions from current emissions are a non-starter in most developing countries, given the relatively low levels of emissions in those countries. This really leaves the question of emission reductions compared to business-as-usual projections. Even here, there seems to be general acceptance that this can only be expected in the context of assistance from higher capacity States.

Another key issue appears to be how to motivate climate change mitigation action in States other than Europe and Japan, the two places that appear to be ahead of the international community on this in terms of domestic action. Much of this likely comes down to the United States. What will it take for the US to take climate change seriously domestically, and to lead on this issue internationally? Short of creating an economic incentive to motivate US leadership, there is no clear answer to this question.[70]

With respect to international assistance, there also appears to be some recognition in line with the capacity principle, that there is some obligations by those with greater capacity to assist those with lower capacity[71] to mitigate and possibly assist with adaptation. The question is whether this takes place in the context of capacity (with some conditions imposed on the recipient of the assistance) or in the context of responsibility (presumably with less or no conditions).

The question of liability for climate change impacts is highly controversial, and is not likely to be solved through negotiations. It has so far largely been avoided, undoubtedly in part because of the potential implications and because it only arises to the extent that efforts to avoid harmful climate change are unsuccessful. What has become clear, however, is that resolving the issue of ultimate liability may be a prerequisite for the motivation needed to bring States to the negotiating table with a willingness to agree to effective measures to mitigate climate change.

---

[69] On the current domestic situation in the United States, see, for example Greg Kahn, "Between Empire and Community: The United States and Multilateralism 2001 – 2003: A Mid-Term Assessment: Environment: The Fate of the Kyoto Protocol under the Bush Administration" (2003) 21 Berkley J. Int'l L. 548, and John C. Dernbach, "Making Sustainable Development Happen: From Johannesburg to Albany" (2004) 8 Alb. L. Envtl. Outlook 173.

[70] See also discussion of state responsibility and liability throughout, and in particular in the Epilogue.

[71] But high mitigating potential.

This may mean that existing international law has to be put to the test to determine whether it can resolve the question of ultimate liability.[72] As discussed in Chapters 5 to 8, international human rights norms, customary international law, and rules established under other regimes such as the WTO and UNCLOS may provide a basis for resolving the liability issue. Other, more general efforts, such as the International Law Commission's work on State responsibility and liability,[73] while showing some promise, is less likely to be effective here because a level of agreement is required that is not likely to be achievable on the climate change issue, given the rising stakes.

The issue of developing countries agreeing to a low emissions path in return for assistance on mitigation and adaptation is also very controversial. In the short term, one of the key questions will be how to engage emerging economic powers in climate change mitigation. The conventional wisdom appears to be that China, India, Indonesia, Brazil and Nigeria will not become full participants in mitigation efforts for decades to come.[74]

Why is it that developed countries are not willing to work with developing countries to achieve reductions in combination with development assistance, essentially assisting in putting those countries on a sustainable development path? The answer appears to be an unwillingness to accept responsibility and acknowledge liability[75] by developed States. There are numerous signs of this throughout the evolution of the climate change regime. Failure to agree on funding mechanisms for climate change adaptation or mitigation in developing States based on responsibility is perhaps the clearest illustration.[76] Another signal is the rejection of responsibility as a guiding principle for developing emission reduction targets under the Kyoto Protocol in 1997.[77]

This further highlights the need to resolve the liability issue outside the negotiations to allow the Parties to bypass the current impasse. If the answer is that there is no liability for climate change resulting from anthropogenic GHG emissions, then the negotiations can proceed on the basis of capacity. Under those

---

[72] For a discussion of the issue of liability for climate change impacts as far back as 1990, see Durwood Zaelke *et al*, "Global Warming and Climate Change: An Overview of the International Legal Process" (1989-1990) 5 Am. U. J. Int'l L. & Pol'y 249, at 261.

[73] For an overview of the work of the International Law Commission on transboundary environmental harm, see Practicia Birnie *et al*, *International Law & the Environment*, 2nd ed. (Oxford: Oxford Univeristy Press, 2002), at 105.

[74] Mueller, *supra* note 12, at 5-1.

[75] *Ibid.* at 5-1.

[76] *Ibid.* at 5-1.

[77] As discussed above, the rejection of the Brazilian Proposal was in large part due to developed State's rejection of historical responsibility as an acceptable basis for allocating emission reduction targets.

circumstances, higher capacity States would be reasonable in expecting some agreement to choose a lower emissions development path in return for the assistance provided.

If the answer is that those responsible for the GHG emissions are liable to those affected by climate change, it is equally clear that any assistance would be a liability mitigation strategy by those with the greatest historical responsibility. In that context, it is less likely that developing countries could be expected to accept conditions in return for assistance. At the same time, the risk of liability would be a motivator to choose a low emissions development path for all, including developing countries.

In the end, it is difficult to see how States will escape some form of liability for future impacts, especially if the response from key developed countries with high emissions continues to lag far behind the scientific evidence of the harm these emissions will cause in the future. Principles such as polluter pays, international rulingson customary law obligations such as the ad hoc arbitral tribunal rulings on the Trail Smelter statements on State responsibility in soft law instruments such as Stockholm and Rio, and the work of the International Law Commission all point to some form of responsibility and liability for extraterritorial impacts of actions taken within a State's boundaries.[78]

There clearly will be considerable debate over issues such as when the science is sufficiently strong to establish the link, and whether responsibility is linked to knowledge of the problem, but ultimately, it is inconceivable that the international community will stand by indefinitely and continue to allow States to suffer the consequences of climate change without holding those responsible for the majority of the GHG concentrations responsible in some form. Any delay on the mitigation side is tantamount to debating the wording of a house insurance policy while the house is burning down, rather than putting out the fire.

Other than the underlying dispute over liability, why is it that developing countries are not willing to accept some constraints on GHG emissions in return for further reductions in Annex I countries and some assistance in achieving their targets? The answer is not a straightforward one. Certainly, any acceptance of a commitment to limit GHG emissions in developing countries is seen as a potential constraint on development. Whether that potential materializes depends on whether the development path of a given country intersects at some point in the future with the GHG emission limit accepted by that country. Given the relative low capacity in these countries, and the priority to meet the basic needs of their populations, their ability to choose a low emission development path depends on the assistance they will receive from developed States.

---

[78] See discussion in Section 4 of the Epilogue.

So far, the negotiations on assistance to developing States and Annex I Party emission reduction efforts have combined to send the message to developing countries not to expect much help and that GHG emission reductions and economic growth are incompatible. Negotiations on capacity-building, technology transfer, and adaptation help for developing countries in the context of the implementation of the Kyoto Protocol have been abysmal.[79] Efforts by Annex I Parties to reduce their GHG emissions while continuing to grow economically have had limited success at best.[80] The overall message from developed to developing States to date, whether it is accurate or not, has been that emission reductions come with a heavy economic price, and that developed States cannot be counted on to pay that price.

It is also important to recognize that for developing States, the issue of responsibility or liability is not limited to mitigation. It is generally accepted that developing States will feel the impacts of climate change disproportionately, that they are disproportionately unequipped to adapt, and that their contribution to the problem to date has been disproportionately low. As a result, it is not surprising that developing States have focussed on impacts and adaptation over mitigation. The fact that developed States have to date not shown much willingness to provide funding, build capacity, transfer technology or provide other resources to developing States on either mitigation or adaptation is a further deterrent to developing country engagement on mitigation.

To summarize, to get to an effective global response to climate change, the following needs to happen:

- Developed States need to either demonstrate by example that economic growth and climate change mitigation are compatible, or demonstrate that a high quality of life is possible without continued economic growth and with reduced GHG emissions;
- Developed States need to demonstrate a commitment to assisting developing States in achieving comparable quality of life for its citizens without high levels of GHG emissions; and
- The role of historical responsibility needs to be resolved, as it appears to be at the heart of the current impasse. Once this is done, the question of the terms under which higher capacity States will assist lower capacity States must be resolved.

---

[79] See developing country Section (c)(viii) in Chapter 2.

[80] See Chapter 2. The EU and Japan are generally recognized as the only jurisdictions that have demonstrated any ability to decouple economic growth and GHG emissions. Sweden provides the most promising example to date.

The question is how do we get there? A combination of an acceptance of responsibility and liability by high GHG emitting States and a much higher level of trust by developing States appears to be the combination of change in circumstances required to make meaningful progress. Movement toward compliance with Kyoto obligations is likely to be an essential step in building trust among Member States. Developments in the science of climate change, assuming current trends continue, will increase the motivation for all States to take this issue seriously. Developments in technology can play an important role in demonstrating that lower GHG emissions does not mean lower quality of life, whether or not it means less economic output. However achieved, it is crucial that there be significant development assistance from higher to lower capacity States, and that the assistance be directed effectively to ensure a low emission development path in the lower capacity States.

What about responsibility for impacts and adaptation? Is this an issue that needs to be addressed to make progress on the mitigation side? Clearly, it would greatly assist the negotiations if Annex I countries accepted liability for adaptation cost based on the responsibility principle. For one, it would provide motivation to take action to prevent climate change.[81] However, assuming that this is next to impossible to achieve in the context of the climate change negotiations,[82] can an Annex I commitment to taking responsibility for mitigation in Non-Annex I Parties be enough to break the impasse?

Given the potential net benefit to developing States of mitigation assistance that also helps with development goals, the answer is probably yes. There is every reason to think that a serious commitment to combine mitigation and development assistance will be received favourably by developing States even in the absence of a commitment to adaptation assistance. The more difficult practical question is when and under what circumstances would key developed States such as the US even agree to a meaningful commitment to fund mitigation in developing States?[83]

It seems that a strategy of development assistance toward a low GHG development path can build capacity to deal with impacts and adaptation, and can reduce the need for adaptation. That is not to say that there will not be need for

---

[81] This is not very helpful from a practical perspective, because a voluntary acknowledgement of liability will only take place once there is a preparedness to take responsibility. A finding of liability for impacts and adaptation by a credible source of international law, however, might influence some state's perspective on mitigation responsibility.

[82] At least without some external influences on developed countries, such as a finding of liability for impacts.

[83] Note that Mueller, *supra* note 12, at 6-2, expresses the view that for developing countries the impacts and adaptation side is a critical issue, and that future progress on mitigation is dependent on movement from Annex I Parties on impacts and adaptation.

adaptation help. If, however, mitigation is offered now in the context of a comprehensive sustainable development assistance strategy, it is not clear on what basis developing States would say no.

5. PROGNOSIS FOR THE CLIMATE CHANGE REGIME

The future of the climate change regime remains uncertain. Until the issue of liability for climate change impacts is resolved, it is unlikely that a completely satisfactory and effective way of distributing future obligations can be found. In the meantime, other crucial steps toward an effective regime can be taken. A number of external influences may also have a significant impact on the regime. Developments in the science of climate change are one. Technological advances to assist with mitigation are another, as is the growing gap between energy demand and supply globally. The need for international cooperation, as evidenced by such developments, is likely to put increasing pressure on States to find a way forward.

In the process, the current impasse in the climate change negotiations, if it continues, will put increasing pressures on existing international regimes to apply existing rules to the climate change context to motivate international cooperation. This in turn may motivate States to work to overcome the impasse through a combination of negotiations, the use of existing legal tools, such as those covered in Chapters 5 to 8 and reliance on other mechanisms such as the International Court of Justice to clarify the line between customary rules of State sovereignty and State responsibility.

The various proposals being considered for the climate change regime implicitly either propose one or a combination of the three principles as guiding principles for the long-term targets, or seek to sidestep the issue by proposing allocations without positioning them in the context of these principles. Furthermore, a number of the proposals are preoccupied with the transition period rather than the equity of the final outcome in terms of the allocation of mitigation and adaptation obligations and impact liability. At the same time, there are a number of proposals that should be able to accommodate a principled approach to long-term mitigation and adaptation obligations, once Parties resolve their differences over the underlying principles. The Multi-Track approach appears particularly well suited to provide a framework for the negotiation of future obligations.

Regardless of which of the proposals is used, the analysis carried out in this book points to the following steps to a more effective climate change regime:

- The inequity of the position of a number of key developed countries, most notably Australia and the United States should be exposed through credible international and domestic legal means, such as those explored in Chapters

5 to 8, as well as the International Court of Justice,[84] and domestic tort claims.[85]

- For negotiations on future commitments to succeed in achieving international consensus on targets that are effective in mitigating climate change, the negotiating capacity of developing countries, especially the least developed countries, needs to be enhanced to reduce the influence of OPEC within the G-77, and to allow developing States to formulate positions that will ensure effective mitigation without compromising legitimate development objectives.[86]
- Considerable work needs to take place bilaterally between developed States and emerging economies, such as China, India and Brazil. An agreement between these two groups on mitigation and development assistance in return for a low emissions development path is crucial.

Given the complexity of the issues, the state of negotiations, the position of the current US administration, and the absence of consensus on principles to guide future negotiations it is unlikely that the current impasse will be resolved easily or quickly. At the same time, the pressure on States to move forward with more meaningful mitigation measures globally will increase significantly over the next decade. The reality is that science is telling us that we are running out of time to save the planet from irreversible harm. Re-engagement of the United States, compliance with the first commitment period targets, and better leadership from the EU and developing States are essential prerequisites for the evolution of the regime. Clarification of liability for impacts and State responsibility for climate change under existing rules of international law can also greatly enhance the prognosis for the climate change regime.

---

[84] See Zaleke, *supra* note 72, at 261.

[85] See, for example, R. S. J. Tol, *et al*, "State Responsibility and Compensation for Climate Change Damages: A Legal and Economic Assessment" (2004) 32 Energy Policy 1109; A. Daniel, "Civil Liability Regimes as a Complement to Multilateral Environmental Agreements: Sound International Policy or False Comfort?" (2003) 12 R. E. C. I. E. L. 225; A. Boyle, "Globalising Environmental Liability: The Interplay of National and International Law" (2005) 17 J. Envtl. L. 3; and J. Brunnée, "Of Sense and Sensibility: Reflections on International Liability Regimes as Tools for Environmental Protection" (2004) 53 Int'l. & Comp. L. Q. 354.

[86] For a more detailed discussion of the challenges for developing country participation, see J. Gupta, "Global Environmental Governance: Challenges for the South from a Theoretical Perspective" in Biermann, F., *et al, A World Environment Organization: Solution or Threat for Effective International Environmental Governance?* (Aldershot, Ashgate Publishing Company, 2005), at 57. On pages 66 – 68, the author discusses the following 8 challenges for effective developing country participation in the negotiation of multilateral environmental obligations: 1. The hollow mandate; 2. The defensive negotiating strategy; 3. The handicapped coalition forming power; 4. The handicapped negotiating power; 5. The structural imbalance in bargaining; 6. The competing hypotheses of problem-solving; 7. Decreasing legitimacy in north-south negotiations; and 8. Regulatory competition and late-comers.

Epilogue: The Climate Change Regime and International Environmental Law

1. INTRODUCTION

Given the profile of climate change in international environmental law over the past decade, it is difficult to conclude without some comment on the potential influence of the climate change regime on international environmental law more broadly. The aim here is a modest one; to reflect on the potential role climate change may play in the future evolution of environmental law, and to offer some very preliminary thoughts on some of the key questions identified about the relationship between climate change and environmental law at the international level. Most importantly, the objective here is to pose and explore questions, not to provide answers.

A number of broader questions about international environmental law flow naturally from the preceding assessment of the climate change regime. How will the climate change regime influence whether, when and how MEAs will be used in the future to address global environmental challenges? Does the compliance system under the Kyoto Protocol provide a good model for other MEAs? To the extent that the effort toward global cooperation on climate change is considered inadequate, does the experience suggest ways to make environmental law more effective, and more capable of responding to global threats in a timely manner?

The comments here are, inevitably, preliminary for a number of reasons. The early stage of the development of the climate change regime makes it difficult to draw firm conclusions about its broader impact. A further limiting factor is that climate change represents only one part of the ever increasing volume of international law in the environmental field,[1] suggesting caution when drawing broader conclusions based on an assessment of the climate change regime and its links to a few other international regimes.

These constraints do not in any way diminish the important role of the climate change regime for the future of international environmental law. The stakes are high, and climate change is receiving more attention than most environmental

---

[1] See, for example, S. Obertur, *et al,* "Reforming International Environmental Governance: An Institutional Perspective on Proposals for a World Environment Organization" in F. Biermann, & S. Bauer, *A World Environment Organization: Solution or Threat for Effective International Environmental Governance?* (Aldershot, Ashgate Publishing Company, 2005) at 206.

issues subject to international law.[2] Because of the inevitable scientific and economic uncertainties surrounding the range of policy responses to climate change, it serves as a good illustration of the general challenges. Its success or failure can therefore be expected to offer valuable insights for international environmental law more generally. In the end, while it will be difficult to draw firm conclusions about international environmental law on the basis of the climate change regime alone, the influence is likely to be considerable.[3]

It is for these reasons that the focus of this Chapter is on the need for further work. As we move from the understanding of the climate change regime to a discussion of possible conclusions about international environmental law more generally, preliminary thoughts offered will have to be confirmed through further research involving other environmental regimes and other sources of international environmental law. Only then will there be an appropriate context for a full understanding of what the climate change regime may have to offer to international environmental law.

Within this context, the potential impact of the climate change regime is considered in three categories. Firstly, some predictions are made about the influence of the climate change regime on the use of regimes in the form of MEAs to address environmental challenges. Included in this discussion is the impact of the climate change regime on the effectiveness of compliance systems under MEAs. Secondly, opportunities for integration among international regimes are outlined. Integration here is considered in the same manner as in Chapter 8; to maximize opportunities to avoid conflict and identify co-benefits with the objectives of MEAs and possibly other international regimes and treaties.

Thirdly, the importance of States developing confidence in international institutions as an effective and fair way of resolving disputes is considered. The importance of having rules of state responsibility and liability as clearly established as possible as a way to provide the necessary context for successful negotiations in a specific context such as climate change is explored. In this context, the likelihood that the current impasse in the climate change and liability negotiations may lead to more reliance on formal dispute settlement processes, the potential for more progressive interpretation of treaties, and the potential role of improved access for non-state actors to arbitration and dispute settlement processes are also raised.

---

[2] Events such as heat waves and flooding in Europe, and extreme events such as hurricane Katrina in the US are bound to keep the focus on this issue.

[3] See Chapter 2, Section 3 for a more detailed discussion of the climate change regime.

2. THE FUTURE OF MULTILATERAL ENVIRONMENTAL AGREEMENTS

A successful climate change regime, one that is seen as instrumental in bringing about global cooperation to meet the ultimate UNFCCC objective of preventing dangerous human interference with the climate system, would undoubtedly reinforce the regime based approach[4] to global environmental challenges. Success, however defined, can more specifically be expected to support a continuation of the trend toward MEAs as an international response to global environmental challenges.

Success of the climate change regime will have a considerable impact on the reliance on regimes because of the complexity of the problem and its possible solutions. In particular, the short term economic costs associated with mitigation, the inequitable distribution of emissions and impacts, and the divergence of views on the relative roles of the principles of responsibility, capacity and potential in allocating mitigation and adaptation obligations, all are recognized as significant challenges to an effective response. The message would be that if the MEA approach can work for climate change, it can work for other, less complex, global environmental challenges.

If the regime in the end is perceived to have failed, either because it has not achieved the more narrow objectives of the UNFCCC or because it is seen to have failed more broadly to assist the global community in its efforts toward sustainability, serious questions about the ability of MEAs to respond in a timely and effective manner will have to be answered. A key issue will be why the ozone layer depletion (OLD) regime appears to have worked so well, while others, including the climate change regime, appear incapable of responding to the challenge.[5] Much will depend in this debate on whether the goal is defined narrowly to be about climate change, or more broadly to be about sustainability. The latter would imply an expectation that regimes deal effectively with the challenge of integration considered in Section 3 below.

In comparing climate change and ozone layer depletion, a key issue will be whether the differences in the outcome shed light on the effectiveness of MEAs, or whether they tell us more about the differences between the issues they sought to address. If the different outcomes are more a reflection of the different chal-

---

[4] That is, the approach of building institutions around a particular issue, and develop successive agreements to move from agreements on structures and principles to agreement on substantive obligations to address the issue at hand.

[5] See discussion of OLD regime and its relationship to the climate change regime in Section 5 of Chapter 1. An example of another regime that has so far not been able to demonstrate the level of success of the OLD regime is the CBD, discussed in Chapter 8.

lenges faced, such as the complexity of the science, the public engagement, the position of key actors such as the EU and the US, and the economic consequences of addressing the respective challenges, does this shed any light on how to design regimes, or on when regimes may be an effective way to achieve international cooperation to respond to an environmental challenge?

In answering these questions, it will be important to look below the surface and carefully consider the reasons for progress or failure. As discussed in Section 5 of Chapter 1, there are key differences between ozone layer depletion and climate change that may raise questions about the appropriateness of using the OLD regime as a template, but may also explain the slower progress of the climate change regime without necessarily indicating a problem with the approach. The complexity of the science, the great divergence of national circumstances on all aspects of climate change and particularly the uneven distribution of short term economic costs and benefits of climate change mitigation are some of the key circumstances that have to be considered in drawing any conclusions about whether or not the slow progress on climate change mitigation points to a problem with regimes.[6]

Regardless of whether or not the climate change regime is successful, the analysis in this book has demonstrated a few things about the role of regimes in international environmental law and the relationship between a given regime and other sources of international law. Some of these issues, such as the need for progressive interpretation of treaties, and the benefit of developing an international body of rulings by credible international tribunals are considered in Section 4 below.

The implication of the climate change regime for compliance with MEAs is worth further reflection, given the nature of the obligations and the unique features of the compliance system. As discussed in some detail in Chapters 3 and 4, the Kyoto compliance system has the opportunity to move compliance in the context of MEAs forward in at least two key respects. Firstly, it has the potential to raise the bar in terms of the level of cooperation within the international community on environmental issues. The level of effort expected of Parties is unprecedented for MEAs. This in turn has the potential to assist in developing a level of trust within the international community that deeper cooperation is possible on environmental issues, and that Parties can implement their commitment with confidence that others will do the same.

---

[6] A detailed assessment of these factors is beyond the scope of this discussion. For an overview of current views on these issues, see, for example, Watson, Robert T. *et al*, eds., *Climate Change 2001: Synthesis Report: A contribution of Working Groups I, II and III to the Third Assessment Report of the Intergovernmental Panel on Climate Change* (Cambridge, England: Cambridge University Press, 2002).

Secondly, if successful, the Kyoto compliance system will demonstrate that facilitation and enforcement are not mutually exclusive but can be complementary. This will allow future negotiations on compliance systems to move past the debate between facilitating and enforcing compliance and instead concentrate on implementing both as effectively and as well coordinated as possible. A related question remains under what circumstances stronger compliance systems will lead to deeper or shallower cooperation.[7] An empirically unexplored component of this question is whether proven performance of an incentives-based compliance system will lead to deeper cooperation in the future. If the compliance system proves to be strong and if it gains the trust and confidence of Parties, it may be possible to answer the question of whether or not confidence in a compliance system itself can lead to deeper cooperation.

Because the economic impact of mitigating climate change or a given Party is likely to depend on other nations taking similar steps to promote mitigation, willingness to act on climate change in the end should depend on two things; agreement on how to allocate the responsibility for climate change, and confidence that all States will do their share. The Kyoto compliance system therefore has the potential to dispel the myth that negotiating States will inevitably choose between substantively meaningful commitments and strong enforcement.

In the end, any consideration of the influence of the climate change regime should consider to what extent success or failure is linked to the regime itself, unique features of the climate change issues, or external influences. In other words, careful thought will have to be given to how the success or failure of the regime might be measured. Is success only about reducing GHG emissions so as to prevent dangerous human interference with the climate system as articulated in Article 2? What about adaptation? Furthermore, to what extent do we consider whether, in the process of addressing GHG emissions, we have created and/or solved other environmental, social, and economic problems? In other words, is the ultimate test whether the climate change regime contributes to sustainability generally or is the focus on GHG emission reductions? These issues were considered in Chapter 8. The broader implications of this for international environmental law are briefly outlined in the next Section.

---

[7] See Oona A. Hathaway, "Between Power and Principle: A Political Theory of International Law" (2005) [unpublished paper, on file with author]. Hathaway suggests that stronger compliance mechanisms in international treaties generally result in fewer states committing to the substance. There is, of course, also the potential that stronger compliance will lead to greater willingness to act. The proposition would be that the assurance that the global community is joining forces to address a global environmental problem will motivate individual states to act, and thereby overcome the tragedy of the commons.

3. INTEGRATION IN INTERNATIONAL LAW

The need for integration in responding to climate change was explored in some detail in Chapter 8. The climate change issue, perhaps more than other global environmental challenges, has brought to light the importance of integration, in moving from fighting fires to sustainability. The main challenge at this stage appears to be when, where and how to ensure an integrated response to the challenges ahead. As a starting point, there are opportunities for integration at the international and the national level.

Internationally, integration can take place before MEAs, regime structures and substantive obligations are agreed upon, or it can take place after some structures and possibly some substantive obligations are put in place. The alternative to integration internationally is to leave integration of various international regimes and their objectives, commitments and obligations to individual Parties. These basic choices are briefly explored here, as they have considerable implications for the future of international environmental law.

If integration is left to individual States, internationally things can proceed essentially as they have, with MEAs negotiated with limited consideration of other existing regimes, other than to try to avoid direct conflict and inconsistencies.[8] As discussed in Chapter 8, integration at the national level is possible, but at least in the case of Canada's Kyoto implementation plan, it does not appear to have taken place in any serious way. Experience has shown that even the most developed states tend to favour short term economic consideration over long term integrated planning for sustainability. Furthermore, there is a risk that integration at the national level will lead to choices that are in the national interest, but compromise the global interest in the process. At the same time, there are clearly some potential advantages to integration at a regional, national, or local level, such as the ability to maximize the benefits of integration by taking into account local circumstances more than may be possible if integration only takes place internationally.

At the international level, a relatively un-intrusive way of improving integration would be to improve coordination among regimes once their structures are established,[9] ideally starting before any substantive obligations are negotiated.

---

[8] See Kal Raustiala and David Victor, "The Regime Complex for Plant Genetic Resources" (2004) 58 International Organization 277, at 278-309, pointing to the problem of conflicts among treaties and how they are resolved over time in the context of plant genetic resources.

[9] For a discussion of the need for integration at the international level, see, for example, J. J. Kirton, "Generating Effective Global Governance: The North's Need for a World Environment Organization" in F. Biermann, & S. Bauer, *A World Environment Organization: Solution or Threat for Effective International Environmental Governance?* (Aldershot, Ashgate Publishing Company, 2005) at 149, 154.

The cooperation would have to be ongoing, as obligations under various regimes evolve over time. Such cooperation would also have to deal with differences in membership. A challenge would be how to ensure, given an inherent focus of each regime on addressing its own issues, that Parties as well as the institutions administering and supporting the respective regimes are sufficiently motivated to ensure proper integration at the international level.

An alternative or complementary form of integration at the international level would be to start the process much earlier, as new challenges are identified, before any of the negotiations on the structure or substance of the response are finalized. One way of achieving integration at this level would be through a world governance body, which could range from an enhanced UNEP, to a World Environment Organization (WEO) or a World Sustainability Organization.[10] There is a range of ways of achieving the objective of maximizing integration generally and more specifically before negotiating the substance of MEAs that have been proposed.[11]

The purpose here is not to resolve the debate which form of global governance is the most effective way to ensure appropriate integration at the international level, but to use the proposals for a WEO to illustrate how and when integration could take place more effectively internationally. The WEO is not used because it provides the best opportunity for integration, but because it may be the most realistic option to at least initiate the process of integration among environmental regimes. The potential role of a WEO in enhancing integration is therefore briefly discussed to explore the potential for integration at this early stage of developing international law.

The long term prognosis for the climate change regime is still very much open for debate. It is, however, clear that after more than fifteen years since formal negotiations began on the climate change regime, the impact of the regime on GHG emissions specifically and sustainability more generally has been limited. An interim assessment would suggest that international law has not been able to respond quickly enough to what is clearly a serious global environmental chal-

---

[10] A full analysis of the implications of establishing a WEO goes well beyond the issue of integration and is beyond the scope of this discussion. For a more detailed analysis of the potential for global governance on the environment, see J. Hierlmeier, "UNEP: Retrospect and Prospect – Options for Reforming the Global Environmental Governance Regime" (2002) 14 Geo. Int'l Envtl. L. Rev. 767, Dena Marshall, "An Organization for the World Environment: Three Models and Analysis" (2002) 15 Geo. Int'l Envtl L. Rev. 79, and Nicholas A. Robinson, "Befogged Vision: International Environmental Governance a Decade After Rio" (2002) 27 Wm. & Mary Envtl. L. & Pol'y Rev. 2.

[11] For a proposal toward improved global governance short of a WEO, see, for example, K. von Molke, "Clustering International Environmental Agreements as an Alternative to a World Environment Organization" in F. Biermann, & S. Bauer, *A World Environment Organization: Solution or Threat for Effective International Environmental Governance?* (Aldershot, Ashgate Publishing Company, 2005) at 175.

lenge. Most importantly, for this discussion, there is limited evidence so far of integration either nationally or internationally.[12]

How could a WEO assist with integration where UNEP has not? Proposals for a WEO certainly predate the creation of UNEP.[13] There are key differences between UNEP as it is currently structured and the various proposals for a WEO. One is that UNEP does not have the same status within the UN as other agencies. Another is that its mandate is limited, with its boundaries determined more based on the UN structure in existence in 1972 than any serious effort to determine an appropriate scope for UNEP.[14] As a result, UNEP as it currently exists is not well positioned to set priorities for environmental protection or sustainability, to improve coordination among existing regimes, or ensure integrated treaty implementation. It is even less likely to be able to ensure an integrated approach to treaty development.

What a WEO (or a reformed UNEP, if sufficiently empowered by the United Nations) could bring to international negotiations is a focus on key changes that are needed to put the global community on the track to sustainability. For example, a WEO could be given a mandate to translate climate change, resource depletion, air pollution, the lack of sustainable energy sources in developing countries and other environmental or sustainability issues into an overall focus for progress toward sustainability.

The end product of such a process by a WEO might be a conclusion that the key area in need of global action is renewable energy and energy conservation and efficiency as a way to address climate change in a manner consistent with other global objectives. A World Sustainability Organization might be able to take this integration a step further by ensuring the integration of environmental, social, and economic considerations in a context of global and intergenerational equity. Whether the world community is politically and technically ready for this additional step is a debate for another day.

The WEO would not necessarily prescribe the way to achieve the goals identified as a result of the integration process, but it would set the agenda for global negotiations. In other words, an organization such as the WEO could take the

---

[12] See Chapter 8.

[13] See S. Bauer, & F. Biermann, "The Debate on a World Environment Organization: An Introduction" in F. Biermann, & S. Bauer, *A World Environment Organization: Solution or Threat for Effective International Environmental Governance?* (Aldershot, Ashgate Publishing Company, 2005) at 1.

[14] For a good discussion of the shortcomings of UNEP, and the need for reform, see L. Elliot, "The United Nations' Record on Environmental Governance: An Assessment" in F. Biermann, & S. Bauer, *A World Environment Organization: Solution or Threat for Effective International Environmental Governance?* (Aldershot, Ashgate Publishing Company, 2005) at 27.

various challenges identified and translate them into a proposed focus on energy conservation and renewable energy.[15] It would be up to Parties to determine what they are willing to do to achieve the objectives set, but it would be up to the WEO to set the priorities and prevent the waste of resources on solutions to one symptom that creates other problems elsewhere, such as a switch from coal to nuclear power, reliance on sinks, bio-fuels, deep sea storage, carbon re-injection and so forth, as primary solutions to climate change.

This is not to say a WEO would be needed to provide this focus, as the international community could find other mechanisms to ensure that a key issue for sustainability is to find sustainable sources of energy, and that the best way to do that is to promote conservation, efficiency and certain renewable sources of energy.[16] Having said this, integration at this early stage has, so far, not taken place, and experience suggests that it may not lend itself to resolution through an issue-specific negotiated process.

One possible response, of course, is that it is clear from international efforts on climate change, sustainable development, and energy that the short term self interest of States and concerns over the loss of sovereignty stand in the way of progress, regardless of institutional changes. On the issue of integration, the point of establishing some form of global governance on these issues through an entity such as the WEO, would be to separate the "what" from the "how" by allowing an independent body to set the priorities while at the same time leaving it to individual States to decide how much they are willing to contribute to making progress on the priority issues identified. The point would be to ensure efforts States are willing to make are focused where they are most needed to move toward sustainability globally, not to take away State sovereignty. The issue of the effect of international cooperation on state sovereignty was discussed in section 4 of Chapter 3.

The idea of centralized integration raises numerous further questions. How can States be convinced to participate in a more centralized process for setting international priorities? How can the line between the "what" and the "how" be drawn? More generally, how would one set up an institution with this general mandate to ensure it can effectively serve the intended purposes?[17] Solutions such as building integration into early stages of negotiations clearly only provide a partial answer. What it does not achieve is a coordinated effort to focus international

[15] The example is taken from Chapter 8.

[16] Such as a reformed UNEP.

[17] These issues are beyond the scope of this discussion. For a detailed assessment of the pros and cons of a WEO, see: F. Biermann, & S. Bauer, *A World Environment Organization: Solution or Threat for Effective International Environmental Governance?* (Aldershot, Ashgate Publishing Company, 2005).

negotiations on key parts of the solutions (as opposed to treatment of particular symptoms that happen to catch the attention of the international community).[18]

In the end, the three options for integration discussed here can complement each other to a significant extent. Certainly, integration nationally, integration through better coordination among existing regimes and institutions, and integration through some central mechanism, are not necessarily mutually exclusive. For example, as the debate continues over the utility of the range of global governance options, efforts should continue to improve integration among existing international bodies. At the same time, integration at the national level is likely to be crucial over the long term, regardless of how effectively integration can be facilitated internationally.

4. THE ROLE OF ARBITRATION AND DISPUTE SETTLEMENT

Part II of this book, which deals with the connection between the climate change regime and a number of other international institutions, has already made the point that there are a number of dispute settlement processes that, if utilized, have the potential to influence both compliance with the Kyoto Protocol, and the future evolution of the climate change regime. Furthermore, as discussed in Chapter 9, there are barriers to progress within the climate change regime that will be difficult to resolve through negotiations, such as the issues of historical responsibility and liability.

As suggested in Chapter 9, existing rules of international law can be utilized through arbitration and dispute settlement procedures to resolve some of these differences, particularly given that much of the dispute can be related to existing principles of international law, including treaty obligations and customary principles of State sovereignty and State responsibility. It is not surprising, in this context of stalled negotiations and the existence of some relevant legal obligations, that litigation on responsibility and liability for climate change, both internationally and domestically, is being considered more and more as an alternative way forward.

The progressive interpretation of rules under international regimes such as UNCLOS, the WTO, and international human rights norms clearly has the potential to help resolve issues of responsibility and liability that are proving difficult to negotiate. The longer the negotiations stall, the greater will be the pressure on international tribunals and arbitrators to apply customary law principles and existing treaty obligations to determine responsibility and liability for climate change impacts and adaptation.

---

[18] See *Ibid.* for a detailed discussion of the broader implications and opportunities of a WEO.

In the process, international environmental law is likely to continue its gradual shift from a contract to a more legislative approach to international law, by interpreting long standing treaty obligations in light of new circumstances, exploring connections among regimes, and developing an overall international legislative system through progressive, evolving interpretation of treaty obligations.[19] This means treaties will be more and more interpreted in an evolving broader context, as opposed to being frozen in time. It means that older treaties will be interpreted in light of more recent treaties, in light of changes in customary law, and in light of changes in the relevant context, such as the discovery of new environmental challenges.[20] Given the ability of States to revoke their ratification of or assent to international treaties, a shift to a more progressive interpretation still preserves state sovereignty, and therefore should not discourage agreement or ratification. At the same time, it will make international law more relevant and reflective of current circumstances.

An important gap in international environmental law identified through this assessment of the climate change regime is the absence of a sufficient body of international law to give meaning to key principles such as State responsibility for transboundary harm, commitments to respect basic human rights, and obligations to prevent marine pollution to name a few. These gaps can be filled through more effective negotiations, through a greater willingness by States to use international legal tools to resolve disputes, and through broader access for non-state actors.

In the absence of successful negotiations on state liability for climate change, for example, increased litigation may have the secondary effect of helping to develop a body of rulings that would build some flesh around the bare bones of existing international law principles, whether customary or treaty based. There are areas where this is happening, such as WTO and international human rights, but it has so far not taken place to the same extent with respect to state responsibility and liability for transboundary environmental harm. Diplomatic channels are understandably often relied upon first, but their use comes at a price, as they do not further the development of international law principles in the way formal rulings do.

There are numerous contexts within which the potential benefit of rulings from credible international tribunals has been identified. Chapters 5 to 8 provide a number of examples where tribunal rulings have the potential to influence compliance with Kyoto and the future of the regime in a constructive manner. As

[19] See the introduction to Chapter 3 for a brief discussion on legislative versus contract approaches to treaties.

[20] See Section 3 of Chapter 6 for a discussion of treaty interpretation in light of such changes in circumstances.

early as 1990, some academics suggested that what may be needed to clarify the issue of State liability for climate change impacts is an advisory opinion from the ICJ.[21] As discussed in the previous Chapter, such a ruling may in fact be a critical external influence that could significantly enhance the evolution of the climate change regime.

While there are many areas of international law that could benefit from further clarification, the one of particular importance for the climate change regime and international environmental law generally is clearly the issue of State responsibility and liability for environmental harm.[22] Efforts to clarify rules on State responsibility and liability to date have included both negotiation and resort to dispute settlement processes. Negotiated efforts have included those on a case specific basis between States involved in a dispute over harm caused by the actions of one in the territory of the other, and more general efforts to clarify when States are responsible and how to determine liability.[23] In case of negotiated efforts in the context of a specific dispute between States, a few have resulted in formal dispute settlement before the ICJ or other international tribunals. More general efforts to clarify the rules of responsibility and liability include the Stockholm Conference, the Rio Conference, the efforts of the International Law Commission (ILC),[24] and negotiations under numerous regimes, including the climate change regime.

On State responsibility, the ILC has made considerable progress to codify and clarify the current state of customary international law. With respect to liability, however, its efforts have been less impressive. In either case, the main sources of law are in fact the few disputes that have been taken by the Parties before recognized international tribunals. The first of these was the Trail Smelter dispute between Canada and the United States in 1938 and 1941[25]. Other examples include the 1949 Corfu Channel Case,[26] the 1957 arbitration ruling on the Affaire du Lac Lanoux involving Spain and France,[27] the 1974 nuclear test cases involving

---

[21] See Durwood Zaelke et al, "Global Warming and Climate Change: An Overview of the International Legal Process" (1989-1990) 5 Am. U. J. Int'l L. & Pol'y 249, at 261.

[22] For a general overview of the state of customary international law on state responsibility for extraterritorial environmental harm, see Praticia Birnie & Alan Boyle, *International Law & the Environment*, 2nd ed. (Oxford: Oxford University Press, 2002), at 104 - 152. For a recent update, see A. E. Boyle, "Globalising Environmental Liability: The Interplay of National and International Law" (2005) 17 J. Envtl. L. 3.

[23] See Jutta Brunnée, "Of Sense and Sensibility: Reflections on International Liability Regimes as Tools for Environmental Protection" (2004) 53 Int'l. & Comp. L. Q. 351, at 352.

[24] For an update of the work of the ILC, see Boyle, *supra* note 23, at 5-7, and Brunnee, *ibid.* at 354

[25] *Trail Smelter Arbitration* (1931-1941), 3 R.I.A.A. 1905.

[26] *Corfu Channel Case (United Kindom v. Albania)* [1949] I.C.J. Rep. 4 (Merits).

[27] *Lake Lanoux Arbitration* (1957), 12 R.I.A.A. 281, 24 I.L.R. 101.

Australia, New Zealand and France,[28] and the fisheries jurisdiction cases that same year involving Iceland.[29] More recent cases include the 1996 case on the Legality of the Threat or Use of Nuclear Weapons,[30] and the 1997 case concerning the Gabcikovo-Nagymaros project between Hungary and Slovakia.[31]

Collectively, these cases have arguably done as much as or more to clarify State responsibility and liability for environmental harm than the efforts of the International Law Commission.[32] The state of customary law has been summarized as consisting of the following two principles:

(i) States have a duty to prevent, reduce and control pollution and environmental harm; and
(ii) States have a duty to cooperate in mitigating environmental risks and emergencies, through notification, consultation, negotiation, and in appropriate cases, environmental impact assessment.[33]

Three rulings by the ICJ have been instrumental in providing what little guidance there is on these general principles. In the nuclear test cases,[34] the majority of the ICJ panel sidestepped the substantive issues because France had already indicated an intention to stop nuclear testing. Some of the dissenting judgments, however, did address the substance and concluded that there is a *prima facie* obligation to conduct an environmental impact assessment to ensure that the proposed activities do not cause harm to the marine environment. Other minority judgments confirmed the obligation not to cause harm in accordance with principle 21 of the Stockholm Declaration.[35]

---

[28] *Nuclear Tests (Australia v. France)* [1973] I.C.J. Rep. 99 (Interim Measures), *Nuclear Tests (Australia v. France)* [1974] I.C.J. Rep. 253 (Merits), *Nuclear Tests (New Zealand v. France)* [1973] I.C.J. Rep. 135 (Interim Measures), *Nuclear Tests (New Zealand v. France)* [1974] I.C.J. Rep. 457 (Merits).

[29] *Icelandic Fisheries Case (UK v. Iceland)*, [1974] I.C.J. Rep. 3, *Icelandic Fisheries Case (Germany v. Iceland)*, [1974] I.C.J. Rep. 175.

[30] *Legality of the Use by a State of Nuclear Weapons in Armed Conflict*, Advisory Opinion, [1996] I.C.J. Rep. 68.

[31] *Case Concerning the Gabcíkovo-Nagymaros Project* [1997] I.C.J. Rep. 1.

[32] The ILC has been working on state responsibility since the 1950s and on liability for environmental harm since 1978. It has released draft articles on state responsibility for acts prohibited by international law. These articles have been finalized and sent to the UN, but have not been implemented by treaty to date. The ILC is still working on liability for acts not prohibited by international law. The ILC is currently exploring facilitating liability claims against non-state actors as an alternative to strict liability of states. See Birnie, *supra* note 22, at 105, and Boyle, *supra* note 22, at 5.

[33] See Birnie, *supra note 22,* at 104.

[34] *Supra* note 28.

[35] See Birnie, *supra* note 22, at 107.

In its advisory opinion on the threat or use of nuclear weapons,[36] the ICJ confirmed the obligation of States "to ensure that activities within their jurisdiction and control respect the environment of other States or areas beyond national control is now part of the corpus of international law relating to the environment".[37] Finally, in the Nagymaros Dam case,[38] the ICJ required the Parties to "co-operate in the joint management of the project, and to institute a continuing process of environmental protection and monitoring".[39]

Based on these rulings of the ICJ and other international tribunals, the bare bones of principles for the allocation of responsibility and liability appear to be in place. What is missing is a sufficient body of rulings from credible international tribunals to flesh out the detail to establish the boundaries for State responsibility and liability. In other words, we know that States have some responsibility for transboundary harm and can be liable, but we don't know where to draw the line and how to award damages.

These developments will not take place over night. However, with greater pressure on international tribunals to resolve disputes over extraterritorial environmental harm, we can expect to gradually develop a body of international law that will provide guidance and certainty on this issue. In the absence of a major breakthrough on the regime development side, the pressure on customary international law to resolve the issue of liability is likely to grow significantly. Only time will tell if and how international tribunals will respond to this challenge.[40]

Based on efforts to date, including those in the context of the climate change regime, it is difficult to see how the issue of State responsibility and liability for environmental harm will be resolved through negotiation and consensus. As much as a negotiated solution may be preferable, time for agreement on general principles to guide allocation of liability may be running out[41] given our increased understanding of the potential liability associated with the expected impacts. This means the pressure on existing dispute resolution mechanisms (both domestic and international) to apply existing rules of law to resolve these issues is likely to increase over the coming years.

---

[36] *Supra* note 30.

[37] *Ibid.* at Par. 29.

[38] *Supra* note 31.

[39] Birnie, *supra* note 22, at 108.

[40] See Laurence R. Helfer & Anne-Marie Slaughter, "Toward a Theory of Effective Supranational Adjudication" (1997) 107 Yale L. J. 273, at 314 on the importance of incrementalism to protect the credibility of the tribunal while allowing the rules of international law to develop.

[41] In particular on the issue of climate change, where the stakes are so clearly high in terms of liability for future harm.

With such an increase in litigation one can expect a greater body of international rulings to flesh out the rules on State responsibility and liability. This raises the question whether such litigation should be facilitated to expedite the development of this area of international law. As discussed, the current impasse in the climate change negotiations itself is likely to encourage some States concerned or affected by inaction on climate change to resort to international arbitration or dispute settlement to encourage action. In fact, the impasse has already started to encourage non-state actors such as the Inuit Circumpolar Conference (ICC) to ask an international body to rule on the relationship between climate change mitigation and their human rights under the IAHR regime. Small Island States have been considering legal action to hold GHG emitting States responsible for climate change impacts.[42] Within the EU, the use of WTO rules to protect domestic industries from unfair competition from States who have not take action on climate change must be expected. There are numerous examples of climate change related litigation domestically in the United States and other States.

It is important to note that to date, the only clear intention of referring a dispute to international arbitration or dispute settlement identified in the climate change context is that by a non-state actor, the ICC. This not by accident. States have been and likely will continue to be more reluctant than non-state actors to initiate formal dispute settlement processes for a wide variety of reasons, including a fear of damaging diplomatic relationships with other States.[43] To date, opportunities for non-state actors to participate in let alone initiate dispute settlement procedures are rare.

Allowing non-state actors access to arbitration and dispute settlement procedures could therefore be a constructive step forward, given the historical pattern of most States to rely on diplomatic channels rather than formal arbitration or dispute settlement. This would eliminate the concern States often have over the implication of dispute settlement procedures on their diplomatic relationships with other States. It is important to consider that similar access is already being provided to non-state investors in the context of the North American Free Trade Agreement and numerous bilateral agreements on investor protection.[44] What is suggested here is an application of the same principle to non-state actors affected

[42] P. Barton, "State Responsibility and Climate Change: Could Canada Be Liable to Small Island States?" (2002) 11 Dal. J. Leg. Stud. 65.

[43] For a discussion of State preference for diplomacy over dispute settlement, see, for example, Brunnee, *supra* note 23, at 353.

[44] For a more detailed discussion of investor protection under NAFTA, see Chris Tollefson, "Games Without Frontiers: Investor Claims and Citizen Submissions Under the NAFTA Regime" (2002) 27 Yale J. Int'l L. 141. The article also assesses the citizen complain procedure, a much weaker tool offered to non-state actors concerned about the failure of a NAFTA Party to enforce its own environmental laws.

by the failure of a State to properly exercise its responsibility for transboundary harm.[45]

What is not clear is whether the development of relevant existing international rules can take place in time for the climate change regime. On the one hand, climate change is a very difficult issue to serve as a mechanism for clarifying these existing rules, given its complexity and the stakes involved. On the other hand, time to let these principles develop through case law on less controversial issues or through efforts of the ILC may also be running out. For as long as the current impasse in the development of the climate change regime continues, the pressure on both is likely to continue to increase.

5. FINAL THOUGHTS

The climate change regime is poised to have a profound impact on international environmental law. Given the complexity of the problem as well as potential solutions and the high stakes in terms of the cost of impacts and mitigation, the climate change regime will likely continue to be at the centre of attention for a long time to come. The three issues briefly raised here only serve to illustrate the potential impact of the climate change regime on international environmental law.

There are numerous other implications that deserve further study. They include the impact of the regime on the role of Europe as a leader in the development of environmental law, the role of environmental issues in dealing with development issues in less developed States and the importance of capacity building to allow developing countries to participate as equal participants in future negotiations of any MEAs, not just those dealing with climate change. The potential of the climate change regime to challenge the WTO to rethink the relationship between trade and environment also deserves further study[46].

Other areas not explored here include the role of domestic enforcement action to motivate States to take climate change seriously and the possible ramifications for international environmental law.[47] Another unanswered question is whether effective mitigation of climate change is compatible with the western way of life, or whether continuing economic growth is sustainable. The international com-

---

45 A full consideration of this proposal is not possible. Rather, the intention here is to generally identify possible avenues forward in international environmental law. With respect to access for non-state actors, issues such as who damages should be awarded to would have to be resolved. Chapter 11 of NAFTA certainly allows non-state actors to receive awards.

46 Discussed in more detail in Chapter 5.

47 See also conclusion to Chapter 9, and Hathaway, *supra* note 7, where the author identifies domestic implementation as one of the key influences on the effectiveness of international law.

munity will continue to be challenged to rethink how it defines progress and quality of life generally, not just in the context of climate change mitigation. In the process, climate change will challenge the global community to rethink sustainability. Climate Change will also continue to challenge developed States to facilitate the shift to equity in international environmental law, and to change the role of developing countries. Only time will tell to what extent the potential climate change represents to the global community to reshape international environmental law will be realized.

It may turn out to be our last chance to get it right.

Table of Abbreviations

AAU	Assigned Amount Units
AOSIS	Association of Small Island States
CAN	Climate Action Network
CBD	Convention on Biological Diversity
CCSBT	Convention for the Conservation of Southern Bluefin Tuna
CDM	Clean Development Mechanism
CER	Certified Emission Reductions
CIEL	Center for International Environmental Law
CITES	Convention on International Trade of Endangered Species of Wild Flora and Fauna
CO2	Carbon Dioxide
COP	Conference of the Parties
CPR	Commitment Period Reserve
CTE	Committee on Trade and Environment
ENGO	Environmental Non Governmental Organization
ERT	Expert Review Team
ERU	Emissions Reduction Unit
ET	Emissions Trading
EU	European Union
FAO	Food and Agriculture Organization
G-77	Group of 77
GATT	General Agreement on Tariffs and Trade
UNESCO	United Nations Educational, Scientific and Cultural Organization
GDP	Gross Domestic Product
GEF	Global Environmental Facility
GESAMP	Joint Group of Experts on the Scientific Aspects of Marine Pollution
GHG	Greenhouse Gas Emissions
IACHR	Inter American Commission on Human Rights
IAHR	Inter American Human Rights
ICC	Inuit Circumpolar Conference
ICJ	International Court of Justice
ILO	International Labour Organization
IMO	International Maritime Organization
IPCC	Intergovernmental Panel on Climate Change
ITLOS	International Tribunal for the Law of the Sea
IUCN	World Conservation Union
JI	Joint Implementation
LDC	Least Developed Countries
LULUCF	Land Use, Land Use Change and Forestry

MCP	Multilateral Consultative Process
MEA	Multilateral Environmental Agreement
MOP	Meeting of the Parties
MOU	Memorandum of Understanding
MT	Megatonnes
NGO	Non Governmental Organization
OAS	Organization of American States
OLD	Ozone Layer Depletion
ODS	Official Development Assistance
OPEC	Organization of Oil Producing Countries
PPB	Parts per Billion
PPM	Parts per Million
PPM	Process and Production Method
RMU	Removal Unit
SD-PAM	Sustainable Development Policies and Measures
SPS	Sanitary and Phytosanitary Measures
TAR	Third Assessment Report
TBT	Technical Barriers to Trade
TED	Turtle Excluder Device
UG	Umbrella Group
UN	United Nations
UNCED	United Nations Conference on Environment and Development
UNCLOS	United Nations Convention on the Law of the Sea (1982)
UNEP	United Nations Environment Programme
UNFCCC	United Nations Framework Convention on Climate Change
US	United States
WEO	World Environment Organization
WHO	World Health Organization
WSSD	World Summit on Sustainable Development
WTO	World Trade Organization

Bibliography

Treaties

Additional Protocol to the American Convention on Human Rights, 17 November 1988, O.A.S.T.S. 69 (1988), 28 I.L.M. 156 reprinted in *Basic Documents Pertaining to Human Rights in the Inter-American System*, OEA/Ser.L.V/II.82 doc.6 rev.1 at 67 (1992) (entered into force 16 November 1999) [hereinafter San Salvador Protocol].

African Charter on Human Rights and Peoples Rights, 27 June 1981, OAU Doc. CAB/LEG/67/3 rev. 5, 21 I.L.M. 58 (entered into force 21 October 1986).

Agreement for Cooperation in Dealing with Pollution of the North Sea by Oil, 9 June 1969, 704 U.N.T.S. 3.

Agreement on Sanitary and Phytosanitary Measures (SPS), reprinted in *The Results of the Uruguay Round of Multilateral Trade Negotiations: The Legal Texts* (Geneva: GATT Secretariat, 1994) 69.

Agreement on Technical Barriers to Trade (TBT), reprinted in *The Results of the Uruguay Round of Multilateral Trade Negotiations: The Legal Texts* (Geneva: GATT Secretariat, 1994) 138.

American Convention on Human Rights, 22 November1969, 1144 U.N.T.S. 123, 9 I.L.M. 673 (entered into force July 18 1978).

Agreement on the Conservation of Nature and Natural Resources (ASEAN) 1985, 15 Envtl. Pol'y & L. 64 (not yet in force).

Barcelona Convention for the Protection of the Mediterranean Sea Against Pollution, 16 February 1976, 15 I.L.M. 290.

Basel Convention on the Transboundary Movements of Hazardous Wastes and their Disposal, 22 March 1989, 28 I.L.M. 657.

Charter of the Organisation of American States, 119 U.N.T.S. 3, (entered into force December 13 1951) reprinted in B. H. Weston *et al*, *Basic Documents in International Law and World Order,* 2nd ed, (St. Paul, MN: West Pub. Co., 1990) *50.*

Conference of the Parties to the Framework Convention on Climate Change: Kyoto Protocol, 10 December 1997, U.N. Doc. FCCC/CP/1997/L.7/add. 1, 37 I.L.M. 22 (1998), [hereinafter the Kyoto Protocol].

Convention between the United States of America and other Powers for the Regulation of Whaling, 24 September 1931, 49 U.S. Stat. 3079, 155 L.N.T.S. 349

Convention for the Conservation of Southern Bluefin Tuna (Adopted in Canberra, Australia 10 May 1993) 26 Law of the Sea Bulletin 57, online: Commission for the Conservation of the Southern Bluefin Tuna <http://www.ccsbt.org/docs/pdf/about_the_commission/convention.pdf>.

Convention for the Protection of Birds Useful to Agriculture, 19 March 1902, 102 B.F.S.P. 969, 191 Cons. T.S. 91(entered into force May 11, 1907).

Convention for the Protection of the Ozone Layer, 22 March 1985, S. TREATY DOC. NO. 9, 99th Cong., 1st Sess. 22 (1985), 26 I.L.M. 1529 (entered into force 22 September 1988), online: UNEP <http://www.unep.ch/ozone/pdfs/vienna-convention2002.pdf>.

Convention on Biological Diversity of the United Nations Conference on the Environment and Development, June 5, 1992, U.N. Doc. DPI/1307, reprinted in 31 I.L.M. 818 (CBD).

Convention on Civil Liability for Oil Pollution Damage, 29 November 1969, 12 U.S.T. 2989, 3 U.N.T.S. 3.

Convention on International Trade in Endangered Species of Wild Fauna and Flora (CITES), 3 March 1973, 27 U.S.T. 1087.

Convention on Long-Range Transboundary Air Pollution, 13 November 1979, 18 I.L.M. 1442 (entered into force 16 March 1983).

Convention on the Prevention of Marine Pollution by Dumping of Wastes and Other Matter, 29 December 1972, 26 U.S.T. 2403, 1046 U.N.T.S. 120.

Convention on the Prevention of Marine Pollution by Dumping of Wastes and Other Matter, London, 29 December 1972, 1046 U.N.T.S. 120 as amended by the *Protocol to the Convention on the Prevention of Marine Pollution by Dumping of Wastes and Other Matter*, London, 7 November 1996, 36 I.L.M. 1, [hereinafter London Convention].

Convention on Wetlands of International Importance Especially as Waterfowl Habitat (RAMSAR), 2 February 1971, 996 U.N.T.S. 245, 11 I.L.M. 963.

Covenant on the Rights of the Child, 20 November 1989, 28 I.L.M. 1448 (entered into force 2 September 1990).

General Agreement on Tariffs and Trade (GATT 1947), reprinted in *The Results of the Uruguay Round of Multilateral Trade Negotiations: The Legal Texts* (Geneva: GATT Secretariat, 1994) 485.

Helsinki Convention on the Protection and Use of Transboundary Watercourses and International Lakes, 17 March 1992, 31 I.L.M. 1312.

International Convention for the Prevention of Pollution from Ships, 2 November 1973, 1340 U.N.T.S. 184, as amended by *Protocol of 1978 relating to the International Convention for the Prevention of Pollution from Ships of 1973*, 17 February 1978, 1340 U.N.T.S. 61, [hereinafter MARPOL].

International Convention on the Elimination of All Forms of Racial Discrimination, 21 December 1965, 5 I.L.M. 350 (entered into force 4 January 1969).

International Covenant on Civil and Political Rights, 16 December 1966, 6 I.L.M. 368 (entered into force 23 March 1976).

International Covenant on Economic, Social and Cultural Rights, 16 December 1966, 6 I.L.M. 360 (entered into force 3 January 1976).

International Labour Organization Convention(no. 169) Concerning Indigenous and Tribal Peoples in Independent Countries, 27 June 1989, 72 I.L.O. Official Bulletin 59, (entered into force 5 September 1991), online: University of Minnesota <http://www1.umn.edu/humanrts/instree/r1citp.htm>, reprinted in B. H. Weston *et al*, *Basic Documents in International Law and World Order*, 2nd ed, (St. Paul, MN: West Pub. Co., 1990) 489.

Montreal Protocol on Substances that Deplete the Ozone Layer, 16 September 1987, amended at London on 29 June 1990, amended at Copenhagen on 25 November 1992, amended at Vienna in 1995, amended at Montreal on 17 September 1997, and amended at Beijing on 3 December 1999, 1522 U.N.T.S. 3, Can. T.S. 1989 No. 42, 26 I.L.M. 1550 (entered into force 1 January 1989), online: United Nations Environment Programme <http://www.unep.org/ozone/pdf/Montreal-Protocol2000.pdf>.

Multilateral Trade Negotiations Final Act Embodying the Results of the Uruguay Round of Multilateral Trade Negotiations, 15 April 1994, 33 I.L.M. 1125 reprinted in *The Results of the Uruguay Round of Multilateral Trade Negotiations: The Legal Texts* (Geneva: GATT Secretariat, 1994)1, (entered into force Jan. 1,

1995), online: World Trade Organization <http://www.wto.org/english/docs_e/legal_e/final_e.htm> [hereinafter WTO Agreement].

Protocol on Environmental Protection under the Antarctic Treaty, 21 June 1991, 30 I.L.M. 1455.

The *American Declaration of the Rights and Duties of Man*, adopted at the 9th International Conference of American States, Bogotá, Colombia, 1948, online: The Inter-American Commission on Human Rights <http://www.cidh.org/Basicos/basic2.htm>, reprinted in Inter-American Commission on Human Rights, *Handbook of Existing Rules Pertaining to Human Rights*, OAS Off. Rec., OEA/Ser.L/V/II.50 (1980) 17.

Treaty Between the United States and Great Britain providing for the Preservation and Protection of Fur Seals, 7 February 1911, United States-Great Britain, 37 Stat. 1538.

Treaty on the Non-Proliferation of Nuclear Weapons, 1 July 1968, 21 U.S.T. 483, (entered into force 5 March 1970).

United Nations Conference on Straddling Fish Stocks and Highly Migratory Fish Stocks: Agreement for the Implementation of the Provisions of the United Nations Convention of the Law of the Sea of 10 December 1982, Relating to the Conservation and Management of Straddling Fish Stocks (4 August 1995), 34 I.L.M. 1542. (1995 UN Fish Stock Agreement)

United Nations Convention on the Law of the Sea (UNCLOS), 10 December 1982, 21 I.L.M. 1261.

United Nations Convention to Combat Desertification (17 June 1994), 33 I.L.M. 1328.

United Nations Framework Convention on Climate Change, Intergovernmental Negotiating Committee for a Framework Convention on Climate Change OR, 5th Sess., Annex, UN Doc. A/AC.237/18 (PartII)/Add.1 (1992), 31 I.L.M. 849, online: UNFCCC <http://unfccc.int/resource/docs/a/18p2a01.pdf> [UNFCCC or The Framework Convention].

Vienna Convention on the Protection of the Ozone Layer, 22 March 1985, 26 I.L.M. 1529 (entered into force 22 September 1988).

WTO/GATT Rulings

United States – Prohibition of Imports of Tuna and Tuna Products from Canada (1982) BISD 29S/91 (Panel Report). The US import prohibition on Canadian tuna was found to violate Article XI:1 of GATT 1947, and not justified under Articles XI:2 or Article XX(g).

Canada – Measures Affecting Exports of Unprocessed Herring and Salmon (1988) BISD 35S/98 (Panel Report) (hereinafter Herring and Salmon Case). Canadian export restrictions were found to violate Article XI:1 of GATT 1947, and not justified under Articles XI:2(b) or Article XX(g).

Thailand – Restrictions on Importation of and Internal Taxes on Cigarettes (1990) BISD 37S/200 (Panel Report).

United States – Restrictions on Imports of Tuna (1991) BISD 39S/155 (Panel Report), (1994) DS29/R (Panel Report) (hereinafter Tuna-Dolphin Case).

United States – Taxes on Automobiles (1994) WTO Doc. DS31/R (Panel Report).

United States – Standards for Reformulated and Conventional Gasoline (1996) WTO Doc. WT/DS2/R, WT/DS2/AB/R (Panel Reports), online: <http://www.wto.org>.

European Communities – Measures Affecting Meat and Meat Products, (1998) WTO Doc. WT/DS26/R/USA, WT/DS26/AB/R, WT/DS48/R/CAN, WT/DS48/AB/R (Panel Report, as modified by the Appellate Body Report).

United States – Import Prohibition of Certain Shrimp and Shrimp Products, (1998) WTO Doc. WT/DS58/AB/R, DSR 1998:VII (Appellate Body Report) (hereinafter Shrimp Turtle (No.1)), online: <http://www.wto.org/english/tratop_e/dispu_e/disp_settlement_cbt_e/a1s1p1_e.htm>.

United States – Import Prohibition of Certain Shrimp and Shrimp Products, Recourse to Article 21.5 of the DSU by Malaysia (2001) WTO Doc. WT/DS58/AB/RW (hereinafter 'Shrimp Turtle (No. 2)'), online: <http://www.wto.org/english/tratop_e/dispu_e/disp_settlement_cbt_e/a1s1p1_e.htm>.

Australia – Measures Affecting Importation of Salmon (1998) WTO Doc. WT/DS18/R (Panel Report), WT/DS18/AB/R (Appellate Body Report), online: <http://www.wto.org/english/tratop_e/dispu_e/disp_settlement_cbt_e/a1s1p1_e.htm>.

Japan – Measures Affecting Agricultural Products (1999) WTO Doc. WT/DS76/R, WT/DS76/AB/R (hereinafter Agricultural Products Case).

European Communities – Measures Affecting Asbestos and Asbestos-Containing Products, (2001) WTO Doc. WT/DS135/R (Panel Report), WT/DS135/AB/R (Appellate Body Report) (hereinafter Asbestos Case), online: <http://www.wto.org/english/tratop_e/dispu_e/disp_settlement_cbt_e/a1s1p1_e.htm>.

European Communities – Trade Description of Sardines (2002) WTO Doc. WT/DS231/R (Panel Report), WT/DS231/AB/R (Appellate Body Report) (hereinafter Sardines Case), online: <http://www.wto.org/english/tratop_e/dispu_e/disp_settlement_cbt_e/a1s1p1_e.htm>.

United States-Taxes on Petroleum and Certain Imported Substances, (1987) WTO Doc. BISD 34S/136 (Panel Report), online: <http://www.wto.org/english/tratop_e/dispu_e/disp_settlement_cbt_e/a1s1p1_e.htm>.

United States – Countervailing Measures Concerning Certain Products from the European Communities, (2002) WTO Doc. WT/DS212/AB/R (Appellate Body Report) (hereinafter US Countervailing Measures), online: <http://www.wto.org/english/tratop_e/dispu_e/disp_settlement_cbt_e/a1s1p1_e.htm>.

United States – Preliminary Determinations With Respect To Certain Softwood Lumber From Canada, Report of the Panel (2002) WTO Doc. WT/DS236/R (Panel Report), online: <http://www.wto.org/english/tratop_e/dispu_e/dispu_status_e.htm>.

Submission of Saudi Arabia, Energy Taxation, Subsidies and Incentives in OECD Countries, (2002) WT/CTE/W/215, TN/TE/W/9 (WTO Committee on Trade and Environment).

European Communities - Provisional Safeguard Measures On Imports Of Certain Steel Products (2003) WTO Doc's; DS248/AB/R, DS249/AB/R, DS251/AB/R, DS252/AB/R , DS2530/AB/R, DS254/AB/R, DS258/AB/R, DS259/AB/R DS260/AB/R.

UNCLOS Rulings

Case concerning Land Reclamation by Singapore in and around the Straits of Johor (Malaysia v. Singapore), International Tribunal for the Law of the Sea, Request for the Prescription of Provisional Measures under Article 290, Paragraph 5, of the UN Convention on the Law of the Sea (10 September, 2003).

The MOX Plant Case (Ireland v. United Kingdom), International Tribunal for the Law of the Sea, Request for the Prescription of Provisional Measures under Article 290, Paragraph 5, of the UN Convention on the Law of the Sea (13 November 2001), reprinted in 41 I.L.M. (2002) 405.

International Court of Justice Ruling on the Delimitation of the Maritime Boundary in the Gulf of Maine Area (Canada v. United Sates of America) (October 21, 1984) Ruling 67, reported in 23 I.L.M. 1197, online: International Court of Justice <http://www.icj-cij.org/icjwww/icases/icigm/icigmframe.htm>.

Southern Bluefin Tuna Cases (New Zealand and Australia v. Japan), International Tribunal for the Law of the Sea, Request for the Prescription of Provisional Measures under Article 290, Paragraph 5, of the UN Convention on the Law of the Sea (27 August, 1999) [hereinafter 'ITLOS Ruling'].

Southern Bluefin Tuna Case, (August 4, 2000), UNCLOS Arbitral Tribunal, 39 ILM 1359 [hereinafter 'Arbitral Tribunal Ruling'].

Human Rights Rulings

Interpretation of the American Declaration of the Rights and Duties of Man Within the Framework of Arcticle 64 of the American Convention on Human Rights, (1989), Advisory Opinion OC-10/89, Inter-Am. Ct. H.R. (Ser. A) No. 10.

James Terry Roach and Jay Pinkerton v. United States, Case 9647, Res. 3/87, 22 September 1987 Annual Report 1986-1987, [46-49];

Rafael Ferrer-Mazorra et al v. United States, Report No. 51/01, Case 9903, April 4, 2001.

Inter-American Commission on Human Rights Res. VIII, Fifth Meeting of Consultation of Ministers of Foreign Affairs, OASOR, August 12-18 1959, Final Act, at 10-11, OEA/ser. C./II.5 (1960).Fifth Meeting of Consultation of Ministers of Foreign Affairs, Santiago, Chile, August 12 to 18, 1959, Final Act, Document OEA/Ser.C/II.5, at 4-11.

Christian B. White and Gary K. Potter v The United States of America and the Commonwealth of Massachusetts, Inter-Am. Ct. H.R. 25, 38, O.E.A./ser.L./V./ II.54, docv.9 rev. 1 (1980-81).

Velasquez Rodriguez Case, Inter-American Court on Human Rights, (1988), (ser. C) No. 4, reprinted in (1989) I.L.M. 291

Report on the Situation of Human Rights in Ecuador (Ecuador Report), (24 April 1997), reprinted in Inter-Am. Ct. H.R. OEA/ser. L. /V./II. 96, doc. 10 rev. 1.

Status of the Yanomami Indians, available online at the American University College of Law, online: Washington College of Law Center for Human Rights and Humanitarian Law <http://www.wcl.american.edu/pub/humright/digest/Inter-American/english/annual/1984_85/res1285.html>.

The Mayagna (Sumo) Awas Tingni Community Case (*Tingni Community Case*), (2001) Case No. 11.557, Inter-Am. Ct. H.R. 79, online at the University of Minnesota <http://www.umn.edu/humanrts/iachr/AwasTingnicase.html>.

Other Primary Sources

Agenda 21, United Nations Conference on Environment and Development, Annex, UN Doc. A/CONF.151/26/Rev.1 (Vol. 1-6) [hereinafter Agenda 21].

Declaration on the Right to Development, GA Res.41/128, UN GAOR, annex 41 UNDocA/41/53 (entered into force 4 December 1986).

Declaration of the United Nations Conference on the Human Environment, 16 June 1972, UN GAOR, U.N. Doc. A/CONF.48/14/Rev.1 (1973), 11 I.L.M. 1416, online: United Nations Environment Programme <http://www.unep.org/Documents/Default.asp?DocumentID=97&ArticleID=1503>, [hereinafter Stockholm Declaration].

Department of Trade and Industry (UK), *Energy White Paper: Our Energy Future–Creating a Low Carbon Economy* (Norwich: The Stationary Office, 2003), online: DTI Energy Group <http://www.dti.gov.uk/energy/whitepaper/ourenergyfuture.pdf>.

Draft Declaration of Principles on Human Rights and the Environment, U.N. Doc. E/CN.4/Sub.2/1994/9 Annex. 1 (1994).

Draft Inter-American Declaration on the Rights of Indigenous Peoples, approved by the Inter-American Commission on Human Rights on February 26, 1997. Reprinted in *Basic Documents Pertaining to Human Rights in the Inter-American System*, OEA/Ser/L/V/II, 90 Doc. 9 rev. 1.

Government of Canada Project Green, *Moving Forward on Climate Change, A Plan for Honouring our Kyoto Commitment*, (Ottawa, Government of Canada, 2005), online: Government of Canada <http://www.climatechange.gc.ca>.

Government of Canada, *Climate Change Plan for Canada*, (Ottawa: Government of Canada, 2002), online: Government of Canada <http://www.climatechange.gc.ca>.

Resolution establishing the *Inter-American Commission on Human Rights*; Res. VIII, Fifth Meeting of Consultation of Ministers of Foreign Affairs, OASOR, 12-18 August 1959, Final Act, at 10-11, OEA/ser. C./II.5 (1960). In 1970, an OAS Charter amendment referred to as the "Protocol of Buenos Aires" opened for signature on 27 February 1967, 21 U.S.T. 607, (entered into force February 27 1970) changed the status of the Commission to that of an official organ of the OAS.

Rio Declaration on Environment and Development, 3-14 June 1992, United Nations Conference on Environment and Development, A/CONF 151/5/Rev.1, 31 I.L.M. 874 online: United Nations <http://www.un.org/documents/ga/conf151/aconf15126-1annex1.htm> [hereinafter Rio Declaration].

Rules of Procedure of the Inter-American Commission on Human Rights, approved by the Commission at its 109th Sess., December 2000, (entered into force 1 May 2001), online: Inter-American Commission on Human Rights <http://www.cidh.org/basicos/basic16.htm>.

Statute of the Inter-American Commission on Human Rights, OAS Res No. 447, 9th Sess., (entered into force October 1979), online: University of Minnesota <http://www1.umn.edu/humanrts/oasinstr/zoas6cts.htm>.

Statute of the International Court of Justice, 26 June 1945, 59 U.S. Stat. 1055.

Stockholm Convention on Persistent Organic Pollutants (22 May 2001) 40 ILM 532.

The American Convention on Human Rights, 22 November 1969, 1144 U.N.T.S. 123, 9 I.L.M. 673 (entered into force July 18 1978), reprinted in Inter-American Commission on Human Rights, *Handbook of Existing Rules Pertaining to Human Rights*, OAS Off. Rec., OEA/Ser.L/V/II.50 (1980) 27.

Understanding on Rules and Procedures Governing the Settlement of Disputes, Annex 2 to World Trade Organization Agreement, 15 April 1994, 33 I.L.M. 1226.

Universal Declaration of Human Rights GA Res 217 (III), UN GAOR, 3d Sess., Supp. No. 13, UN Doc. A/810 (1948) 71, reprinted in B. H. Weston *et al*, *Basic Documents in International Law and World Order,* 2nd ed, (St. Paul, MN: West Pub. Co., 1990) 298.

World Charter for Nature, G.A. Res. 37/7, UN GAOR 37th Sess., U.N. Doc. A/RES/37/7 (1982), online: United Nations < http://www.un.org/documents/ga/res/37/a37r007.htm>.

Reports

Bruce, James P., *et al*, eds., *Climate Change 1995: Economic and Social Dimensions of Climate Change: Contribution of Working Group III to the Second Assessment of the Intergovernmental Panel on Climate Change* (Cambridge, England: Cambridge University Press, 1996)

Climate Change Review (2001), online: U.S. EPA <http://yosemite.epa.gov/oar/globalwarming.nsf/UniqueKeyLookup/SHSU5BNM7H/$File/bush_ccpol_061101.pdf>.

Houghton, John T. *et al*, eds., *Climate Change 1995: The Science of Climate Change: Contribution of Working Group I to the Second Assessment of the Intergovernmental Panel on Climate Change* (Cambridge, England: Cambridge University Press, 1996).

Houghton, John T. *et al*, eds., *Climate Change 2001: The Scientific Basis: Contribution of Working Group I to the Third Assessment Report of the Intergovernmental Panel on Climate Change* (IPCC) (Cambridge, England: Cambridge University Press, 2002).

IMCO/FAO/UNESCO/WMO/WHO/IAEA/UN/UNEP Joint Group of Experts on the Scientific Aspects of Marine Pollution (GESAMP), *Environmental Capacity: An Approach to Marine Pollution Prevention*, No. 80 Reports and Studies (Nairobi: United Nations Environment Programme, 1986).

IMCO/FAO/UNESCO/WMO/WHO/IAEA/UN/UNEP Joint Group of Experts on the Scientific Aspects of Marine Pollution (GESAMP), *Interchange of Pollutants Between the Atmosphere and the Oceans*, No.13 Reports and Studies (Geneva: World Meteorological Organization, 1980).

IMCO/FAO/UNESCO/WMO/WHO/IAEA/UN/UNEP Joint Group of Experts on the Scientific Aspects of Marine Pollution (GESAMP), *Thermal Discharges in the Marine Environment*, No. 24 Reports and Studies (Rome: Food and Agriculture Organization of the United Nations, 1984).

Intergovernmental Panel on Climate Change, *Climate Change and Biodiversity* (Technical Paper, IPCC Working Group II, Technical Support Unit) (IPCC, April 2002).

IPCC, *Second Assessment Climate Change 1995: A Report of the Intergovernmental Panel on Climate Change* (Cambridge, England: Cambridge University Press, 1996).

McCarthy, James J. *et al*, eds., *Climate Change 2001: Impacts, Adaptation & Vulnerability: Contribution of Working Group II to the Third Assessment Report of the Intergovernmental Panel on Climate Change* (IPCC) (Cambridge, England: Cambridge University Press, 2002).

Metz, Bert *et al*, eds., *Climate Change 2001: Mitigation: Contribution of Working Group III to the Third Assessment Report of the Intergovernmental Panel on Climate Change* (IPCC) (Cambridge, England: Cambridge University Press, 2002).

Options for Enhanced Cooperation among the three Rio Conventions, FCCC/SBSTA/2004/INF.19 (November 2, 2004).

Report of the Conference of the Parties on its Seventh Session, Conference of the Parties, United Nations Framework Convention on Climate Change (UN FCCC), 29 October – 10 November 2001, FCCC/CP/2001/13/Add.1(Decisions 1/CP.7 - 14/CP.7), FCCC/CP/2001/13/Add.2(Decisions 15/CP.7 - 19/CP.7), FCCC/CP/2001/13/Add.3(Decisions 20/CP.7 - 24/CP.7), FCCC/CP/2001/13/Add.4(Decisions 25/CP.7 - 39/CP.7 & Resolution 1/CP.7 – 2/CP.7), online: UN FCCC <http://unfccc.int/2860.php> [hereinafter COP 7 or Marrakesh Accords].

Report of the Conference of the Parties on its Eighth Session, Conference of the Parties, United Nations Framework Convention on Climate Change (UN FCCC), 23 October – 1 November 2002, FCCC/CP/2002/7/Add.1(Decisions 1/CP.8 - 20/CP.8), FCCC/CP/2002/7/Add.2(Decisions 21/CP.8 - 25/CP.8 & Resolution 1/CP.8), online: UN FCCC <http://unfccc.int/2860.php> [hereinafter COP 8].

Report of the Conference of the Parties on its Fourth Session, held at Buenos Aires from 2 to 14 November 1998, Addendum, Part Two: Action Taken by the Conference of the Parties at its Fourth Session, UNFCCCOR, UN Doc. FCCC/CP/1998/16/Add.1 (1999) [hereinafter Buenos Aires Plan of Action].

Report of the Conference of the Parties on its Ninth Session, Conference of the Parties, United Nations Framework Convention on Climate Change (UN FCCC), 1-12 December 2003, FCCC/CP/2003/6/Add.1(Decisions 1/CP.9 - 16/CP.9), FCCC/CP/2003/6/Add.2(Decisions 17/CP.9 - 22/CP.9 & Resolution 1/CP.9), online: UN FCCC <http://unfccc.int/2860.php> [hereinafter COP 9].

Report of the Conference of the Parties on its Tenth Session, Conference of the Parties, United Nations Framework Convention on Climate Change (UN FCCC),

6-18 December 2004, FCCC/CP/2004/10/Add.1(Decisions 1/CP.10 - 11/CP.10), FCCC/CP/2004/10/Add.2(Decisions 12/CP.10 - 18/CP.10), online: UN FCCC <http://unfccc.int/2860.php> [hereinafter COP 10].

Report on Biological Diversity and Climate Change, Subsidiary Body on Scientific, Technical and Technological Advice under the Convention on Biological Diversity, 30 September 2003, UNEP/CBD/SBSTTA/9/INF/12, Annex III, Page 11 (Ad Hoc Expert Report).

Report of the United Nations Conference on Environment and Development, (Rio de Janeiro, 3-14 June 1992) ACONF 151/26 vol. 1, online: United Nations <http://www.un.org/documents/ga/conf151/aconf15126-1annex1.htm>.

Report on the United Nations Conference on the Human Environment at Stockholm: Final Documents, 16 June 1972, 11 I.L.M. 1416.

Watson, Robert T. *et al*, eds., *Climate Change 2001: Synthesis Report: A contribution of Working Groups I, II and III to the Third Assessment Report of the Intergovernmental Panel on Climate Change* (Cambridge, England: Cambridge University Press, 2002).

Watson, Robert T., *et al*, eds., *Climate Change 1995: Impacts, Adaptations and Mitigation of Climate Change: Scientific-Technical Analyses: Contribution of Working Group II to the Second Assessment of the Intergovernmental Panel on Climate Change* (Cambridge, England: Cambridge University Press, 1996).

"Winning the Battle Against Global Climate Change", Commission of the European Communities, Brussels, 9 February 2005 {SEC(2005) 180} online: <http://72.14.207.104/search?q=cache:crDpXngSfK8J:europa.eu.int/comm/environment/climat/pdf/comm_en_050209.pdf/>.

Books and Articles

Abbott, Kenneth W., "Modern International Relations Theory: A Prospectus for International Lawyers" (1989) 14 Yale J. Int. L. 335.

Abbott, W. *et al*, "The Concept of Legalization" (2000) 54 International Organization 401.

Abdel-khalik, Jasmine, "Prescriptive Treaties in Global Warming: Applying the Factors Leading to the Montreal Protocol" (2001) 22 Mich. J. Int'l L. 489.

Acquatella, Jean, *Private Finance and Investment Issues in GHG Offset Projects* (Geneva: International Academy of the Environment, 1998).

Adams, Deborah, "Greenhouse Gas Controls: The Future of Tradeable Permits" (1997) Financial Times Energy Publishing 31.

Adler, Jonathan H., ed., *The Costs of Kyoto* (Washington: Competitive Enterprise Institute, 1997).

Allenby, Braden, "Global Warning" (2000) 17 The Environmental Forum 30.

Alvarez Jose E., "Why Nations Behave" (1998) 19 Mich. J. Int'l Law 303.

Aman, Alfred C. Jr., "Introduction: the Montreal Protocol and the Future of Global Legislation" (1993) 15 Law & Pol'y 1.

Ambrose, K. A., "Science and the WTO" (2000) 31 Law & Pol'y Int'l Bus. 861.

Amiott, Jennifer A., "Environment, Equality, and Indigenous Peoples' Land Rights in the Inter-American Human Rights System: *Mayagna (Sumo) Indigenous Community of Awas Tingni v. Nicaragua*" (2002) 32 Environmental Law 873.

Anaya, S. J, "Indigenous Rights Norms in Contemporary International Law" (1991) 8 Ariz. J. Int'l & Comp. L. 1.

Anaya, S. J., "The Awas Tingni Petition to the Inter-American Commission on Human Rights: Indigenous Lands, Loggers, and Government Neglect in Nicaragua" (1998) 9 St. Thomas L. Rev. 157.

Anderson, Molly, "Demonstrable Progress on Climate Change, Prospects and Possibilities" in Trevor Findlay and Oliver Meier, eds., *Verification Yearbook 2003* (London: The Verification Research, Training and Information Centre (VERTIC), December 2003) 191.

Anderson, Molly, "Verification under the Kyoto Protocol" in Trevor Findlay and Oliver Meier, eds., *Verification Yearbook 2002* (London: The Verification Research, Training and Information Centre (VERTIC), December 2002) 147.

Andresen, Steinar *et al*, "The Role of Green NGO's in Promoting Climate Compliance" FNI report 4/2003 (Lysaker: The Fridtjof Nansen Insitute, 2003).

Angelova, Anastasia A., "Compelling Compliance with International Regimes: China and the Missile Technology Control Regime" (1999) 38 Colum. J. Transnat'l L. 419.

Annan, Kofi, "Containing Global Climate Change: a Global Challenge" (2001) 25:2 The Fletcher Forum of World Affairs 5.

Ardia, David S., "Does the Emperor Have No Clothes? Enforcement of International Laws Protecting the Marine Environment" (1998) 19 Mich. J. Int'l L. 497.

Arnold, Frank S., "What Are We Buying with Kyoto?" (1999) 16:4 The Environmental Forum 14.

Atapattu, S., "The Right to a Healthy Life or the Right to Die Polluted?: The Emergence of a Human Right to a Healthy Environment under International Law" (2002) 16 Tul. Envtl. L.J. 65.

Athanasiou, Tom, *et al*, "Cutting the Gordian Knot: Adequacy, Realisms and Equity" December 7, 2004 (on file with author).

Ausubel, J.H., *et al*, "Verification of International Environmental Agreements" (1992) 17 Annual Review of Energy and Environment 1.

Ayres, I. *et al*, *Responsive Regulation: Transcending the Deregulation Debate* (New York: Oxford University Press, 1992)

Bachelder, Andrew, "Using Credit Trading to Reduce Greenhouse Gas Emissions" (2000) 9 J. Envtl. L. & Prac. 281.

Bailey, Zoya E., "The Ship that Sank the Hague: a Comment on the Kyoto Protocol" (2002) 16 Temp. Int'l & Comp. L. J. 103.

Ballentine, Roger, "Kyoto an Important First Step in Fighting Warming (Global Warming)" (2000) 17:3 The Environmental Forum 39.

Banuri, Tariq *et al*, "Equity and Social Considerations" in James P. Bruce *et al*, eds., *Climate Change 1995 - Economic and Social Dimensions of Climate Change:* Contribution of Working Group III to the Second Assessment Report of the Intergovernmental Panel on Climate Change (Cambridge: Cambridge University Press, 1995) 1.

Barnes, James N., "Pollution from Deep Ocean Mining" in Douglas Johnston ed., *The Environmental Law of the Sea* (Gland, Switzerland, IUCN, 1981), at 259

Barratt-Brown, Elizabeth P., "Building a Monitoring and Compliance Regime under the Montreal Protocol" (1991) 16 Yale J. Int'l L. 519.

Barrett, Scott, "International Environmental Agreements: Compliance and Enforcement: International Cooperation and the International Commons" (1999) 10 Duke Envtl. L. & Pol'y F. 131.

Barton, P., "State Responsibility and Climate Change: Could Canada Be Liable to Small Island States?" (2002) 11 Dal. J. Leg. Stud. 65.

Barton, P., "Economic Instruments and the Kyoto Protocol: Can Parliament Implement Emissions Trading without Provincial Co-operation?" (2002) 40 Alta. L. Rev. 417.

Basset, C., *et al*, "Implementing the UNCCD: Towards a Recipe for Success" (2003) 12 R.E.C.I.E.L. 133.

Batruch, Christine, "Hot Air as Precedent for Developing Countries? Equity Considerations" (1999)17 UCLA J. Envtl. L. & Pol'y 45.

Baumert, Kevin A. *et al*, eds., *Building on the Kyoto Protocol: Options for Protecting the Climate* (Washington, D.C.: World Resources Institute, 2002), online: WRI <http://pdf.wri.org/opc_full.pdf>.

Bederman, David J., "Constructivism, Positivism and Empiricism in International Law" (2001) 89 Geo. L. J. 469.

Bello, Judith H., *et al*, "The Post-Uruguay Round Future of Section 301" (1994) 25 Law & Pol'y Int'l Bus. 1297.

Bello, Judith Hippler, "The WTO Dispute Settlement Understanding: Less Is More" (1996) 90 Am. J. Int'l L. 416.

Benedick, Richard E., *Ozone Diplomacy: New Directions in Safeguarding the Planet* (Cambridge: Harvard University Press 1991).

Benedick, Richard E., *Ozone Diplomacy: New Directions in Safeguarding the Planet* 2d ed. (Cambridge: Harvard University Press 1998).

Benedick, Richard Elliot, "The Montreal Ozone Treaty: Implications for Global Warming" (1990) 5 Am. U. J. Int'l L. & Pol'y 227.

Benedickson, Jamie, "The Great Lakes and the Mediterranean Sea: Ecosystem-Management and Sustainability in the Context of Economic Integration" (2004) 14 J. Envtl. L. & P. 107

Betsill, Michele, "Environmental NGOs meet the Sovereign State: the Kyoto Protocol Negotiations on Global Climate Change" (2002) 13 Colo. J. Int'l Envtl. L. & Pol'y 49.

Bhaksar, V., "Distributive Justice and the Control of Global Warming" in V. Bhaskar and Andrew Glyn, eds., *The North the South and the Environment: Ecological Constraints and the Global Economy* (London: Earthscan Publications Ltd., 1995) 102.

Biermann, F., & Bauer, S., *A World Environment Organization: Solution or Threat for Effective International Environmental Governance?* (Aldershot, Ashgate Publishing Company, 2005)

Birnie, Practicia Z., & Boyle, Alan, *International Law & the Environment*, 2nd ed. (Oxford: Oxford University Press, 2002).

Bishop, Kirsten, "Liberalized Trade and International Environmental Law and Policy: Australia's Negotiations under the Kyoto Protocol" (1998) 36 Can. Y.B. Int'l L. 181.

Blaustein, Richard, "Global Warming to be Focus of '96 Conference: Parties to the Framework Convention on Climate Change will Report on Preventive Efforts" (1995) 18:13 Nat'l L. J. C9.

Blegen, Bryce, "International Cooperation in Protection of Atmospheric Ozone: the Montreal Protocol on Substances that Deplete the Ozone Layer" (1988) 16 Denv. J. Int'l L. & Pol'y 413.

Bodansky, Daniel, "The United Nations Framework Convention on Climate Change: a Commentary on a Commentary. (In Commentaries on Daniel Bodansky, The United Nations Framework Convention on Climate Change: A Commentary)" (2000) 25 Yale J. Int'l L. 315.

Bodansky, Daniel, "The United Nations Framework Convention on Climate Change: A Commentary" 18 Yale J. Int'l L. 451.

Boer, B., *et al*, "Legal Aspects of sustainable Soils: International and National" (2003) 12 R.E.C.I.E.L. 149.

Booncharoen, Charlotte *et al*, "International Commitment Toward Curbing Global Warming: the Kyoto Protocol" (1998) 4 Envtl. Law 917.

Borgstrom, Robert E., *et al*, "U.S. Gas Pipelines: the Challenge of Global Warming" (1999) 137 Public Utilities Fortnightly 66.

Bosselmann, Klaus, "Power, Plants and Power Plants: New Zealand's Implementation of the Climate Change Convention" (1995) 12 Envtl. & Planning L. J. 423.

Boyd, David R., *Unnatural Law: Rethinking Canadian Environmental Law and Policy* (Vancouver: UBC Press, 2003).

Boyle, Alan E., "Some Reflections on the Relationship of Treaties and Soft Law" (1999) 48 I.C.L.Q. 901.

Boyle, Alan E., "Globalising Environmental Liability: The Interplay of National and International Law" (2005) 17 J. Envtl. L. 3.

Boyle, Alan E., *et al*, eds., *Human Rights Approaches to Environmental Protection* (Oxford: Clarendon Press, 1996)

Boynton, J., "Issue Salience in Climate Change and Biodiversity Discourses" (Paper prepared for presentation at the 45th Annual Meeting of the International Studies Association, Montreal, Quebec, Canada March 17-20, 2004) [unpublished].

Brack, D., International Environmental Disputes, (Royal Institute of International Affairs, 2001)

Bradbrook, A, "Energy Law: The Neglected Aspect of Environmental Law" (1993) 19 Melbourne U. L. Rev. 1.

Bradbrook, Adrian J., "Energy efficiency and the Energy Charter Treaty" (1997) 14 Envtl. & Planning L. J. 327.

Braithwaite J, *To Punish or Persuade: Enforcement of Coal Mine Safely* (Albany: State University of New York Press, 1985).

Bredhauer, Jacqueline, "Tree Clearing in Western Queensland—a Cost Benefit Analysis of Carbon Sequestration" (2000) 17 Environmental and Planning Law Journal 383.

Breidenich, Clare *et al*, "The Kyoto Protocol to the United Nations Framework Convention on Climate Change" (1998) 92 Am. J. Int'l. L. 315.

Brethour, Patrick, *et al*, "Kyoto not binding, Manley says" The Globe and Mail [Toronto], (14 November 2002) A7.

Brierly, J.L., *The Outlook for International Law* (Oxford: Clarendon Press, 1944) 1-2, cited in H. J. Morgenthau, *Politics Among Nations: the Struggle for Power and Peace* (Brief ed.), revised by K. W. Thompson (New York, NY: McGraw-Hill Inc., 1993) 253.

Brimeyer, Benjamin L., "Bananas, Beef, and Compliance in the World Trade Organization: the Inability of the WTO Dispute Settlement Process to Achieve Compliance from Superpower Nations" (2001) 10 Minn. J. Global Trade 133.

Brooks, Gerald, "Environmental Economics and International Trade: An Adaptive Approach" (1993) 5 Geo. Int'l Envtl. L. Rev. 277.

Brotmann, Matthew, "The Clash Between the WTO and the ESA: Drowning a Turtle to Eat a Shrimp" (1999) 16 Pace Envtl. L. Rev. 321.

Brown, Chester, "Facilitating Joint Implementation under the Framework Convention on Climate Change: Toward a Greenhouse Gas Emission Reduction Protocol" (1997) 14 Environmental and Planning Law Journal 356.

Brown, Chester, *et al*, "People in Greenhouses ... the Kyoto Protocol and its Impact on Australian Industry and Legal Practice" (2000) 74:8 Law Institute Journal 54.

Brown Weiss, Edith *et al*, *International Environmental Law: Basic Instruments and References* (Salem, NH: Butterworth, 1992).

Brown Weiss, Edith, "International Environmental Law, Contemporary Issues and the Emergence of a New World Order" (1993) 81 Geo. L. J. 675.

Brown Weiss, Edith, *et al*, *Engaging Countries' Strengthening Compliance with International Environmental Accords* (Cambridge: The MIT Press, 1998).

Brownlie, Ian, "The Reality and Efficacy of International Law" [1981] Brit. Y.B. Int'l L. 1.

Bruch, Carl *et al*, "The Road from Johannesburg: Type II Partnerships, International Law, and the Commons" (2003) 15 Geo. Int'l Envtl. L. Rev. 855.

Brunnée, Jutta, "A Fine Balance: Facilitation and Enforcement in the Design of a Compliance Regime for the Kyoto Protocol" (2000) 13 Tul. Envtl L. J. 223.

Brunnée, Jutta, "Of Sense and Sensibility: Reflections on International Liability Regimes as Tools for Environmental Protection" (2004) 53 Int'l. & Comp. L. Q. 351.

Brunnée, Jutta, "Review: Legalization and World Politics, Edited by Judith L. Goldstein, Miles Kahler, Robert O. Keohane, and Anne-Marie Slaughter (2001)" (2003) 1/1 Perspectives on Politics (American Political Science Association) 231.

Brunnée, Jutta, "The Kyoto Protocol: A Testing Ground for Compliance Theories?" (2003) 63:2 Heidelberg Journal of International Law 255.

Brunnée, Jutta, *et al*, "Environmental Security and Freshwater Resources: Ecosystem Regime Building" (1997) 91 Am. J. Int'l L. 26.

Brunnée, Jutta, *et al*, "Interactional International Law" (2001) 3 FORUM 186.

Brunnée, Jutta, *et al*, "International Law and Constructivism: Elements of an Interactional Theory of International Law" (2000) 39 Colum. J. Transnat'l L. 19.

Brunnée, Jutta, *et al*, "Persuasion and Enforcement: Explaining Compliance with International Law" (2002) 13 Finnish Yearbook of International Law 1.

Brunnée, Jutta, *et al*, "The Changing Nile Basin Regime: Does Law Matter?" (2002) 43 Harv. Int'l L. J. 105.

Brunner, Annick Emmenegger, "Conflicts Between International Trade and Multilateral Environmental Agreements" (1997) 4 Ann. Surv. Int'l & Comp. L. 74.

Bryce, James T., "Controlling the Temperature: an Analysis of the Kyoto Protocol" (1999) 62 Sask. L. Rev. 379.

Bryk, Dale S., "The Montreal Protocol and Recent Developments to Protect the Ozone Layer" (1991) 15 Harv. Envtl. L. Rev. 275.

Bucholtz, Barbara K., "Coase and the Control of Transboundary Pollution: the Sale of Hydroelectricity Under the United States-Canada Free Trade Agreement of 1988" (1991) 18 B. C. Envtl. Aff. L. Rev. 279.

Buergenthal, Thomas, *International Human Rights*, 2nd ed. (St. Paul, MN: West Pub. Co., 1995) 176

Bugnion, Veronique, *et al*, "A Game of Climate Chicken: Can EPA Regulate Greenhouse Gases before the U.S. Senate Ratifies the Kyoto Protocol?" (2000) 30 Envtl. L. 491.

Burger, M., "Bi-Polar and Polycentric Approaches to Human Rights and the Environment" (2003) 28 Colum. J. Envtl. L. 371.

Burke, Michael E. IV, "China's Stock Markets and the World Trade Organization" (1999) 30 Law & Pol'y Int'l Bus. 321.

Burley, Anne-Marie, "Law Among Liberal States: Liberal Internationalism and the Act of State Doctrine" (1992) 92 Colum. L. Rev. 1907.

Butler, Jo Elizabeth, "The Establishment of a Dispute Resolution / Non-compliance Mechanism in the Climate Change Convention" (1997) 91 Proceedings of the Annual Meeting-American Society of International Law, Annual 250.

Byers, Michael, *Custom, Power and the Power of Rules: International Relations and Customary International Law* (Cambridge: Cambridge University Press, 1999).

Byers, Michael, ed., *The Role of Law in International Politics: Essays in International Relations and International Law* (Oxford: Oxford University Press, 2001).

Caflisch, L.C., "The Settlement of Disputes Relating to the Sea-bed Area" in C. Rozakis & C.A. Stephanou eds., *The New Law Of The Sea: Selected and Edited Papers of the Athens Colloquium on The law of the Sea, September, 1982* (Amsterdam & New York: North Holland, 1983) 279.

Cameron, James, *et al*, "Legal and Regulatory Strategies for GHG Reductions: a Global Survey" (2001) 15 Nat. Resources & Env't 176.

Cameron, Peter, "From Principles to Practice: the Kyoto Protocol.(Protocol to the United Nations Framework Convention on Climate Change)" (2000) 18 Journal of Energy & Natural Resources Law 1.

Canan, Penelope, *et al*, "Ozone Partnerships, the Construction of Regulatory Communities, and the Future of Global Regulatory Power" (1993) 15 Law & Pol'y 61.

Capretta, Annette M., "The future's so bright, I Gotta Wear Shades: Future Impacts of the Montreal Protocol on Substances that Deplete the Ozone Layer" (1988) 29 V. J. Int'l L. 211.

Carr, Donald A., *et al*, "The Kyoto Protocol and U.S. Climate Change Policy: Implications for American Industry" (1998) 7 R.E.C.I.E.L. 191.

Carson, Rachel, *Silent Spring* (London: H. Hamilton, 1963)

Charney, J.I., "Implementing the United Nations Convention on the Law of the Sea: Impact of the Law of the Sea Convention on the Marine Environment" (1995) 7 Geo. Int'l Envtl. L. Rev. 731.

Charney, Jonathan I., "The Implications of Expanding International Dispute Settlement Systems: The 1982 Convention on the Law of the Sea" (1996) 90 Am. J. Int'l L. 69.

Charney, Jonathan I., "The Marine Environment and the 1982 United Nations Convention on the Law of the Sea" (1994) 28 Int'l Law. 879.

Charney, Jonathan, "Third Party Dispute Settlement and International Law" (1997) 36 Colum. J. Transnat'l L. 65.

Charney, Jonathan, "Third State Remedies in International Law" (1989) 10 Mich. J. Int'l L. 57.

Charnovitz, Steve, "Environmental Trade Sanctions and the GATT: An Analysis of the Pelly Amendment on Foreign Environmental Practices" (1994) 9 Am. U. J. Int'l.L & Pol'y 751.

Charnovitz, Steve, "NAFTA and the Expansion of Free Trade: Current Issues and Future Prospects, Critical Guide to the WTO's Report on Trade and Environment" (1997) 14 Ariz. J. Int'l & Comp. L. 341.

Chase, Steven, "Liberals to end Debate on Kyoto, Sources Say" *Globe & Mail* (5 December 2002).

Chayes, Abram, *et al*, *The New Sovereignty: Compliance with International Regulatory Agreements* (Harvard: Harvard University Press, 1995).

Cheyne, Ilona, "Environmental Unilateralism and the WTO GATT System" (1995) 24 Ga. J. Int'l & Comp. L. 433.

Chinn, Lily N., "Can the Market be Fair and Efficient? An Environmental Justice Critique of Emissions Trading" (1999) 26 Ecology L.Q. 80.

Choi, Susan, "Judicial Enforcement of Arbitration Awards Under the ICSID and New York Conventions" (1996) 28 N.Y.U.J. Int'l L. & Pol. 175.

Choudhury, Keya, *et al*, *Integration of Biodiversity Concerns in Climate Change Mitigation Activities* (Berlin: Federal Environmental Agency, 2004).

Churchill, Robin, "The International Tribunal for the Law of the Sea: Survey for 2002" (2003) 18 Int'l J. Mar. & Coast. L. 447.

Chylek, Petr, *et al*, eds., *1st International Conference on Global Warming and the Next Ice Age,* (Conference Proceedings, Halifax, Nova Scotia, Canada, 19-24 August 2001) (Halifax: Dalhousie University, 2001).

Clapp, Jennifer, "The Illegal CFC Trade: An Unexpected Wrinkle in the Ozone Protection Regime" (1997) 9 International Environmental Affairs 259.

Clark, Sean S., *et al*, "Installment Six of the Climate Treaty Debate: a Report on COP-6" (2001) 15 Nat. Resources & Env't 180.

Claussen, Eileen, *et al*, "The Complex Elements of Global Fairness" (Washington, D.C.: Pew Center on Global Climate Change, Oct. 29, 1998) online: Pew Center on Global Climate Change <http://www.pewclimate.org/global-warming-in-depth/all_reports/equity_and_climate_change/index.cfm>.

Climate Action Network (CAN) "A Viable Global Framework for Preventing Dangerous Climate Change" (2003) CAN Discussion Paper, online: Climate Action Network <http://www.climatenetwork.org/pages/publications.html>.

Coghlan, Matthew, "Prospects and Pitfalls of the Kyoto Protocol to the United Nations Framework Convention on Climate Change" (2002) 3 Melbourne Journal of International Law 165.

Cohan, John Alan, "Environmental Rights of Indigenous Peoples under the Alien Tort Claims Act, the Public Trust Doctrine and Corporate Ethics, and Environmental Dispute Resolution" (2001) 20 ULCA J. Envtl. L. & Pol'y 133.

Connor, Steve "The Final Proof: Global Warming is a Man-made Disaster" *The Independent* (19 February 2005).

Cooper, Deborah E., "The Kyoto Protocol and China: Global Warming's Sleeping Giant" (1999) 11 Geo. Int'l Envtl. L. Rev. 401.

Coppock, Rob, "Implementing the Kyoto Protocol" (1998) 13 Issues in Science & Technology 66.

Cranston, Maurice, *What are Human Rights?* (New York: Taplinger Publishing Co.,1973) 36.

Croley, Steven P., *et al*, "WTO Dispute Procedures, Standard of Review, and Deference to National Governments" (1996) 90 Am. J. Int'l L. 193.

Crosby, Hans J., "The World Trade Organization Appellate Body - United States v. Venezuela: Interpreting the Preamble of Article XX - Are Possibilities for

Environmental Protection Under Article XX(g) of GATT Disappearing?" (1998) 9 Vill. Envtl L.J. 283.

Currie, R., "Human Rights and International Mutual Legal Assistance: Resolving the Tension" (2000) 11 Crim. L.F. 143.

Cusack, Vincent, "Perceived Costs versus Benefits of Meeting the Kyoto Target for Greenhouse Gas Emission Reduction: the Australian Perspective" (1999) 16 Environmental and Planning Law Journal 53.

Daniel, A., "Civil Liability Regimes as a Complement to Multilateral Environmental Agreements: Sound International Policy or False Comfort?" (2003) 12 R. E. C. I. E. L. 225.

Danish, Kyle, Book Review of *The New Sovereignty: Compliance with International Agreements* by Abram Chayes and Antonia Handler Chayes (1997) 37 Va. J. Int'l L. 789.

Dannenmaier, Eric *et al*, "Promoting Meaningful Compliance with Climate Change Commitments" (Washington, D.C.: Pew Center on Global Climate Change, 2000), online: Pew Center on Global Climate Change <http://www.pewclimate.org/docUploads/compliance%2Epdf>.

Davey, William J., "Issues of WTO Dispute Settlement" (1997) 91 Am. Soc'y Int'l L. Proc. 279.

Davidson, Christine B., "The Montreal Protocol: the First Step Toward Protecting the Global Ozone Layer" (1988) 20 N.Y.U.J. Int'l L. & Pol. 793.

Davidson, M. G., "Protecting Coral Reefs: The Principal National and International Legal Instruments" (2002) 26 Harv. Envtl. L. Rev. 499.

Davies, Michael, "Poland, the Kyoto Protocol and Opportunities for Emissions Trading" (2003) 5 International Energy Law and Taxation Review 160.

De Klemm, Cyrille, "Living Resources of the Ocean" in Douglas Johnston ed., *The Environmental Law of the Sea* (Gland, Switzerland, IUCN, 1981), at 71

Dejeant-Pons, M. *et al*, *Human Rights and the Environment* (Strasbourg: Council of Europe Publishing, 2002).

Delaume, Georges R., ICSID Arbitration and the Courts, (1983) 77 Am. J. Int'l L. 784.

Dernbach, John C., "Making Sustainable Development Happen: From Johannesburg To Albany" (2004) 8 Alb. L. Envtl. Outlook 173.

Dernbach, John C., "Targets, Timetables and Effective Implementation Mechanisms: Necessary Building Blocks for Sustainable Development" (2002) 27 Wm. & Mary Envtl. L. & Pol'y Rev. 79.

DeSombre, Elizabeth R.., "The Experience of the Montreal Protocol: Particularly Remarkable, and Remarkably Particular" (2001) 19 UCLA J. Envtl. L. & Pol'y 49.

DeSombre, Elizabeth, *et al*, "The Montreal Protocol Multilateral Fund: Partial Success Story" in Robert O. Keohane & Marc A. Levy, eds., *Institutions for Environmental Aid*: Pitfalls and Promise (Cambridge: The MIT Press, 1996) 89.

DiLuigi, Denee A., "Kyoto's So-called Fatal Flaws: a Potential Springboard for Domestic Greenhouse Gas Regulation" (Symposium Comment on Rio's Decade: Reassessing the 1992 Earth Summit) (2002) 32 Golden Gate U. L. Rev. 693.

DiMento, Joseph F. C., "Lessons Learned" (2000/2001), 19 UCLA J. Envtl. L. & Pol'y 281.

DiMento, Joseph F. C., "Process, Norms, Compliance, and International Law" (2003) 18 J. Envtl. L. & Litig. 251.

Dismukes, David E. *et al*, "Clear Skies or Storm Clouds Ahead? The Continuing Debate over Air Pollution and Climate Change" (2003) 51 Oil, Gas & Energy Quarterly 823.

Dismukes, David E. *et al*, "Environmental Costs: Clean Air, Kyoto, and The Boy Who Cried Wolf" (2000) 49 Oil, Gas & Energy Quarterly 529.

Doelle, M., "The Kyoto Protocol; Reflections on its Significance on the Occasion of its Entry into Force" (2005) 27:2 Dal. L.J. 555 [forthcoming in 2005].

Doelle, M., "The Quiet Invasion, Law and Policy Responses to Invasive Species in North America" (2003) 18 Int'l J. Mar. & Coast. L. 261.

Dommen, C., "Claiming Environmental Rights: Some Possibilities Offered by the United Nations Human Rights Mechanisms" (1998) 11 Geo. Int'l Envtl. L. Rev. 1.

Donahue, W. M., "Equivalence: Not Quite Close Enough for the International Harmonization of Environmental Standards" (2000) 30 Envtl. L. 363.

Donnelly, Jack, *International Human Rights*, 2nd ed. (Boulder: Westview Press, 1998).

Doolittle, Diane M., "Underestimating Ozone Depletion: the Meandering Road to the Montreal Protocol and Beyond" (1989) 16 Ecology L.Q. 407.

Dorsey, Michael K., "The Promise and Threat of Climate Justice: Geographies of Resistance in the Context of Uneven Development" (unpublished manuscript, on file with author)

Dowling, A. C., "Un-Locke-ing a Just Right Environmental Regime: Overcoming the Three Bears of International Environmentalism - Sovereignty, Locke, and Compensation" (2002), 26 Wm. & Mary Envtl. L. & Pol'y Rev. 891.

Downs, George W., "Enforcement and the Evolution of Cooperation" (1997 - 1998) 19 Mich. J. Int'l L. 319.

Downs, George W., *et al*, "Is the Good News about Compliance Good News about Cooperation?" (1996) 50 International Organization 379.

Downs, George W., *et al*, "Reputation, Compliance, and International Law" (2002) 31 J. Legal Stud. 95.

Downs, George W., *et al*, "The Transformational Model of International Regime Design: Triumph of Hope or Experience?" (2000) 38 Colum. J. Transnat'l L. 465.

Drennen, Thomas E., "Economic Development and Climate Change: Analyzing the International Response" (1993) (unpublished Ph.D. book, Cornell University) cited in Henry Shue, "After You: May Action by the Rich be Contingent on Action by the Poor?" (1994) 1 Ind. J. Global Legal Stud. 343.

Drumbl, M. A., "Northern Economic Obligation, Southern Moral Entitlement, and International Environmental Governance" (2002) 27 Colum. J Envtl. L. 363.

Dudek, Dan *et al*, "Cooperative Mechanisms under the Kyoto Protocol -The Path Forward" (1 June 1998) online: Environmental Defense <http://www.environmentaldefense.org/pdf.cfm?ContentID=747&FileName=Path-Forward.pdf>

Duffy, Michael F., "Prometheus Re-Bound: how Adoption of the Kyoto Protocol on Climate Change would Devastate the Western U.S. Coal Industry" (1999) 77 Denv. U.L. Rev. 265.

Dunoff, Jeffrey L., "Kyoto Protocol Treaty Means Business - for Attorneys (Environmental Law)" (1998) 21:14 Pennsylvania Law Weekly S11.

Dupuy, Pierre-Marie, "Soft Law and the International Law of the Environment" (1991) 12 Mich. J. Int'l L. 420.

Duruigbo, Emeka, "International Relations, Economics and Compliance with International Law: Harnessing Common Resources to Protect the Environment and Solve Global Problems" (2001) 31 Cal. W. Int'l L. J. 177.

Duval, Lee Anne, "The future of the Montreal Protocol: money and methyl bromide" (1999) 18 Virginia Environmental Law Journal 609.

Ehrenstein, Michael David, "A Moralistic Approach to the Ozone Depletion Crisis" (1990) 21 U. Miami Inter-Am. L. Rev. 611.

Ehrmann, Markus, "Procedures of Compliance Control in International Environmental Treaties" (2002) 13 Colo J. Int'l Envtl L & Pol'y 377.

Eisen, Joel B., "From Stockholm to Kyoto and Back to the United States: International Environmental Law's Effect on Domestic Law" (1999) 32 U. Rich. L. Rev. 1435.

Elder, P.S., "Biological Diversity and Alberta Law" (1996) 34 Alta. L. Rev. 293.

Elrifi, Glenn B. Raiczyk, "Montreal Protocol on Substances that Deplete the Ozone Layer: Conference Calling for Accelerated Phase-out of Ozone-Depleting Chemicals is Planned for 1992" (1991) 5 Temp. Int'l & Comp. L.J. 363.

Elrifi, Ivor, "Protection of the Ozone Layer: a Comment on the Montreal Protocol" (1990) 35 McGill L.J. 387.

Elzen (den), M. G. J., *et al*, "Meeting the EU 2 Degree Climate Target: Global and Regional Emission Implications" (2005) Report 728001031/2005, Netherlands Environmental Assessment Agency, online: <www.mnp.nl>.

Environment Directorate, Organisation for Economic Co-Operation and Development, "Responding to Non-Compliance Under the Climate Change Regime" OECD Information Paper (May 28, 1999) ENV/EPOC(99)21/FINAL online: OECD, Environment Directorate <http://www.oecd.org/LongAbstract/0,2546,en_2649_34359_2385395_1_1_1_1,00.html>.

Erijta, Mars Campins, *et al*, "Compliance Mechanisms in the Framework Convention on Climate Change and the Kyoto Protocol" (2004) 34 R.G.D. 51.

Estapá, Jaume Saura, "Flexibility Mechanisms in the Kyoto Protocol: Constitutive Elements and Challenges Ahead" (2004) 34 R.G.D. 107.

Esty, Daniel C., "An Environmental Perspective on Seattle" (2000) 3 J. Int'l Econ. L. 176.

Esty, Daniel C., "The Fount of Climate Change Scholarship" (In Commentaries on Daniel Bodansky, The United Nations Framework Convention on Climate Change: A Commentary (2000) 25 Yale J. Int'l L. 318.

Evans, M.D., "Decisions of International Tribunals: The Southern Bluefin Tuna Arbitration" (2001) 50 I.C.L.Q. 447.

Farber, Daniel A., "Taking Slippage Seriously: Noncompliance and Creative Compliance in Environmental Law" (1999) 23 Harv. Envtl. L. Rev. 297.

Faye, Andrew A., "APEC and the new regionalism: GATT compliance and prescriptions for the WTO" (1996) 28 Law & Pol'y Int'l Bus. 175.

Ferrante, Alison Raina, "The Dolphin/Tuna Controversy and Environmental Issues: Will the World Trade Organization's Arbitration Court and the International Court of Justice's Chamber for Environmental Matters Assist the United States and the World in Furthering Environmental Goals?" (1996) 5 J. Transnat'l L. & Pol'y 279.

Finnemore, Martha, "Are Legal Norms Distinctive?" (2000) 32 N.Y.U.J. Int'l L. & Pol. 699.

Finnemore, Martha, *et al*, "Alternatives to Legalization: Richer Views of Law and Politics" (2001) 55 International Organization 743.

Fisher, R., *Improving Compliance with International Law* (Charlottesville, VA: University Press of Virginia, 1981).

Fitzgerald, Jack, "The International Panel on Climate Change: Taking the First Steps towards a Global Response" (1990) 14 S. Ill. U.L.J. 231.

Fitzmaurice, Sir Gerald, "The Foundations of the Authority of International Law and the Problem of Enforcement" (1956) 19 The Mod. L. Rev. 1.

Fletcher, Charles R., "Greening World Trade: Reconciling GATT and Multilateral Environmental Agreements Within the Existing World Trade Regime" (1996) 5 J. Transnat'l L. & Pol'y 341.

Forkan, Patricia, "Do not sacrifice dolphins on the altar of free trade" *Pantagraph* (28 April 1997) A9.

Fort, Jeffrey C., *et al*, "Can Emissions Trading Work Beyond a National Program?: Some Practical Observations on the Available Tools" (1997) 18 U. Pa. J. Int'l Econ. L. 463.

Franck, Thomas M., "Legitimacy in the International System" (1998) 82 Am. J. Int'l L. 705.

Franck, Thomas M., "Principles of Fairness in International Law" (Paper presented at the Annual Meeting of the American Political Science Association, 1992).

Franck, Thomas M., *Fairness in International Law and Institutions* (Oxford: Clarendon Press, 1995).

Franck, Thomas M., *The Power of Legitimacy Among Nations* (New York: Oxford University Press, 1990).

Frank, E. Loy, "The United States Policy on the Kyoto Protocol and Climate Change" (2001) 15 Nat. Resources & Env't. 152.

Freeland, Steven, "The Kyoto Protocol: an Agreement without a Future?" (2001) 24 U.N.S.W.L.J. 532.

French, Duncan, "1997 Kyoto Protocol to the 1997 UN Framework Convention on Climate Change" (1998) 10 J. Envtl. L. 227.

Friedland, David M., *et al*, "Worldwide Community Takes Action on Ozone; Amendments to the Montreal Protocol Lead to New Regulations" (1993) 15:41 Nat'l L.J. 30.

Friends of the Earth International, "Tree Trouble: A Compilation of Testimonies on the Negative Impact of Large Scale, Monoculture Tree Plantations" (Released at the sixth Conference of the Parties of the UNFCCC in The Hague, October, 2000) [unpublished, on file with author].

Frischmann, Brett, "A Dynamic Institutional Theory of International Law" (2003) 51 Buffalo L. Rev. 679.

Frischmann, Brett, "Using the Multi-Layered Nature of International Emissions Trading and of International-Domestic Legal Systems to Escape a Multi-State Compliance Dilemma" (2001) 13 Geo. Int'l Envtl. L. Rev. 463.

Fuller, Lon L., *The Morality of Law*, 1969 ed. (Yale University Press: New Haven, 1964).

Gaines, Sanford E., "Foreword: Integrating Environmental Considerations Into the World Trading System" (1994) 13 Stan. Envtl. L.J. vii.

Gallagher, Anne, "The 'New' Montreal Protocol and the Future of International Law for Protection of the Global Environment" (1992) 14 Hous. J. Int'l L. 267.

Gardiner, Caterina, "Trade and Environment in the GATT/WTO" (1998) 16 The Irish Law Times and Solicitors' Journal 20.

Gardner, Allison F., "Environmental Monitoring's Undiscovered Country: Developing a Satellite Remote Monitoring System to Implement the Kyoto Protocol's Global Emissions-Trading Program" (2000) 9 N.Y.U. Envtl. L. J. 152.

George, C., "Testing for Sustainable Development through Environmental Assessment" (1999) 19:2 *Environmental Impact Assessment Review* 175.

Gibbs, W. Wayt, "The Treaty that Worked – Almost" (1995) 273:3 Scientific American 20.

Giorgetti, Chiara, "From Rio to Kyoto: a Study of the Involvement of Non-Governmental Organizations in the Negotiations on Climate Change" (1999) 7 N.Y.U. Envtl. L.J. 201.

Global Commons Institute, online: GCI <http://www.gci.org.uk/>, for technical support and information concerning "Contraction and Convergence" a planning model, "Contraction and Convergence Options," is also available for download.

Goldberg, Donald *et al*, "Building a Compliance Regime under the Kyoto Protocol", (1998) Center for International Environmental Law & EuroNatura (CIEL, 1998), online: CIEL <http://www.ciel.org/Publications/buildingacomplianceregimeunderKP.pdf>.

Goldberg, Donald M. *et al*, "Responsibility for Non-Compliance Under the Kyoto Protocol's Mechanisms for Cooperative Implementation" (Washington, D.C.: The Centre for International Environmental Law and Euronatura-Centre for Environmental Law and Sustainable Development, 1998), online: CIEL <http://www.ciel.org/Publications/pubccp.html>.

Goldberg, Donald, *et al*, "The Compliance Fund: A New Tool for Achieving Compliance under the Kyoto Protocol" (1999) Center for International Environmental Law & EuroNatura (CIEL, 1999) online: CIEL <http://www.ciel.org/Publications/ComplianceFund.pdf>.

Goldschein, Sondra, "Methyl Bromide: the Disparity Between the Pesticide's Phase-Out Dates Under the Clean Air Act and the Montreal Protocol on Substances That Deplete the Ozone Layer" (1998) 4 Envtl. Law. 577.

Gonzalez, Garcia Rogelio, *et al*, "Climate Change and Environmental Policies in Mexico" (1992) 9 Ariz. J. Int'l & Comp. L. 217.

Gosseries, Axel P., "The Legal Architecture of Joint Implementation: What do we Learn from the Pilot Phase?" (1999) 7 N.Y.U. Envtl. L.J. 49.

Goulder, Robert, "Baucus attacks EU gamesmanship on WTO disputes. (Senator Max Baucus; European Union; World Trade Organization) (2000) 87 Tax Notes 1690.

Goulder, Robert, "U.S. Will Comply with WTO's FSC Ruling, but Details Remain Unclear" (2000) 87 Tax Notes 347.

Grabosky, P., "Using Non-Governmental Resources to Foster Regulatory Compliance" (1995) 8 Governance: An International Journal of Policy and Administration 527.

Gray, Mark Allan, "The United Nations Environment Programme: An Assessment" (1990), 20 Envtl. L. 291.

Greene, Owen, "The System for Implementation Review in the Ozone Regime" in David G. Victor *et al* eds., *The Implementation and Effectiveness of International Environmental Commitments: Theory and Practice* (Cambridge, Mass.: MIT Press, 1998) 94.

Grubb, Michael *et al*, The Kyoto Protocol: A Guide and Assessment 90 (London: Earthscan Publications, 1999).

Grubb, Michael J., "Seeking Fair Weather: Ethics and the International Debate on Climate Change" (1995) 71 International Affairs 463.

Grubb, Michael, "International Emissions Trading Under the Kyoto Protocol: Core Issues in Implementation" (1998) 7 R.E.C.I.E.L. 140.

Gumley, Wayne S., "Legal and Economic Responses to Global Warming - an Australian Perspective" (1997) 14 Environmental and Planning Law Journal 341.

Guruswamy, Lakshman, "The Annihilation of Sea Turtles: World Trade Organization Intrasigence and U.S. Equivocation" (2000) 30 Environmental Law Reporter 10261.

Guzman, Andrew, "A Compliance Based Theory of International Law" (2002) 90 Cal. L. Rev. 1823.

Hafetz, J.L., "Fostering Protection of the Marine Environment and Economic Development: Article 121(3) of the Third Law of the Sea Convention" (2000) 15 Am. U. Int'l L. Rev. 583.

Hagem, Catherine *et al*, "Tough Justice for Small Nations: How Strategic Behaviour can Influence the Enforcement of the Kyoto Protocol" (2003) Centre for International Climate and Environmental Research (CICERO Working paper 2003:01), online: CICERO <http://www.cicero.uio.no/media/2186.pdf>.

Hahn, Robert W., *et al*, "The Political Economy of Instrument Choice: an Examination of the U.S. Role in Implementing the Montreal Protocol" (1989) 83 Nw. U.L. Rev. 592.

Hahn, Robert, *et al*, "Where Did All the Markets Go? An Analysis of EPA's Emissions Trading Program" (1989) 6 Yale J. on Reg. 109.

Hallum, V., "International Tribunal for the Law of the Sea: The Mox Nuclear Plant Case" (2002) 11 R.E.C.I.E.L. 372.

Halvorssen, Anita Margrethe, "Climate Change Treaties-New Developments at the Buenos Aires Conference" (1998 Yearbook) Colo. J. Int'l Envtl. L. & Pol'y 1.

Hamwey, Robert, *et al*, "Sizing the Global GHG Offset Market" *Energy Policy* 27:3 (March 1999) 123.

Hanafi, Alex G. "Joint Implementation: Legal and Institutional Issues for an Effective International Program to Combat Climate Change" (1998) 22 Harv. Envt'l. L. Rev. 441.

Handl, Gunther, "Compliance Control Mechanisms and International Environmental Obligations" (1997) 5 Tul. J. Int'l & Comp. L. 29.

Hansen, Patricia Isela, "Transparency, Standards of Review, and the Use of Trade Measures to Protect the Global Environment" (1999) 39 Va. J. Int'l L. 1017.

Hapka, Gerald A., "The Montreal Protocol; a Review of Global Environmental Action" (1991) 9 Del. Law 27.

Harad, George, "The Kyoto Climate Control Protocol" *Pulp & Paper* (1 November, 1998) online: Pulp & Paper <http://www.paperloop.com/db_area/archive/p_p_mag/1998/9811/comment.htm>.

Hardin, R., *Collective Action*, (Baltimore, MD: Johns Hopkins University Press, 1982).

Hare, William, "Australia and Kyoto: In or Out?" (2001) 24 U.N.S.W.L.J. 556.

Hare, William, *et al*, "How Much Warming are we Committed to and How Much can be Avoided?" (2004) PIK Report No. 93, Potsdam Institute for Climate Impact Research, online: <http://www.pik-potsdam.de/publications/pik_reports>.

Harper, M., "Trust but Verify: Innovation in Compliance Monitoring as a Response to the Privatization of Utilities in Developed Nations" (1996) 48 Admin. L. Rev. 593.

Harris, Paul G., "Common but Differentiated Responsibility: the Kyoto Protocol and United States Policy" (1999) 7 N.Y.U. Envtl. L.J. 27.

Harsh, B. A., "Consumerism and Environmental Policy: Moving Past Consumer Culture" (1999) 26 Ecology L.Q. 543.

Harswic, Tamara L., "Developments in Climate Change" [2002] Colo. J. Int'l Envtl. L. & Pol'y 25.

Hassol, Susan Joy, *Arctic Climate Impact Assessment* (Cambridge, Cambridge University Press, 2004).

Hathaway, Oona A., "Between Power and Principle: A Political Theory of International Law" [unpublished paper, on file with author].

Hecht, Joy E., *et al*, "Can the Kyoto Protocol Support Biodiversity Conservation? Legal and Financial Challenges" (1998) 28:9 Environmental Law Reporter 10508.

Helfer, Laurence R., *et al*, "Toward a Theory of Effective Supranational Adjudication" (1997) 107 Yale L. J. 273.

Helfer, Laurence R., "Overlegalizing Human Rights: International Relations Theory and the Commonwealth Caribbean Backlash Against Human Rights Regimes" (2002) 102 Colum. L. Rev. 1832.

Henkin, Louis, *How Nations Behave: Law and Foreign Policy*, 2nd ed. (New York: Columbia University Press, for the Council on Foreign Relations, 1979).

Hierlmeier, J., "UNEP: Retrospect and Prospect – Options for Reforming the Global Environmental Governance Regime" (2002) 14 Geo. Int'l Envtl. L. Rev. 767.

Higgins, Rosalyn, *Problems and Process: International Law and How We Use It* (New York: Oxford University Press, 1994).

Hill, Robert, "The International Climate Change Agreement: an Evolution" (2001) 24 U.N.S.W.L.J. 543.

Hobley, Anthony, "Is Kyoto Dead? Climate Change after Bush" (2002) 10 Environmental Liability 167.

Hoehne, Niklas et al, "Evolution of Commitments under the UNFCCC: Involving Newly Industrialized Economies and Developing Countries" Federal Environmental Agency, Germany, February 2003, Research Report 201 41 255 UBA-FB 000412.

Hoehne, Niklas et al, "Options for the Second Commitment Period of the Kyoto Protocol" Federal Environmental Agency, Germany, February 2005, ISSN 1611-8855, available at http://www.umweltbundesamt.de.

Hoerner, A. J., "The Role of Border Tax Adjustments in Environmental Taxation: Theory and US Experience" (Working Paper, presented at the International Workshop on Market Based Instruments and International Trade of the Institute for Environmental Studies Amsterdam) (19 March, 1998) [unpublished, on file with author].

Hoffman, Andrew J., ed., *Global Climate Change: A Senior Level Debate at the Intersection of Economics, Strategy, Technology, Science, Politics, and International Negotiation* (San Francisco: New Lexington Press, 1998).

Hongju Koh, Harold, "The 1998 Frankel Lecture: Bringing International Law Home" (1998) 35 Hous. L. Rev. 623.

Hongju Koh, Harold, "Transnational Legal Process" (1994) 75 Neb. L. Rev. 181.

Hongju Koh, Harold, "Why Do Nations Obey International Law?" Book Review of *The New Sovereignty: Compliance with International Regulatory Agreements* by A. Chayes & A.

Handler Chayes, and *of Fairness in International Law and Institutions* by T.M. Franck (1997) 106 Yale L. J. 2599.

Horlick, Gary N., "Dispute Resolution Mechanism: Will the United States Play by the Rules?" (1995) 29 J. World Trade 163.

Horn, Laura, "The Kyoto Protocol: Australia's Commitment and Compliance" (2001) 24 U.N.S.W.L.J. 583.

Horn, Laura, "The United Nations Convention on Climatic Change - the First Step" (1993) 10 Environmental and Planning Law Journal 70.

Horsch, Richard A., *et al*, "Does Kyoto Protocol Fall Short of the Mark?" (1998) 219:79 N.Y.L.J. S4.

Hovi, Jon, *et al*, "The Price of Non-compliance with the Kyoto Protocol: The Remarkable Case of Norway" (2004) Centre for International Climate and Environmental Research (CICERO Working Paper 2004:07) online: CICERO <http://www.cicero.uio.no/media/2773.pdf> [submitted to International Environmental Agreements, 2005].

Hubbard, A., "The Convention on Biological Diversity's Fifth Anniversary: A General Overview of the Convention – Where has it been and where is it going?" (1997) 10 Tul. Envtl. L. J. 415.

Hudec, Robert E., *Enforcing International Trade Law: The Evolution of The Modern GATT Legal System* (Salem, NH: Butterworth, 1993).

Hudnall, Shannon, "Towards a Greener International Trade System: Multilateral Environmental Agreements and the World Trade Organization" (1996) 29 Colum. J. L. & Soc. Probs. 175.

Hughes, L., "Limiting the Jurisdiction of Dispute Settlement Panels: The WTO Appellate Body Beef Hormone Decision" (1998) 10 Geo. Int'l Envtl. L. Rev. 915.

Hunter, David *et al*, *International Environmental Law and Policy* (New York: Foundation Press, 1998).

Hurlbut, David, "Beyond the Montreal Protocol: Impact on Non-party States and Lessons for Future Environmental Protection Regimes" (1993) 4 Colo. J. Int'l Envtl. L. & Pol'y 344.

Jackson, John H., "The WTO Dispute Settlement Understanding-Misunderstandings on the Nature of Legal Obligation" (1997) 91 Am. J. Int'l L. 60.

Jacquemont, F., *et al*, "The Convention on Biological Diversity and the Climate Change Convention 10 Years After Rio: Towards a Synergy of the Two Regimes?" (2002) 11 R.E.C.I.E.L. 169.

Jaura, Ramesh, "Environment: U.S.-EU Divide at World Climate Change Talks" INTER PRESS SERVICE, May 31, 1999, available in 1999 WL 5948910.

Jeffrey, Michael, "Where Do We Go From Here? Emissions Trading Under the Kyoto Protocol" (2001) 24 U.N.S.W.L.J. 571.

Jestin, Katya, "International Efforts to Abate the Depletion of the Ozone Layer" (1995) 7 Geo. Int'l Envtl. L. Rev. 829.

Johnson, Pierre Marc, *et al*, *The Environment and Nafta: Understanding and Implementing The New Continental Law* (Washington, DC: Island Press, 1996).

Johnston, Douglas ed., *The Environmental Law of the Sea* (Gland, Switzerland, IUCN, 1981)

Johnston, Douglas M., "International Environmental Law: Recent Developments and Canadian Contributions" in R. St J. MacDonald, Gerald L. Morris, and Douglas M. Johnston, eds., *Canadian Perspectives on International Law and Organization*, (Toronto: University of Toronto Press, 1974).

Jones, David, "The Kyoto Protocol, Carbon Sinks and Integrated Environmental Regulation: an Australian Perspective" (2002) 19 Environmental and Planning Law Journal 109.

Jones, Timothy T., "Implementation of the Montreal Protocol: Barriers, Constraints and Opportunities" (1997) 3 Envtl. Law 813.

Joyner, Christopher C., "Burning International Bridges, Fuelling Global Discontent: the United States and Rejection of the Kyoto Protocol" (2002) 33 V.U.W.L.R. 49.

Kahn, Greg, "Between Empire and Community: The United States and Multilateralism 2001 – 2003: A Mid-Term Assessment: Environment: The Fate of the Kyoto Protocol under the Bush Administration" (2003) 21 Berkley J. Int'l L. 548.

Kahn, Greg, "The Fate of the Kyoto Protocol under the Bush Administration" (2003) 21 Berkeley J. Int'l L. 548.

Kalas, Peggy Rodgers *et al*, "Dispute Resolution under the Kyoto Protocol" (2000) 27 Ecology L.Q. 53.

Kass, Stephen L., *et al*, "Having it All: Trade, Development, Environmental and Human Rights" (2000) 223:87 N.Y.L.J. 3.

Kass, Stephen L., *et al*, "U.S. v. the Marrakesh Accords: Standing Alone on Climate Change" (2002) 227:3 N.Y.L.J. 3.

Katzenstein, Peter J., *et al*, "International Organization and the Study of World Politics" (1998) 52 International Organization 645.

Keeva, Steven, "Storm Clouds: Global Climate Change will Require Unified Response" *ABA Journal*, 86 (September 2000) 107.

Kelly, Claire R., "Realist Theory and Real Constraints" (2004) 44 Va. J. Int'l L. 545.

Keohane Robert *et al*, "Power and Interdependence Revisited" (1987) 41 International Organization 725.

Keohane, Robert, "International Relations and International Law: Two Optics" (1997) 38 Harv. Int'l L. J. 487.

Keohane, Robert, "Rational Choice Theory and International Law: Insights and Limitations" (2002) 31 J. Legal Stud. 307.

Kibel, Paul Stanton, "Justice for the Sea Turtle: Marine Conservation and the Court of International Trade" (1999) 15 UCLA J. Envtl. L. & Pol'y 57.

Kindred, Hugh *et al*, *International Law: Cheifly as Interpreted and Applied in Canada*, 6th ed. (Toronto: Emond Montgomery Publications Ltd., 2000).

King, Gamble John Jr., *et al*, "The 1982 Convention and Customary Law of the Sea: Observations, a Framework, and a Warning" (1984) 21 San Diego L. Rev. 491.

Kingsbury, Benedict, "The Concept of Compliance as a Function of Competing Conceptions of International Law" (1997-1998) 19 Mich. J. Int'l L. 345.

Kirkpatrick, Jeane J., "Law and Reciprocity" (Transcript of Address) [1984] Am. Soc'y Int'l L. Proc. 59.

Kjellen, B., "The Saga of the Convention to Combat Desertification: The Rio/Johannesburg Process and the Global Responsibility for Drylands" (2003) 12 R.E.C.I.E.L. 127.

Klein, N., *Dispute Settlement in the UN Convention on the Law of the Sea,* (Cambridge: Cambridge University Press, 2005).

Knox, John H., "A New Approach to Compliance with International Environmental Law: The Submissions Procedure of the NAFTA Environmental Commission" (2001) 28 Ecology L. Q. 1.

Koivurova, Timo, *et al,* (eds.), *Arctic Governance.* (Juridica Lapponica 29. NIEM, 2004).

Kometani, Kazumochi, "Trade and Environment: How Should WTO Panels Review Environmental Regulations Under GATT Articles III and XX?" (1996) 16 Nw. J. Int'l L. & Bus. 441.

Kornicker Uhlmann, E. M., "State Community Interests, *Jus Cogens* and Protection of the Global Environment: Developing Criteria for Peremptory Norms" (1998) 11 Geo. Int'l Envtl. L. Rev. 101.

Krasner, S. D., ed., *International Regimes,* (Cornell University Press, 1983).

Kratochwil, Friedrich V., "How Do Norms Matter?" in Michael Byers ed., *The Role of Law in International Politics: Essays in International Relations and International Law* (Oxford and New York: Oxford University Press, 2000) 35.

Kratochwil, Friedrich V., *Rules, Norms, and Decisions: On the Conditions of Practical and Legal Reasoning in International Relations and Domestic Affairs* (Cambridge: Cambridge University Press, 1989).

Kuyper, Pieter Jan, "Remedies and Retaliation in the WTO: are They Likely to be Effective? The State Perspective and the Company Perspective" (1997) 91 Am. Soc'y Int'l L. Proc. 282.

Kwiatkowska, B., "The Australia and New Zealand v. Japan Southern Bluefin Tuna (Jurisdiction and Admissibility) Award of the First Law of the Sea Convention Annex VII Arbitral Tribunal" (2001) 16 Int'l J. Mar. & Coast. L. 239.

Kwiatowska, B., "The Southern Bluefin Tuna (New Zealand v Japan; Australia v Japan) Cases" (2000) 15 Int'l J. Mar. & Coast. L. 1.

Lacasta, Nuno S., , *et al*, "Consensus Among Many Voices: Articulating the European Union's Position on Climate Change" (2002) 32 Golden Gate U.L. Rev. 351.

Lanchbery, John, "Verifying Compliance with the Kyoto Protocol" (1998) 7 R.E.C.I.E.L. 170.

Landers, Frederick Pool Jr., "The Black Market Trade in Chlorofluorocarbons: The Montreal Protocol Makes Banned Refrigerants a Hot Commodity" (1997) 26 Ga. J. Int'l & Comp. L. 457.

Lane, Katie A., "Protectionism or Environmental Activism? The WTO as a Means of Reconciling the Conflict Between Global Free Trade and the Environment" (2001) 32 U. Miami Inter-Am. L. Rev. 103.

Lang, Winfried, "Compliance-Control in Respect of the Montreal Protocol" (1995) 89 American Society of International Law Proceedings 206.

Lang, Winfried, "Is the Ozone Depletion Regime a Model for an Emerging Regime on Global Warming?" (1991) 9 UCLA J. Envtl. L. & Pol'y 161.

Larson, Kristin, "Fishing for a Compatible Solution: Toothfish Conservation and the World Trade Organization" (2000) 7 Envtl. Law 123.

Lavorel, Warren, "The World Trade Organization: Looking Ahead" (1997) 91 Am. Soc'y Int'l L. Proc. 20.

Laws, Elliott P., "EC on Target for Kyoto: Why Not U.S.?" (2003) 20 The Environmental Forum 10.

Lee, J., "The Underlying Legal Theory to Support a Well-Defined Human Right to a Healthy Environment as a Principle of Customary International Law" (2000) 25 Colum. J. Envtl. L. 283.

Linden, Henry R., "CO(2) does not pollute: but Kyoto's demise won't end debate" *Public Utilities Fortnightly* 139:10 (15 May 2001) 22.

Llobet, Gabriela, "Trust but Verify: Verification in the Joint Implementation Regime" (1997) 31 Geo. Wash. Int'l L. Rev. 233.

Lodefalk, Magnus, *Climate and Trade Rules – Harmony or Conflict?* (Stockholm: Kommerskollegium, National Board of Trade, 2004).

Loewenberg, Sam, "Chill Hits Global Warming Pact" (1998) 12:18 Legal Times 4.

Lomas, Owen, *et al*, "WTO and the Environment" (1997) 25 Int'l Bus. Law. 120.

Lopez, Todd M., "A look at climate change and the evolution of the Kyoto Protocol" (2003) 43 Natural Resources Journal 285.

Lopina, David A., "The International Centre for Settlement of Investment Disputes: Investment Arbitration for the 1990s" (1988) 4 Ohio St. J. Disp. Resol. 107.

Loy, Frank E., "On a Collision Course? Two Potential Environmental Conflicts between the U.S. and Canada" (Proceedings on the Canada-U.S, Conference on Energy, the Environment and Natural Resources in the Canada/U.S. Context, Cleveland, Ohio, April 19-21 2002) (2002) 28 Can.-U.S.L.J. 11.

Lynn, William S., "Situating the Earth Charter: An Introduction, in Global Ethics and the Earth Charter" (2004) 8 Worldviews 1.

Maggio, G. F., "Inter/Intra-generational Equity: Current Applications under International Law for Promoting the Sustainable Development of Natural Resources" (1997) 4 Buff. Envtl. L. J. 161.

Mallery, David, "Clean Energy and the Kyoto Protocol: Applying Environmental Controls to Grandfathered Power Facilities" (Paper presented in the Internet Symposium: Issues in Modern International Environmental Law) (1999) 10 Colo. J. Int'l Envtl. L. & Pol'y 469.

Malone, L., "Exercising Environmental Human Rights and Remedies in the United Nations System" (2002) 27 Wm. & Mary Envtl. L. & Pol'y Rev. 365.

Mann, Howard, *et al*, *The State of Trade and Environmental Law 2003; Implications for Doha and Beyond* (Winnipeg: International Institute For Sustainable Development, 2003).

Marcich, Marino, "Trade and Environment: What Conflict?" (2000) 31 Law & Pol'y Int'l Bus. 917.

Marin, Fernando Fernandez, "Taxation in Spain and the EU as an Instrument of the Kyoto Protocol" (2001) 19 Journal of Energy & Natural Resources Law 1.

Marong, Alhaji B. M., "From Rio to Johannesburg: Reflection on the Role of International Legal Norms in Sustainable Development" (2003) 16 Geo. Int'l Envtl. L. Rev. 21.

Marshall, Dena, "An Organization for the World Environment: Three Models and Analysis" (2002) 15 Geo. Int'l Envtl L. Rev. 79.

Master, Julie B., "International Trade Trumps Domestic Environmental Protection: Dolphins and Sea Turtles are Sacrificed on the Altar of Free Trade" (1998) 12 Temp. Int'l & Comp. L.J. 423.

McAllister, D. Leigh, "The Climate Change Convention: a Signal of Emerging International Awareness" (1993) 4 Colo. J. Int'l Envtl. L. & Pol'y 484.

McCaskill, K. A., "Dangerous Liasons: The World Trade Organization and the Environmental Agenda" Department of Foreign Affairs and International Trade, Policy Staff Paper No. 94/14 (June 1994).

McConnell, M.L., *et al*, "The Modern Law of the Sea: Framework for the Protection and Preservation of the Marine Environment?" (1991) 23 Case W. Res. J. Int'l L. 83.

McConnell, M.L., *GloBallast Legislative Review: Final Report* (London: Global Ballast Water Management Programme, 2002).

McDougal, Myres, *et al*, "International Law in Policy-Oriented Perspective" in R.St.J. Macdonald and Douglas M. Johnston, eds., *The Structure and Process of International Law: Essays in Legal Philosophy, Doctrine and Theory* (The Hague and Boston: Martinus Nijhoff, 1983) 103.

McFarlane, Amy, "Between Empire and Community: The United States and Multilateralism 2001 – 2003: A Mid-Term Assessment: Development: In the Business of Development: Development Policy in the First Two Years of the Bush Administration" (2003) 21 Berkley J. Int'l. L. 521.

McIntosh, Lee, "An International Investment Agreement and the Environment" (2000) 9 J. Envtl. L. & Prac. 119.

McLaughlin, Richard, "Settling Trade-Related Disputes Over the Protection of Marine Living Resources: UNCLOS or the WTO?" (1997) 10 Geo. Int'l Envtl. L. Rev. 29.

McLaughlin, Robert, "Improving Compliance: Making Non-State International Actors Responsible for Environmental Crimes" (2002) 11 Colo. J. Int'l Envtl. L. & Pol'y 377.

McNeil, J. R., *Something New Under the Sun: An Environmental History of the Twentieth-Century World* (New York: W. W. Norton & Co., 2000).

Meier, Mike, "GATT, WTO, and the Environment: to What Extent do GATT/WTO Rules Permit Member Nations to Protect the Environment when Doing so Adversely Affects Trade?" (1997) 8 Colo. J. Int'l Envtl. L. & Pol'y 241.

Meyer, Susan Boensch, "Is it Safe to Come Out Yet? The Tenth Anniversary of the Montreal Protocol" [1997] Colo. J. Int'l Envtl. L. & Pol'y Y.B. 226.

Miles, E. *et al*, *Environmental Regime Effectiveness: Confronting Theory with Evidence* (Cambridge, Mass.: MIT Press, 2002).

Miller, Carol J., *et al*, "WTO Scrutiny v. Environmental Objectives: Assessment of the International Dolphin Conservation Program Act" (1999) 37 Am. Bus. L. J. 73.

Minchin, Nick, "Responding to Climate Change: Providing a Policy Framework for a Competitive Australia" (2001) 24 U.N.S.W.L.J. 550.

Mintz, Joel A., "Keeping Pandora's Box Shut: a Critical Assessment of the Montreal Protocol on Substances That Deplete the Ozone Layer" (1989) 20 U. Miami Inter-Am. L. Rev. 565.

Mintz, Joel A., "Progress Toward a Healthy Sky: an Assessment of the London Amendments to the Montreal Protocol on Substances that Deplete the Ozone Layer" (1991) 16 Yale J. Int'l L. 571.

Missfeldt, Fanny, "Flexible Mechanisms: Which Path to Take after Kyoto" (1998) 7 R.E.C.I.E.L. 128.

Mitchell, Ronald B., "Compliance Theory: an Overview" in James Cameron, Jacob Werksman, and Peter Roderick, eds., *Improving Compliance with International Environmental Law* (London: Earthscan Publications Ltd., 1996) 3.

Mitchell, Ronald B., "Flexibility, Compliance and Norm Development in the Climate Regime" in Jon Hovi, Olav Schram Stokke, and Geir Ulfstein, eds., *Implementing the Climate Regime: International Compliance*, (London: Earthscan Press, 2005).

Mitchell, Ronald B., "Institutional Aspects of Implementation, Compliance, and Effectiveness" in Urs Luterbacher and Detlef Sprinz, eds., *International Relations and Global Climate Change*, (Cambridge, Mass.: MIT Press, 2001).

Mitchell, Ronald B., "Regime Design Matters: International Oil Pollution and Treaty Compliance" (1994) 48 International Organization 425.

Morgan, D., "Implications of the Proliferation of International Legal Fora: The Example of the Southern Bluefin Tuna Cases" (2002) 43 Harv. Int'l L.J. 541.

Morgan, Jennifer, *et al*, "Compliance Institutions for the Kyoto Protocol: A Joint CIEL/WWF Proposal" (1999) (Discussion draft prepared for the Fifth Conference of the Parties) online: CIEL <http://www.ciel.org/Publications/complianceinstitutions.pdf>.

Morrill, Jackson F., "A Need for Compliance: the Shrimp Turtle Case and the Conflict Between the WTO and the United States Court of International Trade" (2000) 8 Tul. J. Int'l & Comp. L. 413.

Morrisette, Peter M., "The Evolution of Policy Responses to Stratospheric Ozone Depletion" (1989) 29 Natural Resources Journal 793.

Movsesian, Mark L., "Sovereignty, Compliance, and the World Trade Organization: Lessons from the History of Supreme Court Review" (1999) 20 Mich. J. Int'l L. 775.

Mueller, Benito *et al*, "Framing Future Commitments: A Pilot Study on the Evolution of the UNFCCC Greenhouse Gas Mitigation Regime" Oxford Institute for Energy Studies, EV32, June 2003, online: Benito Mueller <http://www.wolfson.ox.ac.uk/~mueller/>.

Murkowski, Frank H., "The Kyoto Protocol is Not the Answer to Climate Change" (2000) 37 Harv. J. on Legis. 345.

Nanda, Ved P., "The Kyoto Protocol on Climate Change and the Challenges to its Implementation: a Commentary" (Paper presented at the Internet Symposium: Issues in Modern International Environmental Law, includes responses) (1999) 10 Colo. J. Int'l Envtl. L. & Pol'y 319.

Neal, Sean Michael, "Bringing Developing Nations on Board the Climate Change Protocol: using Debt-for-Nature Swaps to Implement the Clean Development Mechanism" (1998) 11 Geo. Int'l Envtl. L. Rev. 163.

Neugebauer, R., "Fine-Tuning WTO Jurisprudence and the SPS Agreement: Lessons from the Beef Hormone Case" (2000) 31 Law & Pol'y Int'l Bus. 1255.

Nissen, Jill, "Achieving a Balance Between Trade and the Environment: The Need to Amend the WTO/GATT to Include Multilateral Environmental Agreements" (1997) 28 L. & Pol'y Int'l Bus. 901.

Noble-Allgire, Alice M., "The Ozone Agreements: a Modern Approach to Building Cooperation and Resolving International Environmental Issues" (1990) 14 S. Ill. U.L.J. 265.

Nordhaus, Robert R., *et al*, "A Framework for achieving Environmental Integrity and the Economic Benefits of Emissions Trading under the Kyoto Protocol" (2000) 30 Environmental Law Reporter, 11061.

Nordquist, Myron H., *United Nations Convention on the Law of the Sea 1982: A Commentary, Volume 1* (Dordrecht & Boston: Martinus Nijhof Publishers, 1985).

Nordquist, Myron H., *United Nations Convention on the Law of the Sea 1982: A Commentary, Volume 4* (Dordrecht & Boston: Martinus Nijhof Publishers, 1991).

Nordquist, Myron H., *United Nations Convention on the Law of the Sea 1982: A Commentary, Volume 2* (Dordrecht & Boston: Martinus Nijhof Publishers, 1993).

O'Connell, Mary Ellen, Using Trade to Enforce International Environmental Law: Implications for United States Law, (1994) 1 Ind. J. Global Legal Stud. 273.

O'Connor, Gerard E., "The Pursuit of Justice and Accountability: Why the United States Should Support the Establishment of an International Criminal Court", Note (1999) 27 Hofstra L. Rev. 927.

Ogden, Douglas Hunter, "The Montreal Protocol: Confronting the Threat to Earth's Ozone Layer" (1988) 63 Wash. L. Rev. 997.

Oliva, Louis P., "The International Struggle to Save the Ozone Layer" (1989) 7 Pace Envtl. L. Rev. 213.

Olson, M., *The Logic of Collective Action: Public Goods and the Theory of Groups* (Cambridge, Mass.: Harvard University Press, 1965).

Ott, Hermann E., "The Kyoto Protocol: Unfinished Business" (1998) 40:6 Environment 16.

Ott, Hermann E. et al, "*South-North Dialogue on Equity in the Greenhouse*" May 2004, Wuppertal Institute, Germany, online: Wuppertal Institute <www.wupperinst.org/download/1085_proposal.pdf>.

Ottinger, Richard L., *et al*, "Global Climate Change Kyoto Protocol Implementation: Legal Frameworks for Implementing Clean Energy Solutions" (2000 – 2001) 18 Pace Envtl. L. Rev. 19.

Owen, Scott C., "Might a Future Tuna Embargo Withstand a WTO Challenge in Light of the Recent Shrimp-Turtle Ruling?" (2000) 23 Hous. J. Int'l L. 123.

Oye, Kenneth A., "Explaining Cooperation Under Anarchy: Hypotheses and Strategies, Theories and Method" in Kenneth A. Oye *et al*, eds., *Cooperation Under Anarchy* (Princeton: Princeton University Press, 1996) 1.

Page, Robert, "Kyoto and Emissions Trading: Challenges for the NAFTA Family" (Proceedings on the Canada-U.S, Conference on Energy, the Environment and Natural Resources in the Canada/U.S. Context, Cleveland, Ohio, April 19-21 2002) (2002) 28 Can.-U.S.L.J. 55.

Pappasava, Stella, *et al*, "Adverse Implications of the Montreal Protocol Grace Period for Developing Countries" (1997) 9 International Environmental Affairs 219.

Parker, Christine, "The Greenhouse Challenge: Trivial Pursuit?" (1999) 16 Environmental and Planning Law Journal 63.

Parker, Richard W., "Design for Domestic Carbon Emissions Trading: Comments on WTO Aspects-Summary Memorandum 2" (June 22, 1998) (unpublished manuscript, on file with author).

Parker, Richard W., "The Use and Abuse of Trade Leverage to Protect the Global Commons: What We Can Learn From the Tuna-Dolphin Conflict" (1999) 12 Geo. Int'l, Envtl. L. Rev. 1.

Parks, David M., "GATT and the Environment: Reconciling Liberal Trade Policies with Environmental Preservation" (1997) 15 UCLA J. Envtl. L. & Pol'y 151.

Patlis, Jason M., "The Multilateral Fund of the Montreal Protocol: a Prototype for Financial Mechanisms in Protecting the Global Environment" (1992) 25 Cornell Int'l L.J. 181.

Paul, R. Tourangeau, "The Montreal Protocol on Substances that Deplete the Ozone Layer: Can it Keep us all from Needing Hats, Sunglasses, and Suntan Lotion?" (1988) 11 Hastings Int'l & Comp. L. Rev. 509.

Pei-Jan Tsai, Allen *et al*, "Tracking the Skies: an Airline-based System for Limiting Greenhouse Gas Emissions from International Civil Aviation" (2000) 6 Envt'l. Law. 763.

Peel, J., New State Responsibility Rules and Compliance with Multilateral Environmental Obligations: Some Case Studies of How the New Rules Might Apply in the International Environmental Context" (2001) 10 R.E.C.I.E.L 82.

Perlis, Mark L., "In a Green Squeeze" (1998) 20:46 Legal Times S32.

Perrone, Marissa A., "Fitting the Environmental Piece into the Maastricht Puzzle" (1995) 25 Environmental Law Reporter 10195.

Peters, Margot B., "An International Approach to the Greenhouse Effect: the Problem of Increased Atmospheric Carbon Dioxide can be Approached by an Innovative International Agreement" (1989-1990) 20 Cal. W. Int'l L. J. 67.

Petsonk, Annette, "The Role of the United Nations Environment Programme (UNEP) in the Development of International Environmental Law" (1990) 5 Am. U.J. Int'l L. & Pol'y 351.

Petsonk, Annie *et al*, "The Key to the Success of The Kyoto Protocol: Integrity, Accountability and Compliance" *Linkages Journal* 4:2 (May 28, 1999) online: IISD (International Institute for Sustainable Development, Winnipeg, Manitoba, Canada) <http://www.iisd.ca/journal/petsonkcarpenter.html>.

Petsonk, Annie *et al*, "Market Mechanisms and Global Climate Change: An Analysis of Policy Instruments" (1998) online: Pew Center on Global Climate Change <http://www.pewclimate.org/global-warming-in-depth/all_reports/market_mechanisms/index.cfm>.

Petsonk, Annie, "The Kyoto Protocol & the WTO: Integrating Greenhouse Gas Emissions Allowance Trading into the Global Marketplace" (1999) 10 Duke Envtl. L. & Pol'y F. 185.

Phillipson, Martin, "The United States Withdraws from the Kyoto Protocol" (2001) 36 Ir. Jur. Ann. 288.

Picolotti, R. *et al*, eds., *Linking Human Rights and the Environment* (Tucson: University of Arizona Press, 2003).

Pinkham, Margaret M., "The Montreal Protocol: an Effort to Protect the Ozone Layer" (1991) 15 Suffolk Transnat'l L. Rev. 255.

Pitschas, C., "GATT/WTO Rules for Border Tax Adjustments and the Proposed European Directive Introducing a Tax on Carbon Dioxide Emissions and Energy" (1995) 24 Ga. J. Int'l & Comp. L. 479.

Pitschas, Christian, "GATT/WTO Rules for Border Tax Adjustment and the Proposed European Directive Introducing a Tax on Carbon Dioxide Emissions and Energy" (1995) 24 Ga. J. Int'l & Comp. L. 479.

Plantigna, A.J., *et al*, "An Econometric Analysis of the Costs of Sequestering Carbon in Forests" (1999) 81 American Journal of Agricultural Economics 812.

Pole, Kenneth, "Climate Changes: Emissions Trades of Tax Schemes Cited as Best Options to Meet Kyoto Targets" (2000) 10 Environment Policy & Law 964.

Pole, Kenneth, "House of Commons Ratifies Treaty in 195-77 Vote" (2002) 13 Environment Policy & Law 409.

Popovic, N. A. F., "Human Rights, Environment and Community: A Workshop: Conference held at University at Buffalo Law School" (1998) 7 Buff. Envtl. L.J. 239.

Popovic, N.A.F., "In Pursuit of Environmental Human Rights: Commentary on the Draft Declaration of Principles on Human Rights and the Environment" (1996) 27 Colum. H.R.L. Rev. 487.

Postema, Gerald J., "Implicit Law" (1994) 13 Law and Phil. 361.

Pring George, "The 2002 Johannesburg World Summit On Sustainable Development: International Environmental Law Collides With Reality, Turning Jo'Burg Into 'Joke'Burg'" (2002) 30 Denv. J. Int'l L. & Pol'y 410.

Quick, R. *et al*, "An Appraisal and Criticism of the Ruling in the WTO Hormone Case" 2:3 (1999) J. Int'l Econ. L. 603.

Radford, Bruce W., "Kowtow to Kyoto?" Editorial (1998) 136:4 Public Utilities Fortnightly 6.

Rajamani, Lavanya, "Air and Atmosphere, Re-negotiating Kyoto: A Review of the Sixth Conference of Parties to the Framework Convention on Climate Change" (2000) 11 Colo. J. Int'l Envtl. L. & Pol'y Y.B. 201.

Ratner, Steven, "Does International Law Matter in Preventing Ethnic Conflict?" (2000) 32 N.Y.U.J. Int'l L. & Pol. 591.

Raustiala, Kal *et al*, "The Regime Complex for Plant Genetic Resources" (2004) 58 International Organization 277.

Raustiala, Kal, "The Architecture of International Cooperation: Transgovernmental Networks and the Future of International Law" (2002-2003) 43 Va. J. Int'l L. 1.

Raustiala, Kal "Compliance & Effectiveness in International Regulatory Cooperation" (2000) 32 Case W. Res. J. Int'l L. 387.

Raustiala, Kal, *et al*, "International Law, International Relations and Compliance" in Walter Carlsnaes, Thomas Risse and Beth A. Simmons, eds., *Handbook of International Relations* (London & Thousand Oaks, Calif.: SAGE Publications, 2002).

Reid, Walter V., *et al*, "Are Developing Countries Already Doing as Much as Industrialized Countries to Slow Climate Change?" (1997) 26 Energy Policy 233.

Reilly, William K., "A Climate Policy that Works" *Chicago Daily Law Bulletin* 147:67 (5 April 2001).

Reisman, W. Michael, "The Breakdown of the Control Mechanism in ICSID Arbitration" [1989] Duke L.J. 739. (For a list of pending and concluded cases, see http://www.worldbank.org/icsid>).

Renzulli, Jeffrey J., "The Regulation of Ozone-Depleting Chemicals in the European Community" (1991) 14 B.C. Int'l & Comp. L. Rev. 345.

Rest, A., "Enhanced Implementation of International Environmental Treaties by Judiciary – Access to Justice in International Environmental Law For Individuals and NGOs: Efficacious Enforcement by the Permanent Court of Arbitration" (2004) 1 Macquarie Journal of International and Comparative Environmental Law 1.

Richards, Eric L., *et al*, "The Sea Turtle Dispute: Implications for Sovereignty, the Environment, and International Law" (2000) 71 U. Colo. L. Rev. 295.

Richichi, Thomas, "Although storm clouds threatened throughout the global warming conference in Kyoto, the conferees reached an agreement on greenhouse gas emissions" (1997) 20 Nat'l L.J. B4.

Riebsell, U., "Effects of CO2 Enrichment on Marine Phytoplankton" (2004) 60 Journal of Oceanography 719.

Rinceanu, Johanna, "Enforcement Mechanisms in International Environmental Law: Quo Vadunt?" (2000) 15 J. Envtl. L. & Litig. 147.

Rinkema, Richard A., "Environmental Agreements, Non-State Actors, and the Kyoto Protocol: A 'Third Way' for International Climate Action?" (2003) 24 U. Pa. J. Int'l Econ. L. 729.

Rittberger, Volker, *Regime Theory and International Relations* (Oxford: Clarendon Press 1993).

Robb, C. A. R. *et al*, eds., *Human Rights and Environment, International Environmental Law Reports*, vol. 3 (Cambridge: Cambridge University Press, 2001).

Robinson, Nicholas A., "Befogged Vision: International Environmental Governance a Decade After Rio" (2002) 27 Wm. & Mary Envtl. L. & Pol'y Rev. 2.

Robinson, Nicholas A., "The IUCN Academy of Environmental Law: Seeking Legal Underpinnings for Sustainable Development" (2004) 21 Pace Envtl. L. Rev. 325.

Rochette, A., "Stop the Rape of the World: An Ecofeminist Critique of Sustainable Development" (2002), 51 U.N.B.L.J. 145.

Rodriguez-Rivera, Luis E., "Is the Human Right to Environment Recognized under International Law? It Depends on the Source" (2001) 12 Colo. J. Int'l Envtl. L. & Pol'y 1.

Roessig, J. M. *et al*, "Effects of Global Climate Change on Marine and Estuarine Fisheries and Fisheries" (2004) 14:2 Reviews in Fish Biology and Fisheries 251.

Rolfe, Chris, "Kyoto Protocol to the United Nations Framework Convention on Climate Change: A Guide to the Protocol and Analysis of its Effectiveness" (1998), online: West Coast Environmental Law Association <http:www.wcel.org/wcelpub/1998/12152.html>.

Rolfe, Chris, Sink Solution (2001) online: West Coast Environmental Law Association <http://www.wcel.org/wcelpub/2001/13458.pdf>.

Roman, Andrew J. *et al*, "The Regulatory Framework" in Geoffrey Thompson *et al*, eds., *Environmental Law and Business in Canada* (Auora, Ont.: Canada Law Book Inc., 1993).

Rose, Gregory, "A Compliance System for the Kyoto Protocol" (2001) 24 U.N.S.W.L.J. 588.

Rosenne, Shabtai, "Establishing the International Tribunal for the Law of the Sea" (1995) 89 Am. J. Int'l L. 806.

Royden, Amy, "U.S. Climate Change Policy under President Clinton: a Look Back" (Paper presented to the Symposium: RIO's Decade: Reassessing the 1992 Climate Change Agreement) (2002) 32 Golden Gate U. L. Rev. 415.

Ruggie, John G., "What Makes the World Hang Together? Neo-utilitarianism and the Social Constructivist Challenge" (1998) 52 International Organization 855.

Saab, Saleem S., "Move Over Drugs, There's Something Cooler on the Black Market - Freon: Can the New Licensing System Stop Illegal CFC Trafficking?" (1998) 16 Dick. J. Int'l L. 633.

Sadeleer (de), Nicolas, *Environmental Principles: From Political Slogans to Legal Rules*, (Oxford, Oxford University Press, 2002).

Sakmar, Susan L., "Free Trade and Sea Turtles: The International and Domestic Implications of the Shrimp-Turtles Case" (1999) 10 Colo. J. Int'l Envtl. L. & Pol'y 345.

Salimbene, Franklyn P., "US Business and Technology Transfer in the post-UNCED Environment" (1993) 17 Md. J. Int'l L. & Trade 31.

Sands, Philippe, ed., *Greening International Law* (New York: New Press, 1994).

Sands, Philippe, *Principles of International Environmental Law I: Frameworks, Standards and Implementation* (Manchester: Manchester University Press, 1995).

Schaefer, Matthew, "National Review of WTO Dispute Settlement Reports: in the Name of Sovereignty or Enhanced WTO Rule Compliance?" (1996) 11 St. John's J. Legal Comment. 307.

Schiano di Pepe, Lorenzo, "The World Trade Organization and the Protection of the Natural Environment: Recent Trends in the Interpretation of G.A.T.T. article XX(b) and (g)" (2000)10 Transnat'l L. & Contemp. Probs. 271.

Schlagenhof, Markus, "Trade Measures Based on Environmental Processes and Production Methods" (1995) 29 J. World Trade 123.

Schoenbaum, Thomas J., "International Trade and Protection of the Environment: the Continuing Search for Reconciliation" (1997) 91 Am. J. Int'l L. 268.

Schram Stokke, Olav, "Trade Measures, WTO, and Climate Compliance: The Interplay of International Regimes" FNI report 5/2003 (Lysaker: Fridtjof Nansen Institute Publications, 2003).

Schuler, Joseph F. Jr., "Kyoto Protocol: Economic Threat or Opportunity" (1998) 136:14 Public Utilities Fortnightly 26.

Scott, Dayna Nadine, "Carbon Sinks and the Preservation of Old-Growth Forests under the Kyoto Protocol" (2001) 10 J. Envtl. L. & Prac. 105.

Scott, Inara K., "The Inter-American System of Human Rights: An Effective Means of Environmental Protection?" (2000) 19 Va. Envtl. L.J. 197.

Sedjo, RA, "Temperate Forest Ecosystems in the Global Carbon Cycle" (1992) 21 Ambrio 274.

Setear, John K., "Learning to Live with Losing: International Environmental Law in the New Millennium" (2001) 20 Va. Envtl. L.J. 139.

Setear, John K., "Ozone, Iteration, and International Law" (1999) 40 Va. J. Int'l L. 193.

Setear, John K., "Responses to Breach of a Treaty and Rationalist International Relations Theory: The Rules of Release and Remediation in the Law of State Responsibility" (1997) 83 Va. L. Rev. 1.

Shaffer, Gregory, "The Law and Politics of the Treatment of Trade and Environment Measures in the WTO" (1999) 93 Am. Soc'y Int'l L. Proc. 218.

Shany, Y., *The Competing Jurisdictions of International Courts and Tribunals* (Oxford: Oxford University Press, 2003).

Shelton, Dinah (ed.), *Commitment and Compliance: The Role of Non-Binding Norms in the International Legal System* (Oxford and New York: Oxford University Press, 2000).

Shelton, Dinah, "What Happened in Rio to Human Rights?" (1992) 3 Y.B. Int'l Env. L. 75.

Shihata, Ibrahim F.I., "The Settlement of Disputes Regarding Foreign Investment: The Role of the World Bank, with Particular Reference to ICSID and MIGA" (1986) 1 Am. U. J. Int'l L. & Pol'y 97.

Shim, Sangmin, "Korea's Leading Role in Joining the Kyoto Protocol with the Flexibility Mechanisms as Side-Payments" (2003) 15 Geo. Int'l Envtl. L. Rev. 203.

Shue, Henry, Subsistence Emissions and Luxury Emissions, in Richard L. Revesz ed., *Foundations of Environmental Law and Policy* (New York & Oxford: Oxford University Press, 1997) 322.

Sierra Club of Canada, *Forests, Climate Change and Carbon Reservoirs, Opportunities for Forest Conservation* (Ottawa: Sierra Club of Canada, September 2003).

Sierra Legal Defence Fund, *A Guide to Canada's Species at Risk Act* (Vancouver: Sierra Legal Defence Fund, May 2003).

Sikkink, Kathryn, *et al*, "International Norm Dynamics and Political Change" (1998) 52 International Organization 887.

Silecchia, Lucia Ann, "Ounces of Prevention and Pounds of Cure: Developing Sound Policies for Environmental Compliance Programs" (1996) 7 Fordham Envtl. L. J. 583.

Silverman, Joel L., "The 'Giant Sucking Sound' Revisited: a Blueprint to Prevent Pollution Havens by Extending NAFTA's Unheralded 'Eco-dumping' Provisions to the New World Trade Organization" (1994) 24 Ga. J. Int'l & Comp. L. 347.

Simma, Bruno, *et al*, "The Responsibility of Individuals for Human Rights Abuses in Internal Conflict: A Positivist View" (1999) 93 Am. J. Int'l L. 302.

Simmons, Beth, "Compliance with International Agreements" (1998) 1 Annual Review of Political Science 75.

Simpson, Gerry, "Book Review: Is International Law Fair?" (1996) 17 Mich. J. Int'l L. 615.

Simpson, John, "U.S. Global Warming Action Plan: a Clarion Call to ... What?" (1993) 131:15 Public Utilities Fortnightly 40.

Sinclair, Christine Moran, "Global Warming or not: the Global Climate is Changing and the United States should too", Note (2000) 28 Ga. J. Int'l & Comp. L. 555.

Slaughter, Anne-Marie, "International Law and International Relations" (2000) 277 Rec. des Cours 285.

Slaughter, Anne-Marie, "A Liberal Theory of International Law" (2000) 94 Am. Soc'y Int'l L. Proc. 240.

Smith, K.R., "The Natural Debt: North and South" in T.W. Giambelluca and A. Henderon-Sellers, eds., *Climate Change: Developing Southern Hemisphere Perspectives* (Sussex: John Wiley & Sons Ltd., 1996) 423.

Snoderly, Anna Beth, "Clearing the Air: Environmental Regulation, Dispute Resolution, and Domestic Sovereignty under the World Trade Organization" (1996) 22 N.C.J. Int'l L. & Com. Reg. 241.

Somerset, Margaret E., "An Attempt to Stop the Sky from Falling: the Montreal Protocol to Protect Against Atmospheric Ozone Reduction" (1989) 15 Syracuse J. Int'l L. & Com. 391.

Staffin, Elliot B., "Trade Barrier or Trade Boon? A Critical Evaluation of Environmental Labeling and Its Role in the 'Greening' of World Trade" (1996) 21 Colum. J. Envtl. L. 205.

Stanton, Kibel Paul, "UNCED's Uncertain Legacy: An Introduction to the Issue" (2002) 32 Golden Gate U. L. Rev. 345.

Stasinopoulos, Dinos, "New Energy Policy for the United States" (2002) 20 Journal of Energy & Natural Resources Law 50.

Steinberg, Richard H., "Trade-Environment Negotiations in the EU, NAFTA, and WTO: Regional Trajectories of Rule Development" (1997) 91 Am. J. Int'l L. 231.

Steinmiller, Heather A., "Steel Industry Watch Out! The Kyoto Protocol Is Lurking" (2000) 11 Vill. Envtl. L. J. 161.

Stephens, Tim, "The Limits of International Adjudication in International Environmental Law: Another Perspective on the Southern Bluefin Tuna Case" (2004) 19 Int'l J. of Mar. &Coast. L. 177.

Stephenson, Dale E., "Greenhouse Gas Emissions and Emissions Trading in North America: Kyoto Treaty and U.S. Initiatives" (Proceedings on the Canada-U.S, Conference on Energy, the Environment and Natural Resources in the Canada/ U.S. Context, Cleveland, Ohio, April 19-21 2002) (2002) 28 Can-U.S.L.J. 43.

Stewart, Richard B. *et al*, "Legal Issues Presented by a Pilot International Greenhouse Gas Trading System" 39-40 (United Nations Conference on Trade and Development Pub. No. UNCTAD/GDS/GFSB/Misc. 1, 1996).

Stewart, Richard B., *et al*, "Designing an International Greenhouse Gas Emissions Trading System" (2001) 15 Nat. Resources & Env't 160.

Stewart, Richard *et al*, "The Clean Development Mechanism: Building International Public-Private Partnerships Under the Kyoto Protocol" UN Doc. UNCTAD/GD5/GF5B/Misc.7 (2000).

Stone, C., "Environment 2000 – New Issues for a New Century: Land Use, Biodiversity, and Ecosystem Integrity: Land Use and Biodiversity" (2001) 27 Ecology L. Q. 967.

Strom, Torsten H., "Another Kick at the Can: Tuna/Dolphin II" (1995) 33 Can. Y.B. Int'l Law 149.

Sturtz, L., "Southern Bluefin Tuna Case: Australia and New Zealand v. Japan" (2001) 28 Ecology L. Q. 455.

Sullivan, Mary Anne, "Kyoto Wasn't Last Gasp; Even Without the Protocol, Greenhouse Gas Regulation is Coming here" (2002) 25:23 *Legal Times* 30.

Sussman, Robert M., "Climate Change" (2003) 20:1 The Environmental Forum 19.

Taillant, J., "Environmental Advocacy and the Inter-American Human Rights System" in R.

Talbot, Lori B., "Recent Developments in the Montreal Protocol on Substances that Deplete the Ozone Layer: the June 1990 Meeting and Beyond" (1992) 26 Int'l Law. Spring 145.

Taylor, Gray E., "Global Climate Change Agreements - do the Storm Clouds have a Silver Lining?" (1999) 45 Proceedings of the Forty-Fifth Annual Rocky Mountain Mineral Law Institute 2-1.

Taylor, P. E., "From Environmental to Ecological Human Rights: A New Dynamic in International Law?" (1998) 10 Geo. Int'l Envtl. L. Rev. 309.

Telesetsky, Anastasia, "THE Kyoto Protocol" (1999) 26 Ecology L. Q. 797.

Thacher, Peter S., "Equity under Change" (1987) 81 Am. Soc'y Int'l L. Proc. 133.

The United Nations University Institute for Advanced Studies, *Global Climate Governance; Inter-linkages between the Kyoto Protocol and other Multilateral Regimes*, Final Report (Tokyo, Japan 1999).

Thieme, Dominik, *et al*, "European Community - Subsidies - Free Movement of Goods - Economic Incentives for Renewable Energy - UN Convention on Climate Change" (2002) 96 Am. J. Int'l L. 225.

Thomas, Chris D. *et al*, "Extinction Risk from Climate Change" (2004) 427 *Nature* 145.

Thomas, William L., "The Kyoto Protocol: History, Facts, Figures, and Projections" (1999) 137:8 Public Utilities Fortnightly 48.

Thomas, William L., *et al*, "Creating a Favorable Climate for CDM investment in North America" (2001) 15 Natural Resources & Environment, 172.

Thomas, William L.,"The Kyoto Protocol: a factor in foreign investment in Mexico?" (1999) 137:8 Public Utilities Fortnightly 137:8 40.

Thompson, Alexander "Applying Rational Choice Theory to International Law: The Promise and Pitfalls" (2002) 31 L.S. 285.

Thompson, Michael, *et al*, "Cultural Discourse" in Steve Rayner and Elizabeth Malone, eds., *Human Choice and Climate Change: The Societal Framework,* vol. I (Columbus, Ohio: Battelle National Labs, 1998) 265.

Thompson, Patricia, "The Third Conference of the Parties to the United Nations Framework Convention on Climate Change: the December 1997 Kyoto Protocol" (1997) 9 Colo. J. Int'l Envtl. L. & Pol'y Y.B. 219.

Thoms, Laura, "A Comparative Analysis of International Regimes on Ozone and Climate Change with Implications for Regime Design" (2003) 41 Colum. J. Transnat'l L. 795.

Thorn, C. *et al*, "The Agreement on the Application of Sanitary and Phytosanitary Measures and the Agreement on Technical Barriers to Trade" (2000) 31 Law & Pol'y Int'l Bus. 841.

Tietenberg, Tom *et al*, *International Rules for Greenhouse Gas Emissions Trading* UNCTAD, 1999 UNCTAD/GDS/GFSB/Misc.6 online: <http://r0.unctad.org/ghg/publications/intl_rules.pdf>.

Timmons, Gregory D., *et al*, "Fossil Fuels in the Twenty-First Century" (2001) Energy & Mineral Law Institute Annual 35.

Tinker, C., "The Rio Environmental Treaties Colloquium: A New Breed of Treaty: The United Nations Convention on Biological Diveristy" (1995) 13 Pace Envtl. L. Rev. 191.

Tol, Richard S. J., *et al*, "State Responsibility and Compensation for Climate Change Damages: A Legal and Economic Assessment" (2004) 32 Energy Policy 1109.

Torrie, R., *et al*, "Kyoto and Beyond: The Low-emission Path to Innovation and Efficiency" (October, 2002, David Suzuki Foundation and Climate Action Network Canada), online: David Suzuki Foundation <www.davidsuzuki.org>.

Trask, Jeff, "Montreal Protocol Non-Compliance Procedure: the Best Approach to Resolving International Environmental Disputes?" (1992) 80 Geo. L. J. 1973.

Tripp, James T.B., "The UNEP Montreal Protocol: Industrialized and Developing Countries Sharing the Responsibility for Protecting the Stratospheric Ozone Layer" (1988) 20 N.Y.U.J. Int'l L. & Pol. 733.

Tsai, Sophia, "UNFCCC Technical Workshop on Mechanisms of the Kyoto Protocol" (1999) 10 Colo. J. Int'l Envtl. L. & Pol'y 220.

Tynberg, A., "The Natural Step and its Implications for a Sustainable Future" (2000) 7 Hastings W.-Nw. J. Envtl. L. & Pol'y 73.

Vaughan, Scott, "Trade and Environment: Some North-South Considerations" (1994) 27 Cornell Int'l L.J. 591.

Victor, D. G., *et al*, "The Kyoto Protocol Carbon Bubble: Implications for Russia, Ukraine, and Emission Trading", Interim Report IR-98-094 (1998) 8 International Institute for Applied Systems Analysis.

Victor, David G., "Enforcing International Law: Implications for an Effective Global Warming Regime" (1999) 10 Duke Envtl. L. & Pol'y F. 147.

Victor, David G. *et al* eds., *The Implementation and Effectiveness of International Environmental Commitments: Theory and Practice* (Cambridge, Mass.: MIT Press, 1998).

Victor, David G. *et al* eds., *The Implementation and Effectiveness of International Environmental Commitments: Theory and Practice* (Cambridge, Mass.: MIT Press, 1998).

Victor, David G., "The Early Operation and Effectiveness of the Montreal Protocol's Non-Compliance Procedure" (1996) Executive Report ER-96-2 (International Institute for Applied Systems Analysis).

Victor, David G., *The Collapse of the Kyoto Protocol, and the Struggle to Slow Global Warming*, (Princeton, Princeton University Press, 2001).

Vogt, KA *et al*, "Review of Root Dynamics in Forest Ecosystems Grouped by Climate, Climate Forest Type and Species" (1996) 187 Plant and Soil 159.

Vorlat, Katrien, "The International Ozone regime: Concessions and Loopholes?" (1993) 17 The Fletcher Forum of World Affairs 135.

Wagner, J. M., "The WTO's Interpretation of the SPS Agreement has Undermined the Right of Governments to Establish Appropriate Levels of Protection Against Risk" (2000) 31:3 Law & Pol'y Int'l Bus. 854.

Wang, Xueman, "Towards a System of Compliance: Designing a Mechanism for the Climate Change Convention" (1998)7 R.E.C.I.E.L. 176.

Wang, Xueman, *et al*, "The Implementation and Compliance Regimes under the Climate Change Convention and its Kyoto Protocol" (2002) 11:2 R.E.C.I.E.L. 181.

Ward, Halina, "Common but Differentiated Debates: Environment, Labour and the World Trade Organization" (1996) 45 I.C.L.Q. 592.

Ward, Halina, "Trade and Environment in the Round - and After" (1994) 6 J. Envtl. L. 263.

Ward, Justin R., "Environmental Reform Priorities for the World Trading System" (1995) 35 Santa Clara L. Rev. 1205.

Watson, Robert T., *Millennium Ecosystem Assessment Synthesis Report*, prepublication draft (Washington: Island Press, 23 March 2005) [due for publication in September 2005] online: <http://www.millenniumassessment.org/en/index.aspx>.

Watters, Lawrence, "Indigenous Peoples and the Environment: Convergence from a Nordic Perspective" (2001) 20 UCLA J. Envtl. L. & Pol'y 237.

Wendt, Alexander, "Anarchy is What States Make of It: The Social Construction of Power Politics" (1992) 46 International Organization 391.

Wendt, Alexander, "Collective Identity Formation and the International State" (1994) 88 American Political Science Review 384.

Werksman, Jacob, "Greenhouse Gas Emissions Trading and the WTO" (June 1999) [unpublished manuscript, on file with author].

Werksman, Jacob, "Greenhouse Gas Emissions Trading and the WTO" (June 1999) [unpublished manuscript, on file with author].

Werksman, Jacob, "Procedural and Institutional Aspects of the Emerging Climate Change Regime: Improvised Procedures and Impoverished Rules?" (Paper presented at the concluding workshop for the Project to Enhance Capacity under the Framework Convention on Climate Change and the Kyoto Protocol, London, March 18, 1999) online: <http://www.cserge.ucl.ac.uk/Werksman.pdf>.

Werksman, Jacob, "Responding to Non-compliance under the Climate Change Regime" 1999 OECD Information Paper, (Paris, 1999) online: OECD Documents <http://www.olis.oecd.org/olis/1999doc.nsf/LinkTo/env-epoc(99)21-final>.

Werksman, Jacob, "The Clean Development Mechanism: Unwrapping the Kyoto Surprise" (1998) 7 R.E.C.I.E.L. 147.

Weston, B. H. *et al*, *Basic Documents in International Law and World Order*, 2nd ed, (St. Paul, MN: West Pub. Co., 1990) 298.

Weston, B. H., "Human Rights" (1986) 6 Hum. Rts. Q. 257.

Wettestad, Jørgen, "Enhancing Climate Compliance: What are the Lessons to Learn from Environmental Regimes and the EU?" FNI report 2/2003 (Lysaker: Fridtjof Nansen Institute Publications, 2003).

Wexler, Pamela, "Protecting the Global Atmosphere: Beyond the Montreal Protocol" (1990) 14 Md. J. Int'l L. & Trade 1.

Wiener, Jonathan Baert, "Global Environmental Regulation: Instrument Choice in Legal Context" (1999) 108 Yale L.J. 677.

Wilder, Martijn, "The Kyoto Protocol and Early Action" (2001) 24 U.N.S.W.L.J. 565.

Wilder Martijn, *et al*, "The Clean Development Mechanism" (2001) 24 U.N.S.W.L.J. 577.

Williams, J. Carol, "The Next Frontier: Environmental Law in a Trade-Dominated World" (20th Anniversary Commemorative Issue) (2001) 20 Va. Envtl. L.J. 221.

Winham, G. R., "The World Trade Organization: Institution-Building in the Multilateral Trade System" (1998) 12 The World Economy 349.

Winham, G. R., "The World Trade Organization: Institution-Building in the Multilateral Trade System" (1998) 12 *The World Economy* 349

Winham, G. R., *International Trade and the Tokyo Round of Negotiation* (Princeton: Princeton University Press, 1986).

Winham, G. R., *International Trade and the Tokyo Round of Negotiation* (Princeton University Press, 1986).

Wirth, David A., "The Sixth Session (Part Two) and Seventh Session of the Conference of the Parties to the Framework Convention on Climate Change" (2002) 96 Am. J. Int'l L. 648.

Wiser, Glenn M., "The Clean Development Mechanism v. The World Trade Organization: Can Free-Market Greenhouse Gas Emissions Abatement Survive Free Trade?" (1999) 11 Geo. Int'l Envtl. L. Rev. 531.

Wiser, Glenn M., "The Clean Development Mechanism v. The World Trade Organization: Can Free-Market Greenhouse Gas Emissions Abatement Survive Free Trade?" (1999) 11 Geo. Int'l Envtl. L. Rev. 531.

Wiser, Glenn M., "Compliance Systems under Multilateral Agreements: A Survey for the benefit of Kyoto Policy Makers" (1999) Center for International Environmental Law (CIEL, 1999) online: CIEL <http://www.ciel.org/Publications/SurveyPaper1.pdf>.

Wiser, Glenn M., "Hybrid Liability Revisisted: Bridging the Divide Between Seller and Buyer Liability" (2000) Center for International Environmental Law (CIEL, 2000).

Wiser, Glenn M., "Hybrid Liability Revisited: Bridging the Divide Between Seller and Buyer Liability" (2000) Center for International Environmental Law (CIEL, 2000) online: CIEL <http://www.ciel.org/Publications/Hybrid-LiabilityCOP6.pdf>.

Wiser, Glenn M., "Joint Implementation: Incentives for Private Sector Mitigation of Global Climate Change" (1997) 9 Geo. Int'l Envtl. L. Rev. 747.

Wiser, Glenn M., *et al*, "Restoring the Balance: Using Remedial Measures to Avoid and Cure Non-Compliance Under the Kyoto Protocol" (WWF April 2000) online: <http://www.ciel.org/Publications/restoringbalance.pdf>.

Wiser, Glenn M., *et al*, The Compliance Fund: A New Tool for Achieving Compliance under the Kyoto Protocol (Washington, D.C.: The Center for International Environmental Law, 1999), online: CIEL <http://www.ciel.org/Publications/ComplianceFund.pdf>.

Wofford, C., "A Greener Future at the WTO: The Refinement of WTO Jurisprudence on Environmental Exceptions to GATT" (2000) 24 Harv. Envtl. L. Rev. 563.

Wolfe, Karrie A., "Greening the International Human Rights Sphere? An Examination of Environmental Rights and the Draft Declaration of Principles on Human Rights and the Environment" (2003) 13 J. Envtl. L. & Practice. 109.

Wood, James C., "Intergenerational Equity and Climate Change" (1996) 8 Geo. Int'l Envtl. L. Rev. 293.

Woodward, Jennifer "Turning Down the Heat: What United States Laws Can Do to Help Ease Global Warming" (1989) 39 A.M. U.L. Rev. 203.

Woody, Kristin, "The World Trade Organization's Committee on Trade and Environment" (1996) 8 Geo. Int'l Envtl. L. Rev. 459.

Worika, Ibibia L., *et al*, "Contractual Architecture for the Kyoto Protocol: from Soft and Hard Laws to Concrete Commitments" (2000) 15 J. Land Use & Envtl. L. 489.

World Commission on Environment and Development, *Our Common Future* (Brundtland Report) (Oxford & New York: Oxford University Press, 1987).

Yamin, Farhana, "Principles of Equity in International Environmental Agreements with Special Reference to the Climate Change Convention" in Papers

Presented at the IPCC Working Group III Workshop on Equity and Social Considerations Related to Climate Change, Nairobi, Kenya, (July 1993) 357.

Yamin, Farhana, *The International Climate Change Regime: A Guide to Rules, Institutions and Procedures* (Sussex: Cambridge University Press, 2005).

Yandle, Bruce *et al*, "Bootleggers, Baptists and the Global Warming Battle" (2002) 26 Harv. Envt'l. L. Rev. 177.

Yaninek, Kathleen Doyle, "Turtle Excluder Device Regulations: Laws Sea Turtles Can Live With" (1995) 21 N.C. Centr. L.J. 256.

Yelin-Kefer, Jennifer, "Warming up to an International Greenhouse Gas Market: Lessons from the U.S. Acid Rain Experience" (2001) 20 Stan. Envtl. L.J. 221.

Yoshida, O., "Soft Enforcement of Treaties: the Montreal Protocol's Non-compliance Procedure and the Functions of Internal International Institutions" (1999) 10 Colo. J. Int'l Envtl. L. & Pol'y 95.

Young, O. R., *International Cooperation: Building Regimes for Natural Resources and the Environment* (Ithaca: Cornell University Press, 1989).

Zaelke, Durwood et al, "Global Warming and Climate Change: An Overview of the International Legal Process" (1989-1990) 5 Am. U. J. Int'l L. & Pol'y 249.

Zeidler, J., *et al*, "The Dry and Sub-humid Lands Programme of Work of the Convention on Biological Diversity: Connecting the CBD and the UN Convention to Combat Desertification" (2003) 12 R.E.C.I.E.L. 164.

Zekos, Georgios I., "An Examination of GATT/WTO Arbitration Procedures" (1999) 54:4 Disp. Resol. J. 72.

Zinn, M. D., Policing Environmental Regulatory Enforcement: Cooperation, Capture and Citizen's Suits (2002), 21 Stan. Envtl. L. J. 81.

Index